Microelectronics and Signal Processing

Microelectronics and Signal Processing

Advanced Concepts and Applications

Edited by

Sanket Goel

CRC Press
Taylor & Francis Group
Boca Raton London New York

CRC Press is an imprint of the
Taylor & Francis Group, an **informa** business

First edition published 2021
by CRC Press
6000 Broken Sound Parkway NW, Suite 300, Boca Raton, FL 33487-2742

and by CRC Press
2 Park Square, Milton Park, Abingdon, Oxon, OX14 4RN

Library of Congress Cataloging-in-Publication Data
A catalog record for this title has been requested

ISBN: 9780367640125 (hbk)
ISBN: 9780367767143 (pbk)
ISBN: 9781003168225 (ebk)

Typeset in Times LT Std
by KnowledgeWorks Global Ltd.

Dedication

Dedicated to my parents (Mrs. Pushpalata Goel and Dr. Satish Chandra Goel), who showed me the light and electricity of life. Their doting and everlasting memory always motivates me to rise early and grab the day.

Contents

Preface...ix
Acknowledgments.. xiii
About the Editor... xv
List of Contributors...xvii

Chapter 1   Van der Waals (vdW) Heterostructures Based on Transition
Metal Dichalcogenides (TMD): Current Status and Prospects
in Broadband Photodetector Applications... 1

*Debapriya Som, Srijan Trivedi, Ayantika Chatterjee, and
Sayan Kanungo*

Chapter 2   3D Printing: A State-of-the-Art Approach in Electrochemical
Sensing 17

*Mary Salve, Khairunnisa Amreen, Prasant Kumar Pattnaik,
and Sanket Goel*

Chapter 3   Optimized Electrical Interface for a Vanadium Redox Flow
Battery (VRFB) Storage System: Modeling, Development,
and Implementation...29

Nawin Ra and Ankur Bhattacharjee

Chapter 4   A Review on Recent Advancements in Chamber-Based
Microfluidic PCR Devices 49

Madhusudan B Kulkarni and Sanket Goel

Chapter 5   A Classification and Evaluation of Approximate Multipliers 71

U. Anil Kumar and Syed Ershad Ahmed

Chapter 6   Optical MEMS Accelerometers: A Review 87

Balasubramanian Malayappan and Prasant Kumar Pattnaik

Chapter 7   A 60-GHz SiGe HBT Receiver Front End for Biomedical
Applications.. 109

Puneet Singh and Saroj Mondal

Chapter 8   Gate-Overlap Tunnel Field-Effect Transistors (GOTFETs) for
Ultra-Low-Voltage and Ultra-Low-Power VLSI Applications......... 137

Sanjay Vidhyadharan and Surya Shankar Dan

Chapter 9   The Role of PMU for Frequency Stability in Hybrid Power
Systems.. 165

Renuka Loka, Alivelu M. Parimi, and P. Shambhu Prasad

Chapter 10   A Modular Zigbee-Based IoT Platform for Reliable Health
Monitoring of Industrial Machines Using ReFSA 179

*Amar Kumar Verma, Jaju Vedant Vinod, and Radhika
Sudha*

Chapter 11   A Study on Time-Frequency Analysis of Phonocardiogram
Signals 189

*Samit Kumar Ghosh, Rajesh Kumar Tripathy, and
R. N. Ponnalagu*

Chapter 12   A Study on the Performance of Solar Photovoltaic Systems
in the Underwater Environment 203

*Challa Santhi Durganjali, Sudha Radhika, R. N. Ponnalagu,
and Sanket Goel*

Chapter 13   A Review on Brain Tumor Segmentation Algorithms Using
Recent Deep Neural Network Architectures and a Gentle
Introduction to Deep Neural Network Concepts............................ 227

*B. Dheerendranath, B. V. V. S. N. Prabhakar Rao,
P. Yogeeswari, C. Kesavadas, and Venkateswaran
Rajagopalan*

Chapter 14   LabVIEW-based Simulation Modeling of Building Load
Management for Peak Load Reduction 245

A. Ajitha, Sudha Radhika, and Sanket Goel

Index... 257

Preface

Evidently, electrical and electronic components, processes, and devices have revolutionized the way that the world currently is. In the electrical domain, since the 17th century, various researchers have not only realized efficient renewable energy production but also rationalized its decentralization and deregulation. Likewise, in the electronics domain, since the 19th century, a huge amount of growth has been visible on various fronts, such as semiconductor devices, communication, embedded systems, sensors, etc. In the current century, industrial, mechanical, electrical, and electronics engineering are reliant on parameters measured by processing and control units, sensors and actuators based on electronic components, circuitry, and simulation software, which are used in almost every domain of research area for building prototype models, devices, or systems.

In this context, this book is about general and specific areas involved in electrical and electronics engineering, which comprises broad subjects such as MEMS and Microfluidics, VLSI, and Communication and Signal Processing. This book discusses the recent trends in various aspects of research areas for diverse applications like biomedical, biochemical, and power source systems. It also discusses modeling, simulating, and prototyping of the different electronic-based systems for carrying out varied applications. The main purpose of this work is to bring together technical and general knowledge, distinct ranges of research, reviewing from base level to moderate level of advancement of technology, and the potential ways, standards, and systems that are used in the various scenarios of research fields from the last few years in electrical and electronics engineering, respectively. With this book, readers will understand the multiplatform fundamentals guiding electrical and biomedical devices that form current features such as automation, integration, and miniaturization of a particular device.

This book showcases a unique platform because it covers the different areas of research in this trending era as a benchmark. This book is a link between the electronics and cutting-edge technologies that are being used for numerous applications representing the physical and virtual developments of electronic devices. This link deals with the conditioning and collecting of mechanical and electrical processing signals to build the bridge between various domains to find an advanced and effective study on research areas in recent times. Therefore, this book will mostly uphold the innovation and originality involved in the development of miniaturized devices, and propose new methods, emphasizing with different areas of electrical and electronics engineering. Hence, this book mostly concerns multidomain and multipurpose research areas in the current electrical and electronics scenario.

This book entitles various approaches involved in electrical, biomedical, and electronics for modern distribution of research strategies and covers state-of-the-art research themes such as signal sensing, signal simulators, 3D-printing technology, power systems, data acquisition systems, instrumentation, electrochemical sensing, electromechanical measurements, and signal analysis. Also, it presents approaches involved in the use of virtual and simulation software like MATLAB® and LabVIEW for biomedical applications. The book offers basic and advanced knowledge on design, fabrication, circuit implementation, modeling, and simulation of electronic components, and how the new trends are increasingly associated with growing technological characteristics in electrical and electronics engineering. The scope of this book is to approach electronic instrumentation for the design and development of digital circuitry for interfacing electromechanical measurement systems.

Research in these domains has been carried out by both academia and industry, whereby academic being the backbone whose intellectual outputs have been widely adopted by the industry and implemented for consumer-at-large. This has led to the huge demand for offering an undergraduate (UG) engineering program in Electrical Engineering, which traditionally encompassed Electronics Engineering as well. Even though this UG program has branched out in diverse streams like Electronics Engineering, Communication Engineering, Instrumentation Engineering, etc., the umbrella UG program of Electrical Engineering can still be seen at many universities. Further, several academic research and industry-tailored postgraduate programs covering various subdomains, such as Micro/Nanoelectronics, VLSI, Communications, Embedded Systems, Power Electronics, Control Engineering, Instrumentation, etc., have evolved and become popular among various stakeholders. Undoubtedly, the PhD programs in Electrical and Electronics (E & E) areas have also left a huge impact on the collaborative research and development (R&D) from industry and public-funded research institutes. In this book, the authors intend to discuss the drifting changes associated with the new technology and advancement toward the design and development of the protocol with both by simulation and prototype module that are used for various electrical and biomedical applications with key features like low-cost, easy-to-use, fast response, etc. Besides, we hope that this book will be a good reference for undergraduate and postgraduates, research scholars, and for academic fellows who want to comprehend the foundations of this significant theme in electrical and electronics engineering.

Such huge intervention from Electrical and Electronics Engineering has paved the way for constant streamlined involvement from the academic world to make these programs contemporary, keeping the next generation of students and their career pathway in mind. To this end, since its inception in 1964, the E & E at BITS-Pilani has contributed tremendously to offer industry-focused programs while incorporating the best practices from across the globe. The consistent involvement of R&D and industrial assignment has proven to be a gold standard among peers. Keeping this in view, there is a requirement to have a forum to showcase the current trends in E & E domains to various stakeholders, from freshmen college students to the professionals, in the form of a book. This will provide the academic perspectives of the cutting-edge R&D outputs from the faculty members and PhD students, amalgamating the newer cross-dimensional areas, such as cyber-physical systems, nanoelectronics, smart-sensors, point-of-need devices, etc. The book will become a benchmark for readers to understand the academic aspects of the contemporary work in E & E domains and the way forward on how this will help society-at-large.

The book comprises of 14 chapters and the structure of this book is as follows. Chapter 1 summarizes the recent development on the transition metal dichalcogenides (TMD) based van der Waals (vdW) heterostructures presented for broadband photodetection application. In this chapter, the emphasis has been given on the underlying physics and design aspects of such TMD-based vdW heterostructure photodetectors. Chapter 2 discusses 3D-printing technology, which is emerging as an efficient tool to revolutionize the fabrication of miniaturized devices for biochemical and biomedical applications. Further, it summarizes the recent state-of-the-art approach for advances in the use of 3D printing in versatile analytical and electrochemical sensing applications. Chapter 3 is about the control algorithm to design the optimized electrical interface for the VRFB storage system with a solar photovoltaic source. The maximum power point tracking (MPPT) has also been incorporated in a specific charging phase to enhance the power transfer efficiency. Chapter 4 summarizes various types of microchannel designs, fabrication techniques, processes, and other associated approaches to create chamber-based microfluidic PCR devices for the quantification and detection of nucleic acids. Chapter 5 presents a classification and implementation of state-of-the-art approximate multipliers in an image processing application. To quantify the multiplier designs, exhaustive error analysis is carried out using MATLAB®. Chapter 6 discusses a brief review of various state-of-the-art sensing techniques and interrogation methods used in optical MEMS and nano-photonic accelerometers. Chapter 7 is devoted to the design of a fully integrated RF front end for a low-cost, low DC power consumption, high-sensitivity millimeter-wave receiver operating in the V-band around 60 GHz. Chapter 8 discusses that static power consumption in GOTFET-based basic digital gates such as inverter, NAND, NOR, and XOR circuits is 90–95% lower than that of corresponding conventional CMOS gates.

Chapter 9 discusses the advent of microgrids and increased utilization of renewable energy sources (RESs) contributed to the uncertainty of power generation in hybrid power systems (HPSs), which necessitated stringent monitoring and control of grid conditions. Chapter 10 proposes a modular Zigbee-based IoT platform for reliable health monitoring of industrial machines using a remote fault signature analyzer (ReFSA). Chapter 11 shows a study on the different time-frequency analysis (TFA) approaches used for the analysis of PCG signals. Chapter 12 mainly focuses on detailing the different technologies available, by decreasing the surface temperature of the cells, to improve the photovoltaic cell's efficiency. It also emphasizes the recently developed technology of using solar panels in the underwater environment for improved efficiency. Chapter 13 discusses a review of brain tumor detection algorithms and a gentle introduction to deep neural network concepts are presented in this work. Chapter 14 discusses demand side management (DSM) and demand response (DR) and provides scope for the development of new innovative methods to attain effective building load management by considering consumer comfort. Also, this chapter shows how a substantial amount of energy-saving is obtained with the proposed system.

Acknowledgments

"Coming together is a beginning. Keeping together is progress. Working together is success."

—Henry Ford

With the diversity of contemporary research work even in the same domain, like Engineering, a huge overlap exists that needs to be tapped to have a holistic understanding to impart learning more efficiently. In this context, it is important to have a platform where such an "overlapping" knowledge-base can be amalgamated. This has been the backbone to conceptualize, implement, and accomplish this book project. To build and nurture such a platform, one does not really need to look far to explore its inseparable constituents. Especially for me, the variety in the "Electronics" field led to an opportunity to bring the best research output from our department, Electrical and Electronics Engineering at BITS-Pilani, Hyderabad Campus. I must mention that all the distinguished authors and coauthors have rendered to full, timely, and endearing support to create this platform.

The journey began when, during the COVID-19 lockdown, I received an invitation note from the publisher. After sharing this idea with departmental colleagues, I received a very positive response from the authors and subsequently from the publisher. Everything followed, and all the stakeholders, such as the publisher, authors, and assistant editors, rendered their prompt and treasured support to accomplish this book project.

First and foremost, I am extremely grateful to all the authors and coauthors, all my valued colleagues, who contributed to this project by not only sharing some of their best and cutting-edge research output but also having so much patience to be coherent during the process to develop this book. Without the constant and intelligible support from the 33 authors, it was certainly impossible to even think about realizing this book. I am really lost for words to thank all the authors for being outstanding on-stage players.

"What you don't see backstage is what really controls the show," as someone quoted, has been true in this book project as well. I am grateful to my friend, Mr. Amit Balooni for his invaluable suggestions on articulating various parts of the book. Three of my PhD students, Mr. Prakash Rewatkar, Mr. Madhusudan Kulkarni, and Mr. Sohan Dudala, played the role of Assistant Editors to ensure that the all the book chapters were as per the guidelines set by the publisher. Prakash helped me in performing technical check pertaining to all the chapter in a speedy and professional manner. Madhusudan supported me in checking the formatting and creating the structure to efficiently communicate with the authors. Sohan helped in designing the cover print-art and other formatting issues.

A journey from creating a home from a house has interwoven and conjoined co-passengers, called family members. Pooja Agarwal Goel, my wife, has always motivated and inspired me to do this project. Her "gentle" push has always kept me encouraged to continue with the pace to accomplish this book. Shashwat Goel and

Prisha Goel, our wonderful children, have always shown a lot of endurance and love during this project, and were my "silent" stimuli.

I also acknowledge and thank all of my friends, colleagues, and students who have always enthused me on various aspects of my professional life. Finally, without acknowledging and thanking the publisher, Taylor & Francis, this story would remain incomplete. I thank the publisher for accepting this unique book project and providing all of the support as and when it was necessary.

Now that the book will be available for the readers, I urge them to be in touch with the authors for any question and concerns. Let's keep Connecting, Collaborating, Creating, and Conceiving for better science leading to diversity toward Applied Engineering with a singular objective—a better world today and tomorrow.

About the Editor

Dr. Sanket Goel has been an Associate Professor with the Department of Electrical and Electronics Engineering BITS-Pilani, Hyderabad campus, since 2015. During his tenure, he headed the EEE Department at BITS-Pilani (2017–2020). Prior to this, he led the R&D department and was an associate professor at the University of Petroleum & Energy Studies (UPES), Dehradun, India (2011–2015).

Dr. Goel did his BSc (H-Physics) from the Ramjas College, Delhi University; MSc (Physics) from IIT Delhi; PhD (Electrical Engineering) from the University of Alberta, Canada, on NSERC fellowship; and an MBA in International Business from Amity University in 1998, 2000, 2006, and 2012, respectively. He has worked with two Indian national labs: Institute of Plasma Research, Gandhinagar (2000–2001) and DEBEL-DRDO, Bangalore (2006). As an NIH fellow, Sanket did his postdoctoral work at the Stanford University, United States (2006–2008), and worked as a Principal Investigator at A*STAR, Singapore (2008–2001).

His current research interests are MEMS, Microfluidics, Nanotechnology, Materials and Devices for Energy, Biochemical and Biomedical Applications, Science Policy, and Innovation and Entrepreneurship. As a Principal Investigator, Sanket has implemented several funded projects (from DRDO, DSIR, DST, ICMR, ISRO, MNRE, Government of India; UNESCO; European Commission) and has collaborated with various groups in India and abroad.

During the course of his career, Dr. Goel has won several awards, including the Dr. C R Mitra Best Faculty Award (2021), the Fulbright-Nehru fellowship (2015), the DST Young Scientist Award (2013), the American Electrochemical Society's Best student's paper award (2005), and the University of Alberta PhD thesis award (2005). As of March 2021, he has more than 210 publications and 12 patents (1 US and 9 Indian) to his credits. He has delivered more than 70 invited talks and guided/guiding 24 PhD and several Masters and Bachelors students.

Dr. Goel is a Senior Member, IEEE; Life Member, Institute of Smart Systems and Structures; Life Member, Indian Society of Electrochemical Chemistry. Currently, he is an Associate Editor of *IEEE Transactions on NanoBioscience*, *IEEE Sensors Journal*, *IEEE Access*, *Applied Nanoscience*, and guest editor of *Special Issue in Sensors*: "3D Printed Microfluidic Devices." He also serves as a Visiting Associate Professor with UiT, The Arctic University of Norway.

List of Contributors

Syed Ershad Ahmed
Birla Institute of Technology and
 Science (BITS) Pilani
Hyderabad Campus
Hyderabad, India

A. Ajitha
Birla Institute of Technology and
 Science (BITS) Pilani
Hyderabad Campus
Hyderabad, India

Khairunnisa Amreen
Birla Institute of Technology and
 Science (BITS) Pilani
Hyderabad Campus
Hyderabad, India

Ankur Bhattacharjee
Birla Institute of Technology and
 Science (BITS) Pilani
Hyderabad Campus
Hyderabad, India

Ayantika Chatterjee
Birla Institute of Technology and
 Science (BITS) Pilani
Hyderabad Campus
Hyderabad, India

Surya Shankar Dan
Birla Institute of Technology and
 Science (BITS) Pilani
Hyderabad Campus
Hyderabad, India

B. Dheerendranath
Birla Institute of Technology and
 Science (BITS) Pilani
Hyderabad Campus
Hyderabad, India

Challa Santhi Durganjali
Birla Institute of Technology and
 Science (BITS) Pilani
Hyderabad Campus
Hyderabad, India

Samit Kumar Ghosh
Birla Institute of Technology and
 Science (BITS) Pilani
Hyderabad Campus
Hyderabad, India

Sanket Goel
Birla Institute of Technology and
 Science (BITS) Pilani
Hyderabad Campus
Hyderabad, India

Sayan Kanungo
Birla Institute of Technology and
 Science (BITS) Pilani
Hyderabad Campus
Hyderabad, India

C. Kesavadas
Birla Institute of Technology and
 Science (BITS) Pilani
Hyderabad Campus
Hyderabad, India

Madhusudan B. Kulkarni
Birla Institute of Technology and
 Science (BITS) Pilani
Hyderabad Campus
Hyderabad, India

U. Anil Kumar
Birla Institute of Technology and
 Science (BITS) Pilani
Hyderabad Campus
Hyderabad, India

Renuka Loka
Birla Institute of Technology and
   Science (BITS) Pilani
Hyderabad Campus
Hyderabad, India

Balasubramanian Malayappan
Birla Institute of Technology and
   Science (BITS) Pilani
Hyderabad Campus
Hyderabad, India

Saroj Mondal
Birla Institute of Technology and
   Science (BITS) Pilani
Hyderabad Campus
Hyderabad, India

Alivelu M Parimi
Birla Institute of Technology and
   Science (BITS) Pilani
Hyderabad Campus
Hyderabad, India

Prasant Kumar Pattnaik
Birla Institute of Technology and
   Science (BITS) Pilani
Hyderabad Campus
Hyderabad, India

R. N. Ponnalagu
Birla Institute of Technology and
   Science (BITS) Pilani
Hyderabad Campus
Hyderabad, India

P Shambhu Prasad
Birla Institute of Technology and
   Science (BITS) Pilani
Hyderabad Campus
Hyderabad, India

Nawin Ra
Birla Institute of Technology and
   Science (BITS) Pilani
Hyderabad Campus
Hyderabad, India

Sudha Radhika
Birla Institute of Technology and
   Science (BITS) Pilani
Hyderabad Campus
Hyderabad, India

Venkateswaran Rajagopalan
Birla Institute of Technology and
   Science (BITS) Pilani
Hyderabad Campus
Hyderabad, India

BVVSN Prabhakar Rao
Birla Institute of Technology and
   Science (BITS) Pilani
Hyderabad Campus
Hyderabad, India

Mary Salve
Birla Institute of Technology and
   Science (BITS) Pilani
Hyderabad Campus
Hyderabad, India

Puneet Singh
Birla Institute of Technology and
   Science (BITS) Pilani
Hyderabad Campus
Hyderabad, India

Debapriya Som
Birla Institute of Technology and
   Science (BITS) Pilani
Hyderabad Campus
Hyderabad, India

Rajesh Kumar Tripathy
Birla Institute of Technology and
   Science (BITS) Pilani
Hyderabad Campus
Hyderabad, India

Srijan Trivedi
Birla Institute of Technology and
   Science (BITS) Pilani
Hyderabad Campus
Hyderabad, India

Amar Kumar Verma
Birla Institute of Technology and
   Science (BITS) Pilani
Hyderabad Campus
Hyderabad, India

Jaju Vedant Vinod
Birla Institute of Technology and
   Science (BITS) Pilani
Hyderabad Campus
Hyderabad, India

Sanjay Vidhyadharan
Birla Institute of Technology and
   Science (BITS) Pilani
Hyderabad Campus
Hyderabad, India

P. Yogeeswari
Birla Institute of Technology and
   Science (BITS) Pilani
Hyderabad Campus
Hyderabad, India

1 Van der Waals (vdW) Heterostructures Based on Transition Metal Dichalcogenides (TMD)

Current Status and Prospects in Broadband Photodetector Applications

Debapriya Som, Srijan Trivedi, Ayantika Chatterjee, and Sayan Kanungo

CONTENTS

1.1 Introduction: Background 1
1.2 General Overview of Photodetector 2
1.3 TMD/TMD vdW Heterostructure-based Photodetector 4
1.4 TMD/Non-TMD vdW Heterostructure 7
1.5 Summary 11
References 13

1.1 INTRODUCTION: BACKGROUND

The reliable detection of electromagnetic radiations over a broad spectral range is of primary interest in various fields including biomedical imaging, optical communication, environmental monitoring, food and manufacturing process monitoring, and national security and defense applications. The broadband photodetectors convert the incident electromagnetic radiations over a wide range of wavelengths into electrical signals for detection. The detection mechanism is governed by the interactions of electromagnetic radiation with photoabsorbing material and, subsequently, the choice of such material principally determines the performance of photodetectors (Konstantatos 2018; Mu, Xiang, and Liu 2017). This scenario motivated active explorations of different semiconducting materials for broadband photodetector applications. In this context, two-dimensional (2D) materials with few atomic layers thickness have shown particular promise for photodetector applications owing to their large surface area, pristine surfaces, tunable electronic properties, and mechanical flexibility. Typically, the spatial confinements of charge carriers in out-plane directions lead to high optical response and a long lifetime for photogenerated electrons and holes. Subsequently, a wide range of 2D materials is realized and eventually exploited for ultrafast and ultrasensitive detection of lights in terahertz, infrared, visible, and ultraviolet frequency ranges. In this line, transition metal dichalcogenide (TMD) has emerged as one of the potential 2D semiconductors for optoelectronic applications owing to the presence of suitable energy bandgaps (1–2 eV) that can be tuned by changing the layer thickness. Moreover, TMDs exhibit transition from an indirect to direct bandgap between their multilayers and monolayer configurations. However, like other 2D crystalline semiconductors, TMDs also suffer from low optical absorption and subsequent small photocarrier generation owing to their atomic-scale thicknesses. At the same time, the mobility of TMDs is considerably lower than Graphene (Gr), leading to a relatively slower charge transfer (Mu, Xiang, and Liu 2017; Rao et al. 2019). Therefore, significant efforts have been observed for designing novel 2D materials with superior light absorption and mobility to optimize the photoresponse.

The weak van der Waals (vdW) interaction between individual layers of 2D materials allows the realization of designer heterostructure materials by vertically stacking one or a few layers of two dissimilar 2D materials. These structures are usually referred to as vdW hetero structures having atomically sharp interfaces where emergent properties can be observed. These heterojunction materials have offered an alternative approach to enhance the overall photoresponse of the photodetector because it can exploit the combined optoelectronic properties of individual materials as well as can offer effective separation of photo-generated charge carriers due to desirable band alignment and internal electric field at the heterojunction (Cheng et al., 2014; Rao et al., 2019; Sun et al. 2017). The field of 2D vdW heterostructure-based photodetector design has considerably progressed in recent years, where TMDs remain a principal constituent of such reported vdW heterostructures. However, to the best of authors' knowledge, to date, no exclusive review is available on TMD-based vdW heterostructure photodetectors. Subsequently, this chapter presents a comprehensive summary of TMD-based vdW heterostructure photodetectors, emphasizing the design aspects.

1.2 GENERAL OVERVIEW OF PHOTODETECTOR

Photodetectors are solid-state optoelectronic devices that can convert electromagnetic radiation into measurable electrical signals for the reliable detection of the former. Since the light-matter interactions in solids have manifold consequences, the different physical mechanisms lead to the modulation in the electrical properties of solids. Therefore, based on the underlying detection mechanism, the photodetectors can be broadly classified as thermal and photon photodetectors (Mu, Xiang, and Liu 2017). In thermal photodetectors, the change in electrical properties is due to either bolometric or photothermoelectric effects that exploit the thermal effects of light absorption. The absorption of light increases the lattice temperature and thereby the phonons (quanta of lattice vibration). The subsequent change in the mobility of charge carriers modulates the photodetector current, which is termed as bolometric effects and is shown in Figure 1.1A. On the other hand, the photothermoelectric

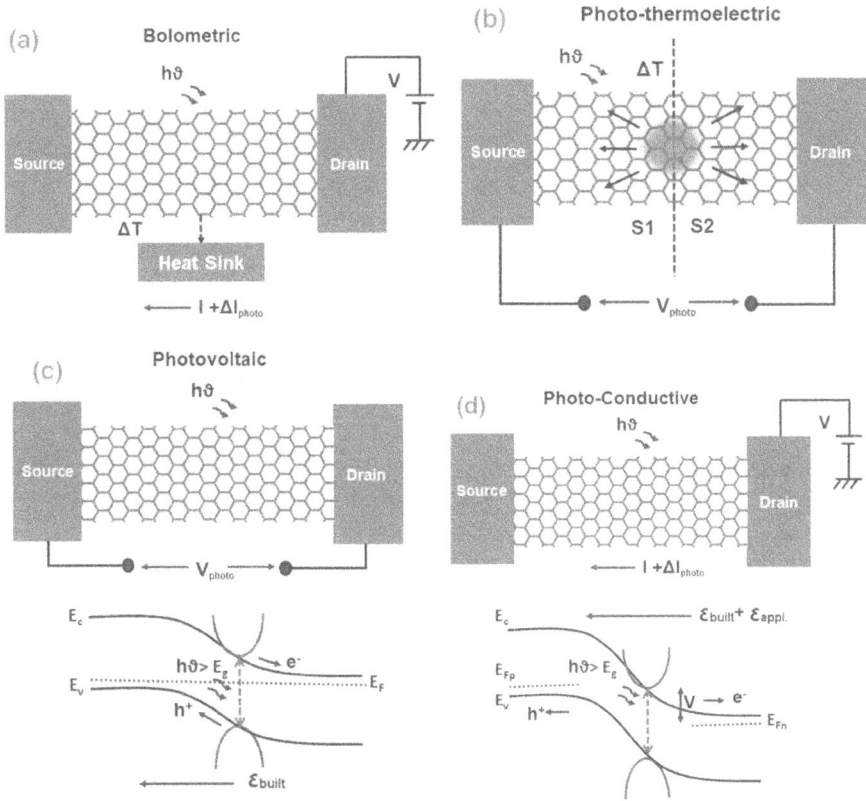

FIGURE 1.1 Different physical mechanisms governing the photodetection: (A) bolometric; (B) photothermoelectric; (C) photovoltaic; and (D) photoconductive.

effect exploits local temperature gradient associated with light absorption to generate a voltage between the electrodes due to the spatial difference in the Seebeck coefficient (represents the temperature difference induced thermoelectric voltage in materials) as depicted in Figure 1.1B. In photon photodetectors, the photogenerated charge carrier changes the electrical properties based on either photovoltaic or photoconductive effects. The photovoltaic effect leads to a photovoltage generation in the contact electrodes of photodetector. If the photon energy of the incident radiation is higher than the bandgap of light-absorbing materials, it generates an electron-hole pair that can move under the influence of a built-in electric field that exists in the junctions and is eventually diffused to the electrodes as shown in Figure 1.1C. However, under the externally applied bias in the electrodes, the photogenerated carriers change the charge carrier density and subsequently, the terminal current. This is known as the photoconductive effect and illustrated in Figure1.1D (Mu, Xiang, and Liu 2017; Rao et al. 2019).

The performance of different photodetectors is benchmarked on the basis of a well-defined figure of merits. The key figure of merits of photodetections is as follows (Rao et al. 2019):

Responsivity (**R**) [unit A/W]: Defined as the ratio between the photogenerated current (I_{ph}) and the incident power (P_0) of the electromagnetic radiation

$$R = I_{ph} / P_0 \tag{1.1}$$

External Quantum Efficiency (EQE) [unitless]: Defined as the ratio of number of photogenerated electron-hole pairs contributed to the photocurrent (I_{ph}) to the incident photon flux (φ_{in})

$$R = I_{ph} / q\,\phi_{in} \tag{1.2}$$

Bandwidth (BW) [unit Hz]: Defined as the frequency of modulation for incident electromagnetic radiation that corresponds to a signal intensity typically 3 dB lower than that of continuous illumination.

Noise Equivalent Power (NEP) [unit W. (Hz)$^{-1/2}$]: Defined as the minimum illumination power to achieve a unity signal-to-noise ratio at 1 Hz bandwidth.

Detectivity (**D***) [unit cm. (Hz)$^{1/2}$/W]: Definition of detectivity involves NEP, area (A), and BW

$$D^* = \left(A.BW\right)^{1/2} / NEP \tag{1.3}$$

Response Time (RT) [unit sec]: This is defined as the time required to change the signal from 10% to 90% in response to incident light intensity modulation.

1.3   TMD/TMD vdW HETEROSTRUCTURE-BASED PHOTODETECTOR

TMDs are represented as MX_2, where one transition metal atom, M (Mo, W, Zr, Hf, Re) is sandwiched covalently between two same chalcogen atoms, X (S, Se, Te) and are arranged in a single atomic layer as illustrated in Figure 1.2A and 1.2B. The VIB MX_2 (M = Mo, W) exhibits natural bandgaps in 2H-phase, whereas VIIB MX_2 (M = Re) and IVB MX_2 (M = Zr, Hf) shows natural bandgaps in 1T-phase. In general, the energy bandgaps of TMDs gradually increases with reducing layer numbers. The VIB MX_2 shows direct bandgaps in their monolayer configurations. However, the IVB MX_2 and VIIB MX_2 remain indirect and direct bandgap semiconductors, respectively, at all layer thicknesses (Mu, Xiang, and Liu 2017).

The vdW heterostructures based on dissimilar TMDs have emerged as a potential aspirant for photodetector applications. The distinctive difference in energy bandgaps and work functions of different TMDs lead to the designer type II band alignments and suitable built-in electric fields at the hetero-interfaces. Since the bandgap and work function of individual TMDs change with the number of layers, the band alignment at the heterointerface can be further tuned by carefully considering the number of layers for individual TMDs in vdW heterostructures. For a well-designed type-II heterojunction, the photogenerated electrons and holes are transferred at the conduction band minima (CBM) and valance band maxima (VBM) of two n-type and p-type TMDs, respectively. This leads to spatial separation of the photogenerated

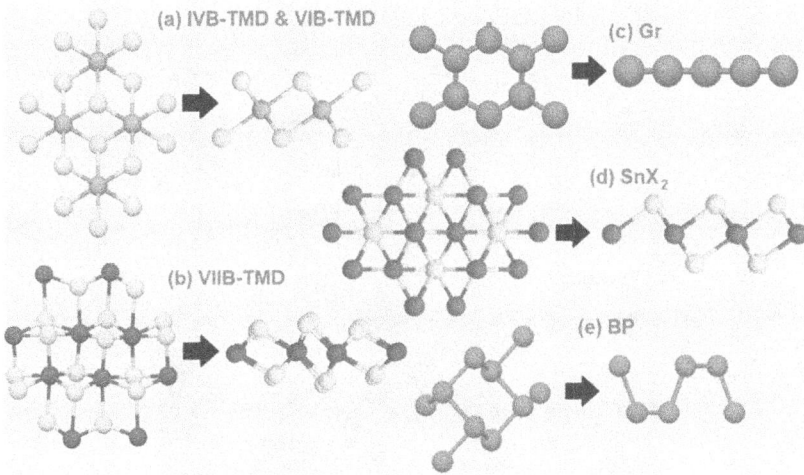

FIGURE 1.2   The crystal structures of different 2D materials: (A) IVB-TMD and VIB-TMD; (B) VIIB-TMD; (C) Gr; (D) SnX$_2$; and (E) BP.

electrons and holes, resulting in a longer lifetime (smaller recombination rate) as well as a larger intrinsic electric field. Subsequently, a superior photoresponse in both photovoltaic (without applied bias) and photoconductive (under applied bias) modes can be observed. At the same time, the light-matter interaction at the vdW interface exhibits a unique feature, where absorption of incident radiations corresponding to a bandgap lesser than that of constituent TMDs can be observed. This is attributed to the interlayer photoexcitation that involves reduced energy differences between electrons and holes situating at the spatially separated CBM and VBM in type-II heterointerfaces. The vdW heterostructures also demonstrate strong photoresponse for the incident radiations corresponding to the natural bandgaps of the constituting TMDs. Subsequently, the TMD-based vdW heterostructures can detect electromagnetic radiation over a broad range of spectrum making them a highly suitable candidate for broadband photodetector application as indicated in Figure 1.3.

In this context, photodiode and field effect phototransistor architectures are primarily adopted for TMD-based vdW photodetectors. The photodiode exploits a rectifying heterojunction formation between two dissimilar TMDs and often introduce metal electrodes with specific work-functions for inducing electronic doping in TMDs by creating a suitable electron or hole blocking barrier at the metal-TMD interfaces. The phototransistor architecture usually incorporates a back-gate in the photodiode architecture. This offers additional flexibility of modulating the band alignment (electrostatics) of the heterojunction. Another notable design strategy for a TMD-based vdW photodetector exploits light absorption at the semiconductor substrates and subsequent charge transfer into the TMDs for enhancing photogeneration of the carrier as well as broadening the operating range of the spectrum.

Cheng et al. (2014) have realized a photodiode using vdW heterojunction of monolayer WSe$_2$ (p-type) and few-layer MoS$_2$ (n-type). A vdW heterojunction of few-layer WSe$_2$ and monolayer MoS$_2$ is reported by Sun et al. (2017). The phototransistor demonstrates responsivity of 17.8 A/W in 660 nm wavelength at zero applied gate bias

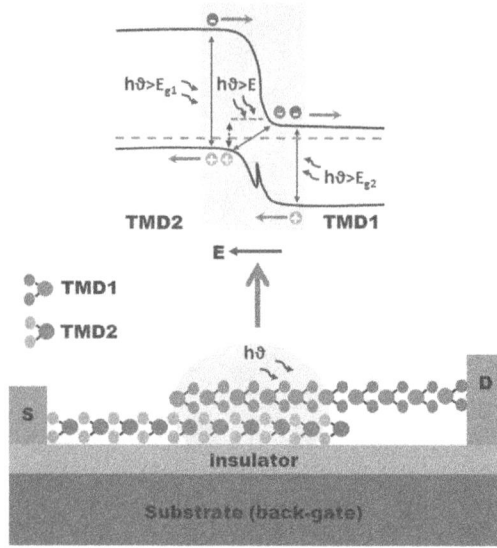

FIGURE 1.3 Schematics of TMD/TMD vdW Photodetectors and the associated photocarrier generation and separation mechanisms.

and responsivity of 0.3 A/W in 850 nm wavelength for a large applied gate bias. Lee et al. (2018) have improvised a few-layer WSe_2- and MoS_2-based vdW photodiode by introducing a platinum electrode underneath WSe_2 to reduce the series resistance contributed by the quasi-neutral region.

Huo et al. (2014) demonstrated few-layer WS_2 (p-type) and MoS_2 (n-type) for realizing a vdW heterojunction phototransistor with strong photoresponse at the photovoltaic mode. Subsequently, the phototransistor exhibits 23% and 75% improvements in responsivity over their WS_2 and MoS_2 counterparts, respectively. Similar few-layer WS_2 and MoS_2-based vdW heterojunction photodiode have been realized by Xue et al. (2016) that exhibits a responsivity of 2.3 A/W at a 450 nm wavelength.

K. Zhang et al. (2016) reported a vdW heterojunction photodiode based on monolayer $MoTe_2$ (p-type) and MoS_2 (n-type) with an effective interlayer photoexcitation bandgap around 0.66 eV. This offers strong photoresponse at a 1,550 nm wavelength. Chen et al. (2018) introduced a few-layer $MoTe_2$ and MoS_2 vdW photodiode capable of detecting a broad electromagnetic spectrum of 750–1,200 nm. The larger thickness of the overlapped region between $MoTe_2$ and MoS_2 contributed to a larger internal electric field that enhances the performance in photovoltaic mode.

Patel et al. (2020) devised a photodiode based on nanocrystal thin films of $MoSe_2$ (n-type) and WSe_2 (p-type) embedded on p-doped silicon substrate, where silver electrodes are deposited on top of $MoSe_2$ and the bottom of silicon. The subsequent Si/WSe_2 and WSe_2/$MoSe_2$ heterointerfaces offer effective charge separations between the electrodes. Xue et al. (2018) put forward a phototransistor based on a vdW heterojunction of few-layer $MoSe_2$ and WSe_2 that can operate on 532 nm, 980 nm, and 1,550 nm wavelengths. Gao et al. (2019) introduced a novel phototransistor architecture based on a few-layer vdW heterostructure of $MoSe_2$ and WSe_2 with n+

and n stacked 4H-SiC substrate. The 4H-SiC exhibits deep energy-level near the mid-gap, and the position of this level is in between the CBM of $MoSe_2$ and CBM of WSe_2. Subsequently, in a WSe_2 (p-type)/4H-SiC (n-type) junction, the photogenerated electrons in WSe_2 are transferred into the mid-gap level of 4H-SiC. Whereas, in $MoSe_2$ (n+-type)/4H-SiC (n-type) junction, the photogenerated and transferred electrons in the mid-gap level of 4H-SiC are transferred into the $MoSe_2$. This electron injection mechanism is exploited to achieve a responsivity of 7.17 A/W for a 532 nm wavelength in photoconductive mode.

Chen et al. (2020) introduced few-layer WSe_2 (p-type)/$MoTe_2$ (n-type) based vdW phototransistor that exhibits strong photoresponse in both photovoltaic and photoconductive modes. The phototransistor detects electromagnetic radiation from 400 nm to 1,700 nm wavelengths with strong photoresponse between 850 nm and 1,200 nm. H. Luo et al. (2019) reported a vdW phototransistor of few-layer $MoTe_2$ (p-type) and $MoSe_2$ (n-type). The photodetector exhibits a responsivity of 1.5 A/W in photovoltaic mode with a suitable back-gate bias.

Cho et al. (2017) demonstrated a vdW heterojunction-based phototransistor using few-layer ReS_2 (n-type) and $ReSe_2$ (n+-type). The phototransistor exhibits a good photoresponse between 400 nm and 550 nm wavelengths with an optimum responsivity of 0.021 A/W. A similar approach has been adopted by Jo et al. (2018) and successfully detected incident radiation between 410 nm and 1,310 nm wavelengths, using a ReS_2/$ReSe_2$ vdW phototransistor. The phototransistor demonstrated an extremely high photoresponsivity ($>10^3$ A/W) in photoconductive mode. M. Luo et al. (2019) have introduced a vdW phototransistor based on few-layer ReS_2 and $MoTe_2$ operating at a 520 nm wavelength. Varghese et al. (2020) reported successful photodetection between 950 nm and 1,250 nm wavelengths using a vdW photodiode based on monolayer WSe_2 and few-layer ReS_2.

On the other hand, it should be noted that different TMD/TMD vdW heterostructures are theoretically investigated and their potential for photodetector applications are assessed from their optoelectronic properties. In this context, the different promising type-II vdW heterointerfaces are proposed based on $MoSe_2$/$HfSe_2$ (Aretouli et al. 2015), ReS_2/MoS_2 (Saeed et al. 2020), ReS_2/$MoSe_2$, ReS_2/WSe_2, ReS_2/$MoTe_2$ (Saha, Varghese, and Lodha 2020). The experimental realizations of these proposed heterojunctions are expected to further widen the horizon of vdW photodetectors.

1.4 TMD/NON-TMD vdW HETEROSTRUCTURE

The exploration of photodetectors based on vdW heterostructures between two dissimilar TMDs also encouraged vdW photodetector designs based on TMD and non-TMD 2D materials. The incorporation of non-TMD 2D materials with TMD materials offers a vast array of type-II vdW heterointerfaces with diversified optoelectronic properties that can be effectively exploited for broadband photodiode and phototransistor designs. Among such photodetectors, the TMD/Gr, TMD/Black Phosphorous (BP), and TMD/SnX_2 have shown particular promise.

The Gr is a single layer of graphite arranged in a hexagonal honeycomb structure as shown in Figure 1.2C. Gr shows a semi-metallic nature because the CBM and VBM touch each other at the Dirac point. Also, the linear dispersion relation at CBM/VBM indicates that the electrons and holes behave like massless quasi-relativistic particles leading to an ultrafast charge transfer in Gr. Due to the absence of a natural bandgap in Gr, it exhibits an ultrabroadband light absorption from the ultraviolet to terahertz range of the spectrum. However, the atomic layer thickness of Gr limits the light absorption and at the same time, the absence of a bandgap drastically reduces the lifetime of photo generated carriers (Konstantatos 2018; Rao et al. 2019). In this context, the Gr/TMD vdW heterostructures have emerged as a potential alternative to mitigate the limitations of Gr while preserving its advantages. In this context, the bolometric mechanism that is prominent in Gr can act in unison with the photoconductive mechanism and thereby increases the overall photodetection capability of Gr/TMD vdW photodetectors. Furthermore, the semi-metallic nature of Gr often leads to Schottky barrier formation with TMDs, where the photodetection is enhanced by the internal photoemission process. In the internal photoemission, the incident light energy is higher than the Schottky barrier height and subsequently, the majority carriers near the Schottky junction can have adequate energy to overcome the barrier as shown in Figure 1.4.

Zhang et al. (2014) introduced a vdW phototransistor based on monolayer Gr/MoS$_2$, where the Gr is acting as a charge transport layer and subsequently deposited on MoS$_2$ and contacted. The photogenerated charge carriers in MoS$_2$ enter in Gr under the combined influence of a built-in and applied (back-gate bias) electric field in the vertical direction across the heterojunction, and efficiently separated by the lateral electric field due to the applied contact potential. The phototransistor exhibits an extremely high responsivity in the order of 10^7 A/W. De Fazio et al. (2016) presented a monolayer Gr/MoS$_2$-based vdW phototransistor on a flexible polyethylene terephthalate substrate that exhibits a significantly higher responsivity of 570 A/W at

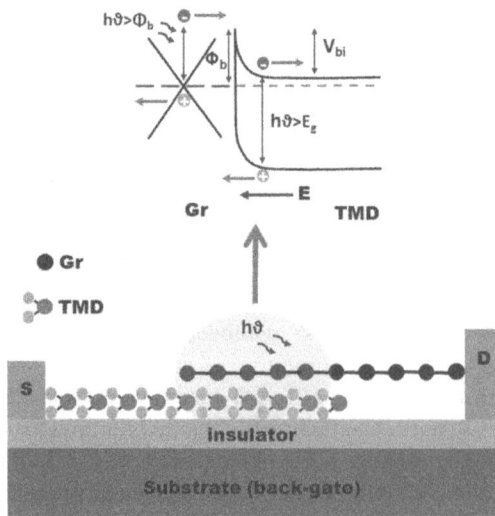

FIGURE 1.4 Schematics of Gr/TMD Schottky vdW Photodetectors and the associated photocarrier generation and separation mechanisms.

a 642 nm wavelength. Vabbina et al. (2015) devised a Schottky photodiode based on a vdW heterojunction between a few layers of MoS_2 (p-type) and Gr, where extremely fast hole transfer can be observed from Gr to MoS_2. The photodiode demonstrates a strong photoresponse between 400 nm and 1,500 nm wavelengths with a maximum responsivity of 1.26 A/W at 1,440 nm. Lan et al. (2017) demonstrated a monolayer Gr (p-type)/WS_2 based vdW phototransistor, where the built-in field-induced fast transfer of photogenerated electrons from WS_2 to Gr can be observed. This leads to a photogating effect contributed by the partially trapped holes in the WS_2 layer. Subsequently, the phototransistor shows photodetection over the 340 nm to 680 nm range of the spectrum with a maximum responsivity of 950 A/W at a 405 nm wavelength. Liu et al. (2019) introduced a vdW photodiode based on a few-layer $MoSe_2$ and monolayer Gr operating in photovoltaic mode. The photodiode can successfully detect electromagnetic radiation between 450 nm and 900 nm wavelengths. Kuiri et al. (2016) demonstrated a vdW heterojunction of bilayer Gr and few-layer $MoTe_2$ for phototransistor design with a floating top-gate introduced through the ionic liquid. The vdW phototransistor shows superior photoresponse over its $MoTe_2$-based counterpart. Yu et al. (2017) reported a phototransistor based on a vdW heterojunction between monolayer Gr and few-layer $MoTe_2$ acting as charge transport and photoabsorption layers, respectively. The photogenerated electrons are partially rapped into the localized sates of $MoTe_2$, leading to a photogating effect that eventually contributes to the strong photoresponse (maximum responsivity 971 A/W) operating between 514 nm and 1064 nm wavelengths. Kang et al. (2018) demonstrated a vdW heterojunction between a few-layer Gr of ReS_2 acting as charge transport and photoabsorption layers, respectively. The n-type Schottky junction restricts electron transfer from ReS_2 to Gr and that leads to photogating effects by the trapped electrons. Subsequently, the phototransistor exhibits a responsivity of 7×10^5 A/W.

The incorporation of two vdW heterointerfaces between Gr and dissimilar TMDs or one TMD and one non-TMD leads to novel functionalities in the heterointerfaces that has been successfully exploited to achieve highly promising photoresponse characteristics. In this line, Murali et al. (2019) demonstrated a few-layer Gr/WS_2/MoS_2 vdW heterojunction where the Gr is exploited as both light absorption and carrier transport layers. In this architecture, the photogenerated electrons are partially transferred into MoS_2 and experience a quantum confinement effect due to the conduction band offsets with neighboring WS_2 and SiO_2 (substrate) layers. This leads to a photogating effect with incident radiation that enhances the overall conductivity modulation. In absence of the incident radiation, the photogating effect quickly diminishes as electrons tunnels from MoS_2 to Gr by the influence of the built-in electric field in vdW heterointerface. Correspondingly, the phototransistor demonstrates a responsivity of 4.4×10^6 A/W with a fast transient response and high signal-to-noise ratio. Long et al. (2016) devised a vdW heterojunction by sandwiching Gr between few-layer MoS_2 (n-type) and WSe_2 (p-type), where the Dirac point of Gr lies between the CBM of MoS_2 and VBM of WSe_2. Subsequently, the photogenerated electrons and holes from Gr are transferred to MoS_2 and WSe_2, respectively. The phototransistor based on this vdW heterojunction exhibits a strong photoresponse between 400 nm and 1,500 nm wavelengths with a maximum responsivity of 7×10^4 A/W. Massicotte et al. (2016) introduced a vdW phototransistor based on a few-layer WSe_2 sandwiched between monolayer Gr encapsulated within 2D insulator layers of hexagonal boron nitride (hBN). The structural compatibility of Gr with hBN allows a high-quality interface and thereby an extremely high charge mobility in the encapsulated Gr. For a suitable applied bias, the photogenerated carriers in WSe_2 are separately transferred into two neighboring Gr layers and are transferred into the electrodes. The phototransistor exhibits an ultrafast photoresponse between wavelengths of 500 nm and 750 nm. A similar approach has been reported by Zhang et al. (2017) in their vdW phototransistor of a few-layer $MoTe_2$ sandwiched in monolayer Gr. The suitable tuning of the two Schottky junctions of the phototransistor allows photodetection over a 473 nm to 1,064 nm range of the spectrum. Xiong et al. (2019) demonstrated the all-in-fiber photodetector using a vdW heterostructure of few-layer $Gr/MoS_2/WS_2$ on the end of an optical fiber. The photoconductive and bolometric mechanisms acting in synergy enhances the overall performance of the photodiode. The photodiode exhibits a strong photoresponse between a 400 nm and 1600 nm wavelength range of the spectrum with an optimum responsivity of 6.6×10^7 A/W. Tan et al. (2017) proposed a vdW phototransistor by sandwiching a pair of Gr electrodes between monolayer WS_2 and MoS_2. The efficient interlayer photocarrier separation between WS_2 and MoS_2 and subsequent charge transfer in Gr lead to a maximum responsivity of 2,340 A/W.

Tin dichalcogenides, SnX_2 (X = S, Se) exhibit indirect bandgaps in their 1T phase and have a hexagonal lattice structure as indicated in Figure 1.2D. Owing to the structural symmetry of SnX_2 with TMD, different SnX_2/TMD vdW photodetectors are reported that exhibit type-II band alignments with the underlying photodetection mechanism that is very similar to that of TMD/TMD vdW photodetectors. Zhou et al. (2017) demonstrated a vdW photodiode based on monolayer MoS_2 (p-type) and few-layer $SnSe_2$ (n-type) that exhibits responsivity of 9.1×10^3 A/W at 500 nm wavelength. Murali and Majumdar (2018) reported a vdW heterojunction of few-layer WSe_2 (p-type) and $SnSe_2$ (n-type). The photodiode achieved a responsivity of 1,100 A/W at a 785 nm wavelength. Similar vdW phototransistors of few-layer WSe_2 and $SnSe_2$ are reported by Xue et al. (2019) that can successfully detect between a spectrum range of 532 nm and 1,550 nm. The phototransistor demonstrated a responsivity of 588 A/W at a 532 nm wavelength. A. Li et al. (2018) proposed a vdW heterostructure of $hBN/MoTe_2$ (p-type)$/Gr/SnS_2$ (n-type)/hBN, where the typical band alignment leads to the transfer of photogenerated carriers into SnS_2 and $MoTe_2$, respectively. The vertical built-in electric field offers an efficient special charge separation from few-layer Gr over a 300 nm to 1,600 nm range of the spectrum. The phototransistor exhibits a maximum responsivity of 2,600 A/W.

The BP shows a buckled honeycomb structure as shown in Figure 1.2E. The BP exhibits a small direct bandgap as well as small effective masses in transport direction while maintaining a large density of states. Subsequently, the suitable type-II band alignments of TMD/BP vdW heterojunctions have drawn significant research interest for photodetector design. The vdW phototransistor based on few-layer BP and monolayer MoS_2 was reported by Deng et al. (2014). The phototransistor exhibits a responsivity of 0.42 A/W at the 633 nm wavelength. Ye et al. (2016) demonstrated a vdW heterojunction of few-layer BP and MoS_2. The phototransistor is able to detect at 532 nm and 1,550 nm wavelengths with a maximum responsivity of 22.3 A/W.

Similarly, Zheng et al. (2018) reported a vdW photodiode based on few-layer BP and MoS_2 with a responsivity of 2.17 A/W at a 582 nm wavelength. Similarly, Chen et al. (2016) achieved good photoresponse using a vdW phototransistor based on few-layer BP and WSe_2. Cao et al. (2018) put forward a vdW phototransistor based on few-layer BP (p-type) and ReS_2 (n-type) that exploited strong photoadsorption in the direct bandgap materials and highly efficient photocarrier separations in the heterointerface. Subsequently, a maximum responsivity of 1.2×10^4 A/W has been achieved. Zhu et al. (2019) adopted a similar approach in their few-layer BP (p-type) and ReS_2 (n-type) based vdW photodiode that exhibits strong photoresponse over a 550 nm to 1,750 nm range of the spectrum with notable sensitivity toward polarization of light. In this context, few other TMD-based vdW heterostructures including MoS_2/reduced graphene oxide (rGO) (Kumar et al. 2018); MoS_2/GaTe (Yang et al. 2016); $MoSe_2$/InAs and WSe_2/InAs (Li et al. 2018); $MoSe_2$/Blue Phosphorous, WS_2/Blue Phosphorous, and WSe_2/Blue Phosphorous (Peng et al. 2016); and ZrS_2/g-C_3N_4 (X. Zhang et al. 2016) have shown considerable promise for broadband photodetector applications.

1.5 SUMMARY

To summarize the discussions, the performances of different TMD-based vdW photodetectors are benchmarked against their silicon (Si) and indium gallium arsenide (InGaAs) based counterparts and the detection range of the spectrum are indicated in Figure 1.5. The responsivity for these photodetectors is depicted as a function of detectivity, EQE, and response time as shown in Figure 1.6A, 1.6B, and 1.6C,

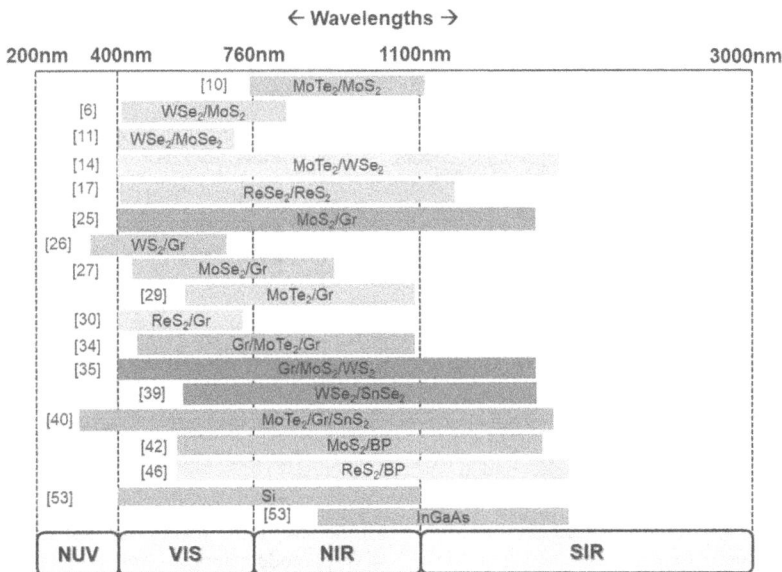

FIGURE 1.5 Photodetection range of different TMD-based vdW and Si/InGaAs photodetectors.

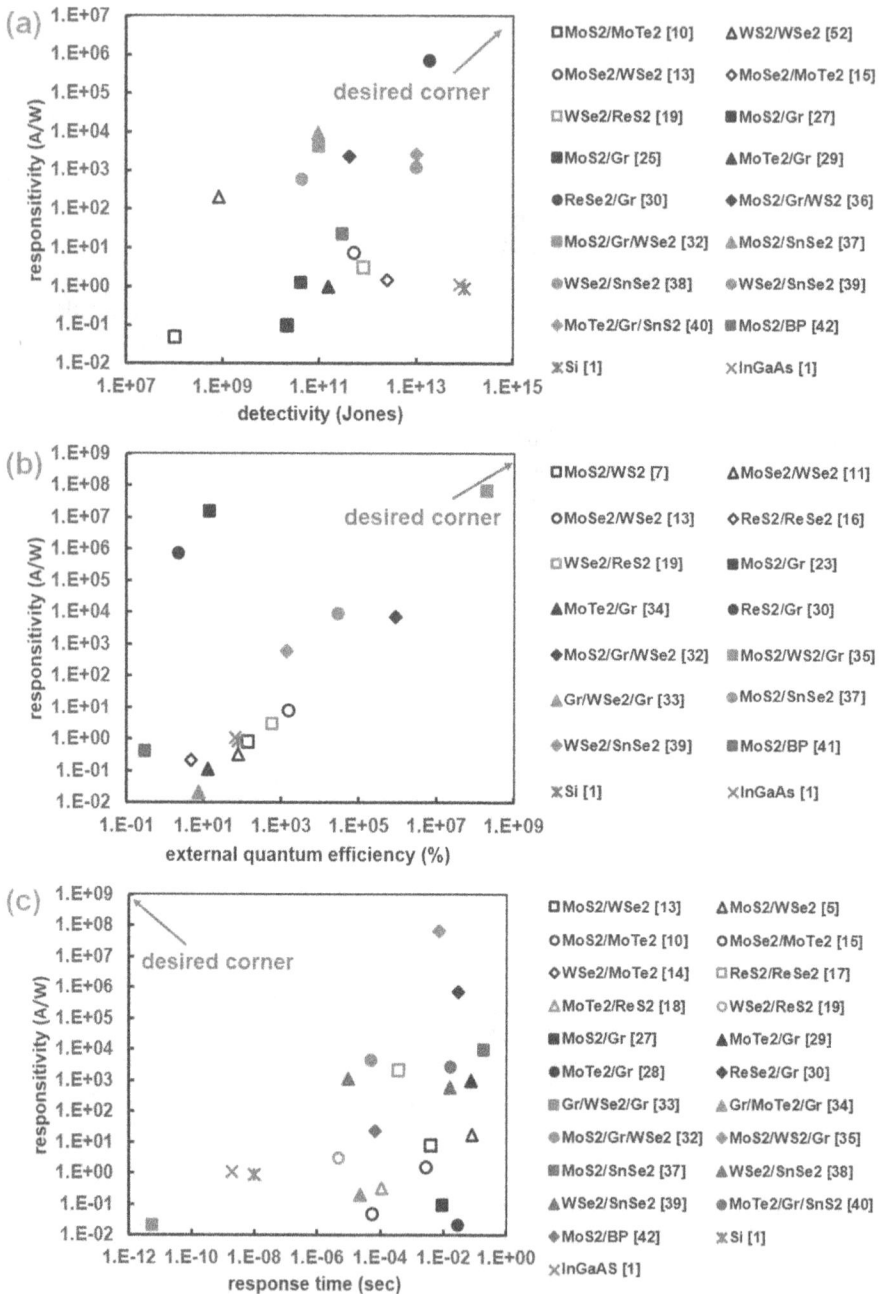

FIGURE 1.6 The reported responsivity as a function of (A) detectivity, (B) EQE, and (C) response time for different TMD-based vdW photodetectors as well as their Si and InGaAs counterparts.

respectively. It should also be noted that apart from the material specifications, the performance of any vdW heterostructure photodetectors is significantly influenced by device architecture, dielectric and electrodes specification, material synthesis, and fabrication techniques. The study indicates that, in general, the TMD-based vdW photodetectors exhibit comparable or superior responsivity, EQE, as well as a detectable wavelength range than that of Si- and InGaAs-based photodetectors. This can be attributed to the interlayer photocarrier generation as well as special charge-separation at the vdW heterointerfaces which is most effectively tuned by stacking different 2D semiconductors or semimetal (Gr) with TMDs. On the other hand, the relatively smaller mobility in TMDs results in lower response time for TMD-based vdW photodetectors. Therefore, the TMD-based vdW broadband photodetectors are typically suitable for highly sensitive photodetection applications without ultrahigh speed requirement. However, substantial research efforts are still necessary to cooptimize the detectivity and response time without hindering the responsivity of TMD-based vdW photodetectors.

REFERENCES

Aretouli, K. E., P. Tsipas, D. Tsoutsou, J. Marquez-Velasco, E. Xenogiannopoulou, S. A Giamini, E. Vassalou, N. Kelaidis, and A. Dimoulas. 2015. "Two-dimensional semiconductor $HfSe_2$ and $MoSe_2/HfSe_2$ van der Waals heterostructures by molecular beam epitaxy." *Applied Physics Letters* 106(14): 143105. https://doi.org/10.1063/1.4917422.

Cao, S., Y. Xing, J. Han, X. Luo, W. Lv, W. Lv, B. Zhang, and Z. Zeng. 2018. "Ultrahigh-photoresponsive UV photodetector based on a BP/ReS_2 heterostructure p-n diode." *Nanoscale* 10(35): 16805–11. https://doi.org/10.1039/c8nr05291c.

Chen, J., Y. Shan, Q. Wang, J. Zhu, and R. Liu. 2020. "P-type laser-doped $WSe_2/MoTe_2$ van der Waals heterostructure photodetector." *Nanotechnology* 31(29): 295201. https://doi.org/10.1088/1361-6528/ab87cd.

Chen, P., T. T. Zhang, J. Zhang, J. Xiang, H. Yu, S. Wu, X. Lu, G. Wang, F. Wen, Z. Liu, et al. 2016. "Gate tunable WSe_2-BP van der Waals heterojunction devices." *Nanoscale* 8(6): 3254–58. https://doi.org/10.1039/c5nr09218c.

Chen, Y., X. Wang, G. Wu, Z. Wang, H. Fang, T. Lin, S. Sun, H. Shen, W. Hu, J. Wang, et al. 2018. "High-performance photovoltaic detector based on $MoTe_2/MoS_2$ van der Waals heterostructure." *Small* 14(9). https://doi.org/10.1002/smll.201703293.

Cheng, R., D. Li, H. Zhou, C. Wang, A. Yin, S. Jiang, Y. Liu, Y. Chen, Y. Huang, and X. Duan. 2014 "Electroluminescence and photocurrent generation from atomically sharp WSe_2/MoS_2 heterojunction p–n diodes." *Nano Letters* 14(10): 5590–97. dx.doi.org/10.1021/nl502075n.

Cho, A.-J., S. D. Namgung, H. Kim, and J.-Y. Kwon. 2017 "Electric and photovoltaic characteristics of a multi-layer $ReS_2/ReSe_2$ heterostructure." *APL Materials* 5(7): 076101. https://doi.org/10.1063/1.4991028.

De Fazio, D., I. Goykhman, D. Yoon, M. Bruna, A. Eiden, S. Milana, U. Sassi, M. Barbone, D. Dumcenko, K. Marinov, et al. 2016. "High responsivity, large-area graphene/MoS_2 flexible photodetectors." *ACS Nano* 10(9): 8252–62. https://doi.org/10.1021/acsnano.6b05109.

Deng, Y., Z. Luo, N. J. Conrad, H. Liu, Y. Gong, S. Najmaei, P. M. Ajayan, J. Lou, X. Xu, and P. D. Ye. 2014. "Black phosphorus-monolayer MoS_2 van der Waals heterojunction p-n diode." *ACS Nano* 8(8): 8292–99. https://doi.org/10.1021/nn5027388.

Gao, W., F. Zhang, Z. Zheng, and J. Li. 2019 "Unique and tunable photodetecting performance for two-dimensional layered $MoSe_2/WSe_2$ p-n junction on the 4H-SiC substrate."

ACS Applied Materials and Interfaces 11(21): 19277–85. https://doi.org/10.1021/acsami.9b03709.

Huo, N., J. Kang, Z. Wei, S.-S. Li, J. Li, and S.-H. Wei. 2014. "Novel and enhanced optoelectronic performances of multi-layer MoS$_2$-WS$_2$ heterostructure transistors." *Advanced Functional Materials* 24(44): 7025–31. https://doi.org/10.1002/adfm.201401504.

Jo, S-H, H-W. Lee, J. Shim, K. Heo, M. Kim, Y-J. Song, and J-H. Park. 2018. "Highly efficient infrared photodetection in a gate-controllable van der Waals heterojunction with staggered bandgap alignment." *Advanced Science* 5(4): 1–9. https://doi.org/10.1002/advs.201700423.

Kang, B., Y. Kim, W. J. Yoo, and C. Lee. 2018. "Ultrahigh photoresponsive device based on ReS2/graphene heterostructure" *Small* 14(35). https://doi.org/10.1002/smll.201802593.

Konstantatos, G. 2018. "Current status and technological prospect of photodetectors based on two-dimensional materials." *Nature Communications* 9(1): 9–11. https://doi.org/10.1038/s41467-018-07643-7.

Kuiri, M., B. Chakraborty, A. Paul, S. Das, A. K. Sood, and A. Das. 2016. "Enhancing photoresponsivity using MoTe2-graphene vertical heterostructures." *Applied Physics Letters* 108: 063506. https://doi.org/10.1063/1.4941996.

Kumar, R., N. Goel, R. Raliya, P. Biswas, and M. Kumar. 2018. "High-performance photodetector based on hybrid of MoS$_2$ and reduced graphene oxide." *Nanotechnology* 29(40): 1–8. https://doi.org/10.1088/1361-6528/aad2f6.

Lan, C., C. Li, S. Wang, T. He, Z. Zhou, D. Wei, H. Guo, H. Yang, and Y. Liu. 2017. "Highly responsive and broadband photodetectors based on WS$_2$-graphene van der Waals epitaxial heterostructures." *Journal of Materials Chemistry C* 5(6): 1–24. https://doi.org/10.1039/c6tc05037a.

Lee, H. S., J. Ahn, W. Shim, S. Im, and D. K. Hwang. 2018. "2D WSe$_2$/MoS$_2$ van Waals heterojunction photodiode for visible-near infrared broadband detection." *Applied Physics Letters* 113(16): 163102. https://doi.org/10.1063/1.5042440.

Li, A., Q. Chen, P. Wang, Y. Gan, T. Qi, P. Wang, F. Tang, J. Z. Wu, R. Chen, L. Zhang, et al. 2018. "Ultrahigh-sensitive broadband photodetectors based on dielectric shielded MoTe$_2$/Graphene/SnS$_2$ p–g–n junctions." *Advanced Materials* 31(6). https://doi.org/10.1002/adma.201805656.

Li, X., G. Jia, J. Du, X. Song, C. Xia, Z. Wei, and J. Li. 2018. "Type-II InSe/MoSe$_2$ (WSe$_2$) van der Waals heterostructures: vertical strain and electric field effects." *Journal of Materials Chemistry C* 6: 10010–19. https://doi.org/10.1039/C8TC03047B.

Liu, B., C. Zhao, X. Chen, L. Zhang, Y. Li, H. Yan, and Y. Zhang. 2019. "Self-powered and fast photodetector based on graphene/MoSe$_2$/Au heterojunction." *Superlattices and Microstructures* 130: 87–92. https://doi.org/10.1016/j.spmi.2019.04.021.

Long, M., E. Liu, P. Wang, A. Gao, H. Xia, W. Luo, B. Wang, J. Zeng, Y. Fu, K. Xu, et al. 2016 "Broadband photovoltaic detectors based on an atomically thin heterostructure." *Nano Letters* 16(4): 2254–59. https://doi.org/10.1021/acs.nanolett.5b04538.

Luo, H., B. Wang, E. Wang, X. Wang, Y. Sun, and K. Liu. 2019. "High-responsivity photovoltaic photodetectors based on MoTe$_2$/MoSe$_2$ van der Waals heterojunctions." *Crystals* 9(6): 315. https://doi.org/10.3390/cryst9060315.

Luo, M., X. Chen, P. Wu, H. Wang, Y. Chen, F. Chen, L. Zhang, and X. Chen. 2019. "Gate-tunable ReS$_2$/MoTe$_2$ heterojunction with high-performance photodetection." *Optical and Quantum Electronics* 51(5): 1–10. https://doi.org/10.1007/s11082-019-1839-3.

Massicotte, M., P. Schmidt, F. Vialla, K. G. Schädler, A. Reserbat-Plantey, K. Watanabe, T. Taniguchi, K.J. Tielrooij, and F. H. L. Koppens. 2016. "Picosecond photoresponse in van der Waals heterostructures." *Nature Nanotechnology* 11(1): 42–46. https://doi.org/10.1038/nnano.2015.227.

Mu, C., J. Xiang, and Z. Liu. 2017. "Photodetectors based on sensitized two-dimensional transition metal dichalcogenides – A review." *Journal of Materials Research* 32(22): 4115–31. https://doi.org/10.1557/jmr.2017.402.

Murali, K., N. Abraham, S. Das, S. Kallatt, and K. Majumdar. 2019 "Highly sensitive, fast graphene photodetector with responsivity >10^6 A/W using a floating quantum well Gate." *ACS Applied Materials & Interfaces* 11(33): 30010–8. https://doi.org/10.1021/acsami.9b06835.

Murali, K., and K. Majumdar. 2018. "Self-powered, highly sensitive, high-speed photodetection using ITO/WSe$_2$/SnSe$_2$ vertical heterojunction." *IEEE Transactions on Electron Devices* 65(10): 4141–48. https://doi.org/10.1109/TED.2018.2864250.

Patel, A. B., P. Chauhan, K. Patel, C. K. Sumesh, S. Narayan, K. D. Patel, G. K. Solanki, V. M. Pathak, P. K. Jha, and V. Patel. 2020. "Solution-processed uniform MoSe$_2$–WSe$_2$ heterojunction thin film on silicon substrate for superior and tunable photodetection." *ACS Sustainable Chemistry & Engineering* 8(12): 4809–17. https://doi.org/10.1021/acssuschemeng.9b07449.

Peng, Q., Z. Wang, B. Sa, B. Wu, and Z. Sun. 2016. "Electronic structures and enhanced optical properties of blue phosphorene/transition metal dichalcogenides van der Waals heterostructures." *Scientific Reports* 6: 2–11. https://doi.org/10.1038/srep31994.

Rao, G., X. Wang, Y. Wang, P. Wangyang, C. Yan, J. Chu, L. Xue, C. Gong, J. Huang, J. Xiong, et al. 2019. "Two-dimensional heterostructure promoted infrared photodetection devices." *InfoMat* 1(3): 272–88. https://doi.org/10.1002/inf2.12018.

Saeed, M., W. Uddin, A. S. Saleemi, M. Hafeez, M. Kamil, I. A. Mir, Sunila, R. Ullah, S. U. Rehman, and Z. Ling. 2020. "Optoelectronic properties of MoS$_2$-ReS$_2$ and ReS$_2$-MoS$_2$ heterostructures." *Physica B: Condensed Matter* 577: 411809. https://doi.org/10.1016/j.physb.2019.411809.

Saha, D., A. Varghese, and S. Lodha. 2020. "Atomistic modeling of van der Waals heterostructures with Group-6 and Group-7 monolayer transition metal dichalcogenides for near infrared/short-wave infrared photodetection." *ACS Applied Nano Materials* 3: 820–29. https://doi.org/10.1021/acsanm.9b02342.

Sun, M., Q. Fang, D. Xie, Y. Sun, J. Xu, C. Teng, R. Dai, P. Yang, Z. Li, W. Li, et al. 2017. "Novel transfer behaviors in 2D MoS$_2$/WSe$_2$ heterotransistor and its applications in visible-near infrared photodetection." *Advanced Electronic Materials* 3(4). https://doi.org/10.1002/aelm.201600502.

Tan, H., W. Xu, Y. Sheng, C. S. Lau, Y. Fan, Q. Chen, M. Tweedie, X. Wang, Y. Zhou, and J. H. Warner. 2017. "Lateral graphene-contacted vertically stacked WS$_2$/MoS$_2$ hybrid photodetectors with large gain." *Advanced Materials* 29(46). https://doi.org/10.1002/adma.201702917.

Vabbina, P., N. Choudhary, A. A. Chowdhury, R. Sinha, M. Karabiyik, S. Das, W. Choi, and N. Pala. 2015. "Highly sensitive wide bandwidth photodetector based on internal photoemission in CVD grown p-type MoS$_2$/graphene Schottky junction." *ACS Applied Materials and Interfaces* 7(28): 15206–13. https://doi.org/10.1021/acsami.5b00887.

Varghese, A., D. Saha, K. Thakar, V. Jindal, S. Ghosh, N. V. Medhekar, S. Ghosh, and S. Lodha. 2020. "Near-direct bandgap WSe$_2$/ReS$_2$ type-II pn heterojunction for enhanced ultrafast photodetection and high-performance photovoltaics." *Nano Letters* 20(3): 1707–17. https://doi.org/10.1021/acs.nanolett.9b04879.

Xiong, Y. F., J. H. Chen, Y. Q. Lu, and F. Xu. 2019. "Broadband optical-fiber-compatible photodetector based on a graphene-MoS$_2$-WS$_2$ heterostructure with a synergetic photogenerating mechanism." *Advanced Electronic Materials* 5(1). https://doi.org/10.1002/aelm.201800562.

Xue, H., Y. Dai, W. Kim, Y. Wang, X. Bai, M. Qi, K. Halonen, H. Lipsanen, and Z. Sun. 2019. "High photoresponsivity and broadband photodetection with a band-engineered WSe$_2$/SnSe$_2$ heterostructure." *Nanoscale* 11(7): 3240–47. https://doi.org/10.1039/c8nr09248f.

Xue, H., Y. Wang, Y. Dai, W. Kim, H. Jussila, M. Qi, J. Susoma, Z. Ren, Q. Dai, J. Zhao, et al. 2018. "A MoSe$_2$/WSe$_2$ heterojunction-based photodetector at telecommunication wavelengths." *Advanced Functional Materials* 28(47): 1–7. https://doi.org/10.1002/adfm.201804388.

Xue, Y., Y. Zhang, Y. Liu, H. Liu, J. Song, J. Sophia, J. Liu, Z. Xu, Q. Xu, Z. Wang, et al. 2016. "Scalable production of a few-layer MoS_2/WS_2 vertical heterojunction array and its application for photodetectors." *ACS Nano* 10(1): 573–80. https://doi.org/10.1021/acsnano.5b05596.

Yang, S., C. Wang, C. Ataca, Y. Li, H. Chen, H. Cai, A. Suslu, J. C. Grossman, C. Jiang, Q. Liu, and S. Tongay. 2016. "Self-driven photodetector and ambipolar transistor in atomically thin GaTe-MoS_2 p-n vdW heterostructure." *ACS Applied Materials and Interfaces* 8(4): 2533–39. https://doi.org/10.1021/acsami.5b10001.

Ye, L., H. Li, Z. Chen, and J. Xu. 2016. "Near-infrared photodetector based on MoS_2/black phosphorus heterojunction." *ACS Photonics* 3(4): 692–99. https://doi.org/10.1021/acsphotonics.6b00079.

Yu, W., S. Li, Y. Zhang, W. Ma, T. Sun, J. Yuan, K. Fu, and Q. Bao. 2017. "Near-infrared photodetectors based on $MoTe_2$/graphene heterostructure with high responsivity and flexibility." *Small* 13(24): 1–8. https://doi.org/10.1002/smll.201700268.

Zhang, K., X. Fang, Y. Wang, Y Wan, Q. Song, W. Zhai, Y. Li, G. Ran, Y. Ye, and L. Dai. 2017. "Ultrasensitive near-infrared photodetectors based on a graphene-$MoTe_2$-graphene vertical van der Waals heterostructure." *ACS Applied Materials and Interfaces* 9(6), 5392–98. https://doi.org/10.1021/acsami.6b14483.

Zhang, K., T. Zhang, G. Cheng, T. Li, S. Wang, W. Wei, X. Zhou, W. Yu, Y. Sun, P. Wang, et al. 2016. "Interlayer transition and infrared photodetection in atomically thin type-II $MoTe_2/MoS_2$ van der Waals heterostructures." *ACS Nano* 10(3): 3852–58. https://doi.org/10.1021/acsnano.6b00980.

Zhang, W., C. P. Chuu, J. K. Huang, C. H. Chen, M. L. Tsai, Y. H. Chang, C. Te. Liang, Y. Z. Chen, Y. L. Chueh, J. H. He, et al. 2014. "Ultrahigh-gain photodetectors based on atomically thin graphene-MoS_2 heterostructures." *Scientific Reports* 4(3826): 1–8. https://doi.org/10.1038/srep03826.

Zhang, X., Z. Meng, D. Rao, Y. Wang, Q. Shi, Y. Liu, H. Wu, K. Deng, H. Liu, and R. Lu. 2016. "Efficient band structure tuning, charge separation, and visible-light response in ZrS_2-based van der Waals heterostructures." *Energy Environmental Science* 9: 841–49. https://doi.org/10.1039/c5ee03490f.

Zheng, S., E. Wu, Z. Feng, R. Zhang, Y. Xie, Y. Yu, R. Zhang, Q. Li, J. Liu, W. Pang, et al.. 2018. "Acoustically enhanced photodetection by a black phosphorus-MoS_2 van der Waals heterojunction p-n diode." *Nanoscale* 10(21): 10148–53. https://doi.org/10.1039/c8nr02022a.

Zhou, X., N. Zhou, C. Li, H. Song, Q. Zhang, X. Hu, L. Gan, H. Li, J. Lü, J. Luo, et al. 2017. "Vertical heterostructures based on $SnSe_2/MoS_2$ for high performance photodetectors." *2D Materials* 4(025048): 1–10. https://doi.org/10.1088/2053-1583/aa6422.

Zhu, W., X. Wei, F. Yan, Q. Lv, C. Hu, and K. Wang. 2019. "Broadband polarized photodetector based on p-BP/n-ReS_2 heterojunction." *Journal of Semiconductors* 40(092001): 1–8. https://doi.org/10.1088/1674-4926/40/9/092001.

2 3D Printing
A State-of-the-Art Approach in Electrochemical Sensing

Mary Salve, Khairunnisa Amreen, Prasant Kumar Pattnaik, and Sanket Goel

CONTENTS

2.1 Introduction 17
2.2 3D-Printed Fluidic Platform.. 18
2.3 3D-Printed Electrode.. 19
2.4 3D-Printed Electrochemical Sensor 21
2.5 Conclusion and Future Perspective...25
References..26

2.1 INTRODUCTION

Three-dimensional (3D) printing, or additive manufacturing, has received remarkable attention from researchers on both the academic laboratory and industrial level. 3D printing facilitates vast opportunities for rapid prototyping. Therefore, it has a broad spectrum use in various research areas stretching from mechanical, electrical, chemical engineering, medical, materials, aerospace engineering, and chemistry. Electrochemistry, basically a bifurcation of analytical chemistry, has recently benefited from the 3D printing methodology. This gives an economic design and fabrication method for a high-performance and extensively available electrochemical sensor. 3D printing technology has been extensively applied to prepare multiplex devices with a microfluidic channel that can be used as a platform to build sensors that are fabricated by traditional technique (Symes et al. 2012; Erkal et al. 2014). However, utilizing the features of 3D printing to develop materials useful as electrodes for an electroanalytical sensor has been recently developed due to the availability of the conductive material to be used during the 3D printing process (Ambrosi and Pumera 2016; Manzanares-Palenzuela and Pumera 2018).

The procedure of 3D printing comprises of designing of 3D solid object built on the deposition of the desired material layer-by-layer via computer-aided design (CAD) software. This model has to be changed to Standard Triangle Language (.stl) file format (i.e., compatible to the 3D printing software). A 3D image is changed by this into a corresponding 2D layer of the respective object giving a G-code file. Finally, the 3D printer is able to print the object in a layer-by-layer manner by depositing the material. Ambrosi and Pumera (2016) in their review article have discussed the different 3D printing approaches with electrochemical applications.

Different types of 3D printers available on the market can be grouped together depending on printing processes such as lamination, powder based, photopolymerization, and extrusion. The photopolymerization incorporates stereolithography (SLA) and material jetting and scattered technique especially in the designing of a microfluidic/miniaturized platform. The process is based on polymerization of resin formulation by ultraviolet (UV) light, in a layer-by-layer way monitored by an optical arrangement until the formation of the desired object. Up until now, using this process, no applications are found for fabrication of an electrochemical sensor as per our knowledge. Fused deposition modeling (FDM) is mostly used in extrusion-based technique, in that the filament, which is thermoplastic in nature, is heated continuously to a partly molten form in a layer-by-layer depositing before extrusion. This process is simple and cost-effective, and used to print multiplexed material objects even though the precision and the quality of the surface may be relatively poor in comparison to powder-based technique (Bikas, Stavropoulos, and Chryssolouris 2016). Selective laser melting (SLM) uses metal powder and a high-cost 3D printer. An SLM-based 3D printer developed electrochemical sensor has been reported here. Inkjet-based 3D printing is another type of powder-based 3D printing, which works on the principles of drop on demand of ink onto any solid or flexible substrate. Powders of different variety can be prepared based on the opted binder to meet the essential need of the resultant ink such as surface tension and density viscosity. Conductivity plays an important role for electrochemical sensing, so conductive reagent needs to be added to polymeric binder (Pang et al. 2020). To the best of our knowledge, no reports are available for the development of an electrochemical sensor using an ink-based 3D printer. Also, no reports are available on a lamination-process-based electrochemical sensor so far.

The electrochemical sensor's 3D-printing provides substantial advantages over traditional fabrication techniques such as rapid prototyping, cost-effective production, increase in manufacturing speed, and fabrication of complex design. Herein, we report the development of an electrochemical sensor for biosensing applications using SLA- and FDM-based 3D printers because these printers are the most affordable with the bench top printer ranging between a few hundred and a thousand dollars. The literature reported here shows the usage of commercially available bench top 3D printers for the fabrication of cost-effective and functional objects. The techniques have been explored for the development of a microfluidics platform, electrodes, and an electrochemical cell for electrochemical and electroanalytical applications.

2.2   3D-PRINTED FLUIDIC PLATFORM

Various biosensors based on a 3D-printed microfluidic device combined with optical detection methodology or with a smartphone camera as an optical readout has gained much attention (Chan et al. 2016). 3D-printed microfluidic devices paired with the electrochemical sensing method has also gained attention in recent years for biosensor development. The combination of a microfluidic device fabricated via 3D printing incorporated with an electrochemical instrumentation can provide a reasonably inexpensive, simple, easy to handle, biosensing compatible sensor and can be used in a point-of-care (POC) application in a critical environment.

In electrochemistry, the pioneer application of 3D-printing for an electrochemical application was demonstrated by Snowden et al. (2010) using the SLA deposition technique. They designed a versatile flow cell that can integrate an electrode of choice and secured it using a cotton thread for electrochemical sensing with a flow rate up to 64 mL/min using a polycrystalline boron doped diamond (pBDD) electrode with gold in a supporting electrolyte of 0.1 M KNO_3. The experiments were conducted to investigate the electrochemical signal obtained by using the redox probe (ferrocenylmethyl)trimethylammonium hexaflorophosphate ($FcTMA^+$) at different flow rates and then comparing the result with numerical simulation (Snowden et al. 2010).

Alabdulwaheed and Bishop utilized an SLA 3D printer to design flow cell devices with a threaded port located at input and output of the channel for the fluid flow using market available fitting and tubing with a threaded port at the center for the electrode incorporation. The electrodes were prepared using a conductive silver wire and pencil graphite rod connected to a 24-gauge copper wire for electrochemical measurements. The working electrode was a gold nanoparticle modified for a biomarker protein, S100B electrochemical immunoassay (Alabdulwaheed and Bishop 2018).

Recently, FDM 3D printing has been used to develop an electrochemical flow cell that was further tested for electrochemical sensing applications. Kizek et al. (2014) fabricated a 3D-printed microfluidic platform to detect influenza hemagglutinin. The 3D-printed device has two steps that include isolation and electrochemical detection. Isolation based on glycan paramagnetic particles and a microfluidic chip was fabricated using polylactic acid (PLA), which consists of glass carbon microelectrode, graphite lead as reference, and platinum as an auxiliary electrode. The 3D-printed chip using a CDS quantum dot was used for electrochemical detection of the influenza virus. Similarly, our group has developed a 3D-printed micro reactor for hydrogen peroxide (H_2O_2) electrochemical sensing with real sample analysis in bleach, medicated H_2O_2 using a pencil graphite electrode modified with a silver nanoparticle (Salve et al. 2020). Dias et al. (2016) made a paper device, enzymatic, for detection of glucose through a batch injection analysis (BIA) via a 3D-printed electrochemical cell.

2.3   3D-PRINTED ELECTRODE

Herein, we have highlighted the use of a 3D-printed metal and polymer electrode in the context of electrochemical sensor development. The 3D printing of metal requires an expensive instrument in comparison to polymer filament printing. The metals used for 3D printing are generally iron, steel, titanium, and aluminum. 3D-printed iron electrodes are the most feasible ones because they do not passivate for creating the reproducible surface. In most cases, an extra step of electroplating with another metal layer is needed to make the electrode suitable for sensing applications. Metal electrodes limit the potential window, which reduces its scope in sensing applications.

Loo, Chua, and Pumera (2017) have used a metal printed electrode that was electroplated using different metal for electrochemical sensing applications. Stainless steel helical electrodes were gold electroplated for single-strand DNA detection. The helical electrode of stainless steel functioned here as a transducing platform with electroactive methylene blue and has been used for analyzing the DNA hybridization process. It gives super selectivity over a noncomplimentary DNA target with a linear detection range from 1 to 1,000 nM. Similarly, Liyarita, Ambrosi, and Pumera (2018) have used a 3D-printed metal electrode for pain reliever acetaminophen (AC) and for neurotransmitter dopamine (DA) using a 3D-printed stainless steel helical shape electrode, which gives better response compared to a glassy carbon electrode and disk gold electrode. Further, the electrode was used for the estimation of AC in commercial tablets. In other work, Ho, Ambrosi, and Pumera (2018) tested the 3D-printed helical shape stainless steel metal electrode both with surface modification and without it for gold film used for detection of ascorbic acid and uric acid. It showed better sensitivity with gold electroplating. The 3D-printed electrode was used for real-time sample testing using a commercial tablet of vitamin C for ascorbic acid concentration variation that gives a variation of 0.5% from the claimed concentration. All the studies highlighted here show the potential of a 3D printer for printing a metal electrode; however, the electrochemical behavior can be achieved after electroplating.

Due to all of the aforementioned issues, carbon-based material acts as an attractive material for the development of a 3D-printed electrode. The thermoplastic polymers mostly used in the 3D printing of a conductive filament in electrochemical devices are acrylonitrile-butadiene-styrene (ABS) and polylactic acid (PLA). Conductive filament based on ABS and PLA obtained are acquired by assimilation of conductive material in the polymer matrix. The conductive material generally uses nanomaterial such as carbon nanotube (CNT), carbon black, carbon nanofiber (CNF), and graphene. All these are used owing to their high surface areas, chemical inertia, electrical conducting nature, and possibility for functionalization.

Carbon particles containing commercially available filaments are used in the development of electrochemical sensors and biosensor applications because the material is of low-cost, highly electrically conductive, biocompatible, and design flexible. Two kinds of conductive filaments such as PLA/graphene (Black Magic) and PLA/Carbon Black (Proto-Pasta) are sold on the market, based on our knowledge, that are being adapted for 3D-printed electrochemical sensor designing. Manzanares-Palenzuela et al. (2018) demonstrated a proof of concept of using a graphene/PLA-based fabricated electrode for sensing of chemicals such as ascorbic acid and picric acid. The electrode response was relatively low as compared to a carbonaceous electrode surface such as glassy carbon or carbon paste.

To enhance the electron transfer and conductivity rate of the electrochemical sensor, the investigation for surface treatment of the 3D-printed electrodes was done by various research groups. To increase the electrochemical activity and to eliminate the nonconductive material partially from the surface of the electrode, different techniques were used such as electrochemical or chemical activation by using different solvents and mechanical polishing biological digestion using enzymes.

Manzanares-Palenzuela et al. (2018) reported a nontreated graphene/PLA-based electrode of ring and disc shape for electrochemical detection. The electrodes were activated by immersing for 10 minutes in a dimethylformamide DMF solution and then cleaned using ethanol and air dried for at least 24 hours. As a proof of concept, ascorbic acid and picric acid sensing was performed. Sometime later, they explored the activity via immersion in solvent DMF and electrochemical sensing in supporting electrolyte 0.1 mol/L potassium chloride of a 3D-printed electrode, which improved the electrochemical sensing of the redox mediator (Browne et al. 2018). The electrochemical activation increases the electrode surface conductivity as the exposed surface of the electrode gets oxidize. Santos et al. (2019) performed an electrochemical pretreatment of oxidation and reduction on a PLA filament coated with graphene (Black Magic 3D). The electrode at varied potential in a pH 7 PBS of 0.1 mol/L concentration for understanding the morphological behavior of an electrode. The oxidation was performed at a potential of +1.8 V and reaction of reduction between 0 V and −1.8 V, leading to an increase in structure defects and the plane edges of surface exposure and oxygen functionalities occurrence. The activity of the modified system was examined toward DA sensing. The graphene 3D-printed electrode, which was electrochemically activated using a Prussian blue for sensitive detection of H_2O_2 in a real-time sample of milk and mouthwash was also reported. In other work, a biocatalytic pretreatment was reported for activating a graphene/PLA electrode (Manzanares-Palenzuela et al. 2019). The biotechnical activation of a 3D-printed electrode depending on partial digestion of the PLA electrode and a proteinase K enzyme in combination with edges of graphene was used for electrochemical detection of 1-naphthol. Different protocols for graphene/PLA surface treatment were compared for DA detection as a target analyte and the best results were obtained using hydroxide activation using sodium hydroxide (NaOH 1 mol/L) for immersion for about 30 minutes, which helps in the saponification process of the PLA surface enhancing graphene sites (Kalinke et al. 2020). After electrochemical activation, great improvement was observed, leading to increased electron transfer activity, surface area, and defects. The electrode was used for detection of DA in human serum samples and synthetic urine.

Cardoso et al. (2020) used a 3D-printed graphene/PLA electrode that gives an appropriate parameter for the immobilization of enzyme by cross linking glutaral aldehyde used for glucose sensing in plasma. The combination of mechanical polishing followed by chemical treatment (immersing in DMF solution for 10 minutes) prior to further modification with the biomolecule was reported for determining uric acid and nitrite, simultaneously. This was performed via differential pulse voltammetry (DPV) in combination with a batch injection analysis for real-sample examination in saliva and urine. Hamzah et al. (2018) used a carbon black/ABS filament for evaluating the electrochemical behavior of a 3D printing electrode based on different printing directions, both in the horizontal and vertical directions, and it was observed that the horizontal direction results in smooth electrode surface in comparison to the vertical direction. It was analyzed using ferrocene carboxylic acid and serotonin hydrochloric acid as electroactive species. All these experiments show that the 3D-printed conductive material is viable to be used as an electrode and its performance is enhanced after surface modification and gives a better result in comparison to the glassy carbon electrode.

2.4   3D-PRINTED ELECTROCHEMICAL SENSOR

3D-printing technology proves to be a great benefit to modern electroanalytical chemistry because it gives feasibility to develop a fully integrated miniaturized electrochemical sensor of desired geometry and is easily reproduced for large-scale production. A dual-extruder 3D printer, based on the FDM technique, has recently been used to develop an integrated miniaturized electrochemical system in one step using two materials simultaneously with one conductive filament as an electrode with a view to develop a low-cost, portable, low-sample volume consumption platform for electrochemical analysis for biosensing applications.

Katseli, Economou, and Kokkinos (2019) demonstrated that a 3D printer is employable for the fabrication of a three electrode system in the same device in one step. A three electrode system was printed using a carbon-deposited PLA filament and the electrode holder was printed using a nonconductive PLA filament in a single stem via a dual-extruder FDM 3D printer. The developed platform was applied for sensing three different target analytes with appropriate modification such as mercury, caffeine, and glucose. In the next work, Katseli et al. (2020) reported the chip-based 3D-printed cell for simultaneous determination of caffeine and paracetamol. The electrochemical cell was printed using PLA and studies were carried out on different conductive filaments and it was observed that the carbon-deposited ABS filament gave more sensitivity and a low background signal.

In a study by Cardoso et al. (2018), a 3D-printed fluidic electrochemical cell was fabricated using ABS and an electrode using a highly conductive PLA with graphene doping and was deployed for diclofenac, dipyrone, DA, and tertbutyl hydroquinone detection. They have used both the flow injection analysis as well as a batch injection analysis with different working electrodes in comparison to a proposed 3D-printed working electrode and it showed appreciable electroanalytical performance with other reported carbon dependent electrodes. However, a 3D-printed electrode gave lesser electron transfer behavior as compared to plain GCE, but better performance for DA and catechol sensing with the help of an oxygenated group present at the graphene/PLA surface. Richter et al. (2019) developed a 3D-printed electrochemical cell by using an ABS filament and the electrodes were printed using a carbon black/PLA filament and the reference electrode was coated with silver/silver chloride (Ag/AgCl) ink and was tested for target analyteuric acid, ascorbic acid, DA, hexaammineruthenium(III) chloride, and ferricyanide/ferrocyanide. The working electrode was later examined by polishing it on sandpaper and then electrochemically activating it in an NaOH solution to improve the electrochemical behavior.

O'Neil et al. (2019) fabricated an electrochemical flow cell with an integrated graphene/PLA electrode using a dual-extruder FDM 3D printer in a single-step process. The working electrode was employed for one electron oxidation of ferrocene methanol for characterizing the unmodified cells under stagnant condition modified by gold electro plating and was used to detect catechol and ferrocene methanol under flow condition FDM 3D printing to provide a great potential to develop a fully integrated electrochemical sensor in a single step. The working electrode can be modified or pretreatment can be done for various biosensing applications.

Recently, our group has worked on developing silver nanoparticles using the green synthesize method and utilized them as a supercapacitor and in electrochemical sensing applications. The entire experimental set up was in a 3D-printed microfluidic electrochemical platform. Figure 2.1 shows the designed platform. A commercial 3D printer based on fused deposition modeling (Flashforge) was employed to fabricate the miniaturized platform using an ABS filament of a diameter of 1.75 mm.

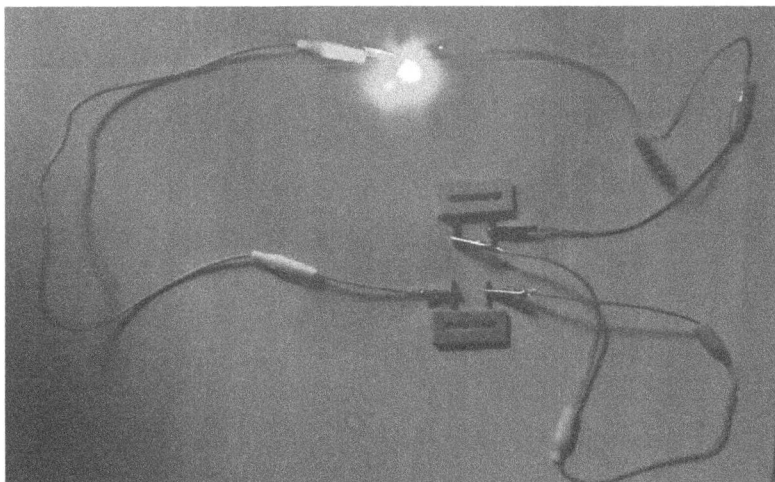

FIGURE 2.1 LED powered using two symmetric electrodes in series connection in a 3D-printed microfluidic channel.

The CAD model was made using 123D design software and then the file was saved in .stl format so that can be it can be used with 3D printer software. The design was then printed with an adjusted extruder temperature of 240°C, temperature of bed as 120°C, infill 100%, and a layer height of 100 mm. Herein, two separate devices were printed for electrochemical sensing and supercapacitor studies. The developed device was of $40 \times 18 \times 5$ mm in size consisting of a microchannel with a rectangular reservoir with $30 \times 4 \times 2$ mm dimensions and inlet regions for the electrodes were designed. The three electrodes were pencil graphite (PGE), platinum (Pt), and Ag/AgCl electrodes. The microchannel was then bonded to a glass slide with a double-sided tape (Figure 2.2). Similarly, we were successful in developing a 3D-printed

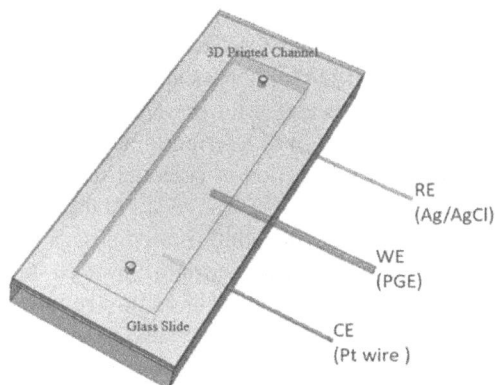

FIGURE 2.2 Schematic representation of a 3D microfluidic channel for supercapacitor and electrochemical sensing.

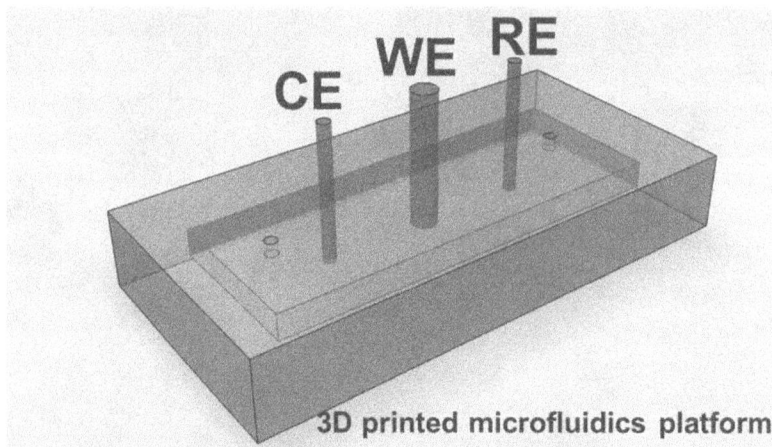

FIGURE 2.3   3D-printed miniaturized three-electrode platform for warfarin sensing.

reusable platform for three-electrode-based electrochemical sensing. The developed platform was fully reusable, and after each experiment, the platform was cleaned using a water gun and oven dried at 40°C for around 30 seconds. The CAD of the miniaturized device was designed in 123D design software (Autodesk) and converted to require 3D printer format (.stl file). The printing material used was an ABS filament of 1.75 mm diameter, which was compatible with the printer. The optimized parameter for the printing of the ABS filament was 240°C extruder temperature, 120°C bed temperature with a layer height of 100 mm, and 100% infill.

Figure 2.3 shows the developed miniaturized device printed with overall dimensions of $40 \times 20 \times 5$ mm (length × breadth × height) with a close rectangular reservoir of $30 \times 8 \times 3$ mm for the fluid. The volume of the consumed reagent was 750 µL for the reservoir to be fully filled. The inlet and outlet ports for the fluid were realized on the top layer of diameter 1 mm and for the three electrode ports for GCE (5 mm diameter), Pt wire (0.8 mm diameter), and Ag/AgCl wire (0.8 mm diameter).

The fabricated platform was used for electrochemical detection of warfarin. Warfarin, is an anticoagulant drug that is basically a coumarin derivative (Rezaei, Rahmanian, and Ensafi 2014). It is useful for preventing venous and arterial thromboembolism. In common language, it is referred to as a "blood thinner." Warfarin acts on the liver to reduce the production of essential proteins in blood that cause blood clotting (Walfisch and Koren 2010). The required dosage of warfarin is about 2.0–5.0 g mL^{-1}. As a result, monitoring administered dosage in a patient is crucial. Owing to the fact that warfarin has a longer half-life of 20–60 hours, dosage has to be carefully maintained. Hence, detection of its dosage in biological samples of a patient is important. Substantial traditional methods such as high performance liquid chromatography (HPLC) with fluorescence detection, UV-Vis, HPLC-MS/MS, liquid chromatography, electrophoresis, and spectrophotometric methods are used. But these have several disadvantages because they are time-consuming and require tedious sample preparations. Also, large sample and reagent volume

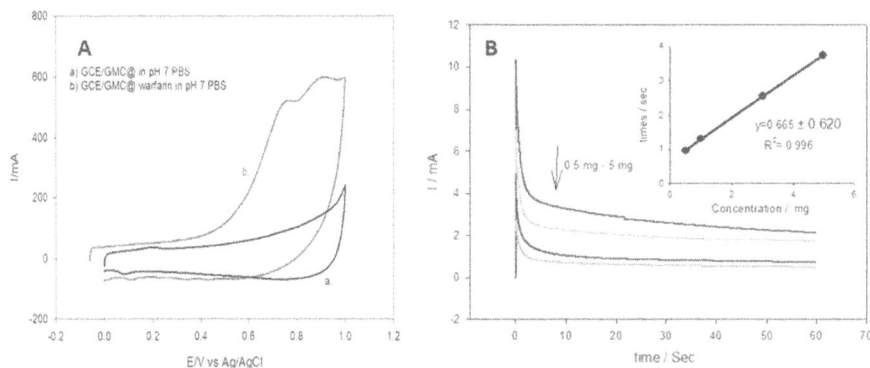

FIGURE 2.4 (A) Comparative CV response of GCE/GMC in pH 7 PBS and 1 mg warfarin at 10 mV s^{-1} and (B) effect of variable concentration of warfarin on GCE/GMC with chrono-amperometry technique.

consumption is another disadvantage. In this context, simple, rapid, and low sample consuming detection methods are in high demand. Therefore, an electrochemical sensing approach is being widely adapted. There are a couple of reports where carbon nanotube, quantum dots, polymer, gold particles, and so on, have been used for bulk volume sensing (Gholivand and Solgi 2017, 2018). Herein, we have designed a 3D-printed platform for warfarin sensing.

A graphitized mesoporous carbon (GMC) chemically modified glassy carbon electrode designated as (GCE/GMC) was a working electrode. The other electrodes were a Pt wire as counter and Ag/AgCl as a reference electrode. A pH 7 phosphate buffer solution (PBS) was used as a supporting electrolyte and 1 mg of a warfarin commercial tablet was used as a real sample. Cyclic voltammetry (CV) was performed using biologic potentiostat in a potential window: 0.5 to 0.6 V at a scan rate of 10 mV s^{-1}. Figure 2.4A is a typical CV wherein a distinctive oxidation peak at 0.3 V corresponding warfarin oxidation is observed. Such type of peak is not observed in the absence warfarin (Figure 2.4A), that is, control (GCE/GMC) in plain PBS. To further authenticate that the peak obtained was due to warfarin, the effect of variable warfarin concentration was studied using the chronoamperometric technique. Figure 2.4B, shows the concentration effect. It is observed that GCE/GMC gave a linear response toward change in concentration from 0.5 mg to 5 mg. Therefore, it was confirmed that the response obtained was due to warfarin only. However, this was just a prototype analysis focusing the use of a 3D-printed platform. A detailed study about warfarin analysis will be conducted in the future.

2.5 CONCLUSION AND FUTURE PERSPECTIVE

This chapter focuses on the use of additive manufacturing technology in the development of electrochemical platforms and sensors. Recently, the use of a 3D-printed platform, 3D-printed electrode, and the fully integrated 3D-printed electrochemical sensor has gained much attention for the development of biosensors. 3D-printing technologies are increasingly adopted by industry and research institutes and laboratories because the materials used in 3D printing are cost-effective and can be easily prepared in a laboratory. 3D printing proves beneficial to electrochemistry from fabrication of novel microfluidic platform, electrochemical cells, through rapid prototyping.

Using, 3D printing in the development of electrochemical sensors is still in initial stages and there is significant potential that can be used in printing sensors of a desired geometrical shape. In the future, the number of issues needs to be investigated for improving the efficiency that can be used in addressing real-world analytical problems. Such as a detailed study on material that can be used in microscale and macroscale for development of electrochemical sensors and also the development of conductive material that does not require pretreatment. In the near future, conductive polymer resin can be developed that can be used with an SLA or DLP printer because these printers have high resolution compared to the FDM printer. Also, the exciting potential opportunity is to integrate 3D-printed electrochemical sensors with 3D bioprinting that is mostly used in tissue engineering applications (Murphy and Atala 2014). By solving this issue, we believe that a tremendous increase can be seen in the use of 3D-printed electrochemical sensors and in electrochemical research.

REFERENCES

Alabdulwaheed, Abdulhameed, and G.W. Bishop. 2018. "3D-printed fluidic devices and incorporated graphite electrodes for electrochemical immunoassay of biomarker proteins." East Tennessee State University, Johnson City, TN.

Ambrosi, Adriano, and Martin Pumera. 2016. "3D-printing technologies for electrochemical applications." *Chemical Society Reviews* 45: 2740–55. doi:10.1039/C5CS00714C.

Bikas, H., P. Stavropoulos, and G. Chryssolouris. 2016. "Additive manufacturing methods and modeling approaches: A critical review." *International Journal of Advanced Manufacturing Technology* 83 (1–4): 389–405. doi:10.1007/s00170-015-7576-2.

Browne, Michelle P., Filip Novotný, Zdeněk Sofer, and Martin Pumera. 2018. "3D printed graphene electrodes' electrochemical activation." *ACS Applied Materials and Interfaces* 10 (46): 40294–301. doi:10.1021/acsami.8b14701.

Cardoso, Rafael M., Dianderson M.H. Mendonça, Weberson P. Silva, Murilo N.T. Silva, Edson Nossol, Rodrigo A.B. da Silva, Eduardo M. Richter, and Rodrigo A.A. Muñoz. 2018. "3D printing for electroanalysis: From multiuse electrochemical cells to sensors." *Analytica Chimica Acta* 1033: 49–57. doi:10.1016/j.aca.2018.06.021.

Cardoso, Rafael M., Pablo R.L. Silva, Ana P. Lima, Diego P. Rocha, Thiago C. Oliveira, Thiago M. do Prado, Elson L. Fava, Orlando Fatibello-Filho, Eduardo M. Richter, and Rodrigo A.A. Muñoz. 2020. "3D-printed graphene/polylactic acid electrode for bioanalysis: Biosensing of glucose and simultaneous determination of uric acid and nitrite in biological fluids." *Sensors and Actuators, B: Chemical* 307. doi:10.1016/j. snb.2019.127621.

Chan, Ho Nam, Yiwei Shu, Bin Xiong, Yangfan Chen, Yin Chen, Qian Tian, Sean A. Michael, Bo Shen, and Hongkai Wu. 2016. "Simple, cost-effective 3D printed microfluidic components for disposable, point-of-care colorimetric analysis." *ACS Sensors* 1 (3): 227–34. doi:10.1021/acssensors.5b00100.

Dias, Anderson A., Thiago M.G. Cardoso, Rafael M. Cardoso, Lucas C. Duarte, Rodrigo A.A. Muñoz, Eduardo M. Richter, and Wendell K.T. Coltro. 2016. "Paper-based enzymatic reactors for batch injection analysis of glucose on 3D printed cell coupled with amperometric detection." *Sensors and Actuators, B: Chemical* 226: 196–203. doi:10.1016/j.snb.2015.11.040.

Erkal, Jayda L., Asmira Selimovic, Bethany C. Gross, Sarah Y. Lockwood, Eric L. Walton, Stephen McNamara, R. Scott Martin, and Dana M. Spence. 2014. "3D printed microfluidic devices with integrated versatile and reusable electrodes." *Lab on a Chip* 14 (12): 2023–32. doi:10.1039/c4lc00171k.

Gholivand, Mohammad Bagher, and Mohammad Solgi. 2017. "Sensitive warfarin sensor based on cobalt oxide nanoparticles electrodeposited at multi-walled carbon nanotubes modified glassy carbon electrode (CoxOyNPs/MWCNTs/GCE)." *Electrochimica Acta* 246: 689–98. doi:10.1016/j.electacta.2017.06.105.

Gholivand, Mohammad Bagher, and Mohammad Solgi. 2018. "Simultaneous electrochemical sensing of warfarin and maycophenolic acid in biological samples." *Analytica Chimica Acta* 1034: 46–55. doi:10.1016/j.aca.2018.06.045.

Hamzah, Hairul Hisham Bin, Oliver Keattch, Derek Covill, and Bhavik Anil Patel. 2018. "The effects of printing orientation on the electrochemical behaviour of 3D printed acrylonitrile butadiene styrene (ABS)/carbon black electrodes." *Scientific Reports* 8 (1): 1–8. doi:10.1038/s41598-018-27188-5.

Ho, Eugene Hong Zhuang, Adriano Ambrosi, and Martin Pumera. 2018. "Additive manufacturing of electrochemical interfaces: Simultaneous detection of biomarkers." *Applied Materials Today* 12: 43–50. doi:10.1016/j.apmt.2018.03.008.

Kalinke, Cristiane, Naile Vacilotto Neumsteir, Gabriel De Oliveira Aparecido, Thiago Vasconcelos De Barros Ferraz, Pãmyla Layene Dos Santos, Bruno Campos Janegitz, and Juliano Alves Bonacin. 2020. "Comparison of activation processes for 3D printed PLA-graphene electrodes: Electrochemical properties and application for sensing of dopamine." *Analyst* 145 (4): 1207–18. doi:10.1039/c9an01926j.

Katseli, Vassiliki, Anastasios Economou, and Christos Kokkinos. 2019. "Single-step fabrication of an integrated 3D-printed device for electrochemical sensing applications." *Electrochemistry Communications* 103 (May): 100–103. doi:10.1016/j.elecom.2019.05.008.

Katseli, Vassiliki, Anastasios Economou, and Christos Kokkinos. 2020. "A novel all-3D-printed cell-on-a-chip device as a useful electroanalytical tool: Application to the simultaneous voltammetric determination of caffeine and paracetamol." *Talanta* 208. doi:10.1016/j.talanta.2019.120388.

Krejcova L, Nejdl L, Rodrigo MA, Zurek M, Matousek M, Hynek D, Zitka O, Kopel P, Adam V, Kizek R. 2014. "3D printed chip for electrochemical detection of influenza virus labeled with CdS quantum dots." *Biosensors and Bioelectronics* 54: 421–7. doi: 10.1016/j.bios.2013.10.031.

Liyarita, Bella Rosa, Adriano Ambrosi, and Martin Pumera. 2018. "3D-printed electrodes for sensing of biologically active molecules." *Electroanalysis* 30 (7): 1319–26. doi:10.1002/elan.201700828.

Loo, Adeline Huiling, Chun Kiang Chua, and Martin Pumera. 2017. "DNA biosensing with 3D printing technology." *Analyst* 142 (2): 279–83. doi:10.1039/c6an02038k.

Manzanares-Palenzuela, Carmen Lorena, Sona Hermanova, Zdenek Sofer, and Martin Pumera. 2019. "Proteinase-sculptured 3D-printed graphene/polylactic acid electrodes as potential biosensing platforms: Towards enzymatic modeling of 3D-printed structures." *Nanoscale* 11 (25): 12124–31. doi:10.1039/c9nr02754h.

Manzanares-Palenzuela, C. Lorena, Filip Novotný, Petr Krupička, Zdeněk Sofer, and Martin Pumera. 2018. "3D-printed graphene/polylactic acid electrodes promise high sensitivity in electroanalysis." *Analytical Chemistry* 90 (9): 5753–57. doi:10.1021/acs.analchem.8b00083.

Manzanares Palenzuela, C. Lorena, and Martin Pumera. 2018. "(Bio)analytical chemistry enabled by 3D printing: Sensors and biosensors." *TrAC - Trends in Analytical Chemistry* 103: 110–18. doi:10.1016/j.trac.2018.03.016.

Murphy, Sean V., and Anthony Atala. 2014. "3D bioprinting of tissues and organs." *Nature Biotechnology* 32 (8): 773–85. doi:10.1038/nbt.2958.

O'Neil, Glen D., Shakir Ahmed, Kevin Halloran, Jordyn N. Janusz, Alexandra Rodríguez, and Irina M. Terrero Rodríguez. 2019. "Single-step fabrication of electrochemical flow cells utilizing multi-material 3D printing." *Electrochemistry Communications* 99 (December 2018): 56–60. doi:10.1016/j.elecom.2018.12.006.

Pang, Yaokun, Yunteng Cao, Yihang Chu, Minghong Liu, Kent Snyder, Devin MacKenzie, and Changyong Cao. 2020. "Additive manufacturing of batteries." *Advanced Functional Materials* 30 (1): 1–22. doi:10.1002/adfm.201906244.

Rezaei, Behzad, Omid Rahmanian, and Ali Asghar Ensafi. 2014. "An electrochemical sensor based on multiwall carbon nanotubes and molecular imprinting strategy for warfarin recognition and determination." *Sensors and Actuators, B: Chemical.* 196: 539–45. doi:10.1016/j.snb.2014.02.037.

Richter, Eduardo M., Diego P. Rocha, Rafael M. Cardoso, Edmund M. Keefe, Christopher W. Foster, Rodrigo A.A. Munoz, and Craig E. Banks. 2019. "Complete additively manufactured (3D-printed) electrochemical sensing platform." *Analytical Chemistry* 91 (20): 12844–51. doi:10.1021/acs.analchem.9b02573.

Salve, Mary, Aurnab Mandal, Khairunnisa Amreen, Prasant Kumar Pattnaik, and Sanket Goel. 2020. "Greenly synthesized silver nanoparticles for supercapacitor and electrochemical sensing applications in a 3D printed microfluidic platform." *Microchemical Journal* 157 (April). doi:10.1016/j.microc.2020.104973.

Santos, Pãmyla L. dos, Vera Katic, Hugo C. Loureiro, Matheus F. dos Santos, Diego P. dos Santos, André L.B. Formiga, and Juliano A. Bonacin. 2019. "Enhanced performance of 3D printed graphene electrodes after electrochemical pre-treatment: Role of exposed graphene sheets." *Sensors and Actuators, B: Chemical* 281: 837–48. doi:10.1016/j.snb.2018.11.013.

Snowden, Michael E., Philip H. King, James A. Covington, Julie V. MacPherson, and Patrick R. Unwin. 2010. "Fabrication of versatile channel flow cells for quantitative electroanalysis using prototyping." *Analytical Chemistry* 82 (8): 3124–31. doi:10.1021/ac100345v.

Symes, Mark D., Philip J. Kitson, Jun Yan, Craig J. Richmond, Geoffrey J.T. Cooper, Richard W. Bowman, Turlif Vilbrandt, and Leroy Cronin. 2012. "Integrated 3D-printed reactionware for chemical synthesis and analysis." *Nature Chemistry* 4 (5): 349–54. doi:10.1038/nchem.1313.

Walfisch, Asnat, and Gideon Koren. 2010. "The 'warfarin window' in pregnancy: The importance of half-life." *Journal of Obstetrics and Gynaecology Canada* 32 (10): 988–89. doi:10.1016/S1701-2163(16)34689-8.

3 Optimized Electrical Interface for a Vanadium Redox Flow Battery (VRFB) Storage System
Modeling, Development, and Implementation

Nawin Ra and Ankur Bhattacharjee

CONTENTS

3.1 Introduction 29
3.2 Schematic of the Optimized Electrical Interface for a VRFB 31
3.3 Modeling of the Proposed System...32
    3.3.1 VRFB System Modeling 32
    3.3.2 Charge Controller Topologies...34
        3.3.2.1 Maximum Power Points Tracking (MPPT) Technique 34
        3.3.2.2 Work Flow of Three Levels CC-CV Charging of a VRFB 35
        3.3.2.3 Modified Three Levels for Solar PV Charging of a VRFB 36
    3.3.3 DC Stage Conversion for Optimized Electrical Interface
        of a VRFB...37
    3.3.4 Optimization of Dynamically Varying Flow Rate 39
3.4 Development of the Optimized VRFB Charge Controller............................40
3.5 Performance Validation 40
    3.5.1 Validation of the Proposed Charge Controller under Practical
        Conditions..42
        3.5.1.1 Condition 1: Sunny Weather Conditions..............................42
        3.5.1.2 Condition 2: Cloudy Weather Conditions 43
3.6 Conclusion 45
References...47

3.1 INTRODUCTION

Renewable energy sources (RES) such as solar photovoltaics (PV) and wind power are highly dependent on environmental factors. To utilize RES effectively, storing electrical energy plays a crucial role. Storage of energy generated from PV technology during the daytime employing energy storage systems (ESS) eases the reliance on the grid during the nighttime peak hours. Redox flow batteries (RFBs) are considered to be promising battery storage solutions compared to other conventional battery storage such as lead-acid, nickel-cadmium (Ni-Cd), lithium-ion (Li-ion), and sodium-sulfide (Na-S) since flow batteries offer several advantages. RFBs offer versatility in designing. According to our suitability, the scaling of the battery energy storage system (BESS) can be varied in the future through the separate control of power density and energy density. In the case of RFBs, the stack size decides the power (kW), and the volume of the electrolyte decides the energy capacity (kWh) of the storage system (Parasuraman et al. 2013; Arenas, Ponce de León, and Walsh 2017; Lourenssen et al. 2019). Among several available RFB technologies, a vanadium redox flow battery (VRFB) is the widely accepted battery technology due to several advantages. Advantages of a VRFB include long cycle life, deep discharge capacity, high coulombic efficiency, and safe operation because of exposed structure fast response time (Franco and Frayret 2015; Guarnieri et al. 2016). In a VRFB, a separate heat dissipation chamber is not needed, since the electrolyte acts as a built-in coolant.

Modeling of the VRFB system demonstrated simplicity in predicting the dynamic behavior of the system. A comprehensive electrical equivalent circuit model was proposed for power grid analysis (Zhang et al. 2015). The inherent features of the VRFB system such as pump energy consumption, ion diffusion and shunt current are included in the model. (Bhattacharjee et al. 2018a). The electrical interfacing of the VRFB system with source end (renewable sources such as solar PV) and the load end (grid) was indicated in the paper through impedance matching including of three subsystems including a VRFB stack, charge controller (source end), and inverter (load end). The charge controller plays a crucial role in the electrical interface of a VRFB in renewable energy applications and there has been a gradual improvement in the Research and Development of the charge controller over the last decade.

The charge controller for a VRFB had not been discussed in detail in the existing literature. For a defined capacity rating, voltage excursion per cell in a VRFB is almost 38%, which is very high in comparison with conventional batteries such as lead acid batteries, whose voltage excursion per cell for the same capacity is 20% (Qiu et al. 2014). Therefore, an MPPT-based charge controller is significant to attain complete charging of a VRFB.

Charging current and electrolyte flow rate are the parameters that need to be simultaneously controlled for the design of an efficient charge controller. During VRFB charging, a sudden increase in the charging current leads to an instantaneous increase of the VRFB stack temperature, which may shoot the VRFB stack temperature over 40%. The surge in temperature can ultimately lead to thermal instability in the VRFB stack (Tang, Bao, and Skyllas-Kazacos 2012; Yan, Skyllas-Kazacos, and Bao 2017). In the proposed system, thermal instability is mitigated by dynamically varying optimum flow rate. The work flow of three different charging topologies is discussed in this chapter to obtain a more suitable charging topology for optimized interfacing of a VRFB with RES. Modified three levels an MPPT-based solar PV charging of a VRFB was found to be the practically feasible charging topology. This chapter experimentally validates the significant improvement in the overall system efficiency of a VRFB after the introduction of the charge controller in the VRFB ESS. The charge controller discussed in this chapter can be adapted for charging a large-scale VRFB system based on a solar PV source.

This chapter is organized into six sections. The schematic of the proposed system is discussed in Section 3.2. The system modeling is described in Section 3.3. The development of the optimized charge controller is detailed in Section 3.4 and its performance validation is analyzed in Section 3.5. The conclusion of this book chapter is given in Section 3.6.

3.2 SCHEMATIC OF THE OPTIMIZED ELECTRICAL INTERFACE FOR A VRFB

The proposed system consist of a solar PV source, DC-DC converter, VRFB system, DC-AC converters, and a 16-bit dsPIC microcontroller (dsPIC33FJ32MC204). The microcontroller performs as the central processing unit (CPU) by executing the simultaneous real-time hydraulic pump control as well as the control of charging algorithms. As shown in Figure 3.1, the DC-DC converter is utilized to ensure the efficient transfer of power from the solar PV source to a VRFB. For the charging of a VRFB, control switches SW1 and SW2 are turned ON. Switches SW3 and SW4 are turned ON for VRFB discharge.

The dynamic flow rate under real-time varying state of charge (SOC) is obtained with the help of a pump controller, which controls the pumps supplying electrolytes to the VRFB stack. The bidirectional DC-AC converter smoothens the power transfer between the VRFB system and the power grid.

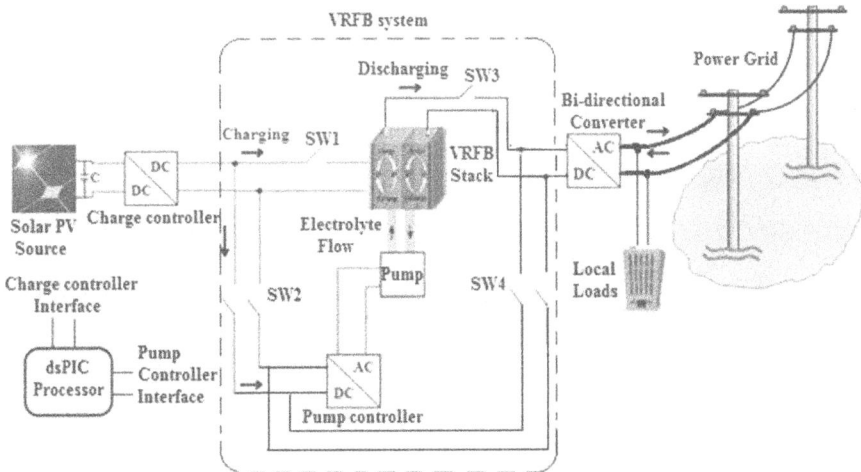

FIGURE 3.1 Schematic representation of the electrical interface for a VRFB storage system with power grid. (Reproduced from Bhattacharjee et al. 2018b. Copyright 2018, with permission from Elsevier.)

3.3 MODELING OF THE PROPOSED SYSTEM

The modeling of the proposed system topology shown in Figure 3.1 is carried out in a MATLAB®/Simulink environment.

3.3.1 VRFB SYSTEM MODELING

The modeling of the VRFB system can be utilized to estimate the accurate stack terminal voltage under the dynamically varying SOC with optimal flow rate control. In Figure 3.2A, the electrical equivalent circuit model of the VRFB system designed in MATLAB®/Simulink is shown (Bhattacharjee et al. 2018a). The specifications of the proposed system is displayed in Table 3.1. The parameters such as self-discharge potential drop, diffusion current, flow rate, and SOC are considered in the development of the VRFB model. In addition to these considerations, the impedance matching among the charge controller, VRFB stack, and inverter is implemented to ensure that the maximum power is transferred from the solar PV source to the power grid as shown in Figure 3.2B.

In the model described in this chapter, the effect of shunt current is neglected because it contributes the negligible fraction of less than 1% to the total terminal current magnitude (Zhang et al. 2015).

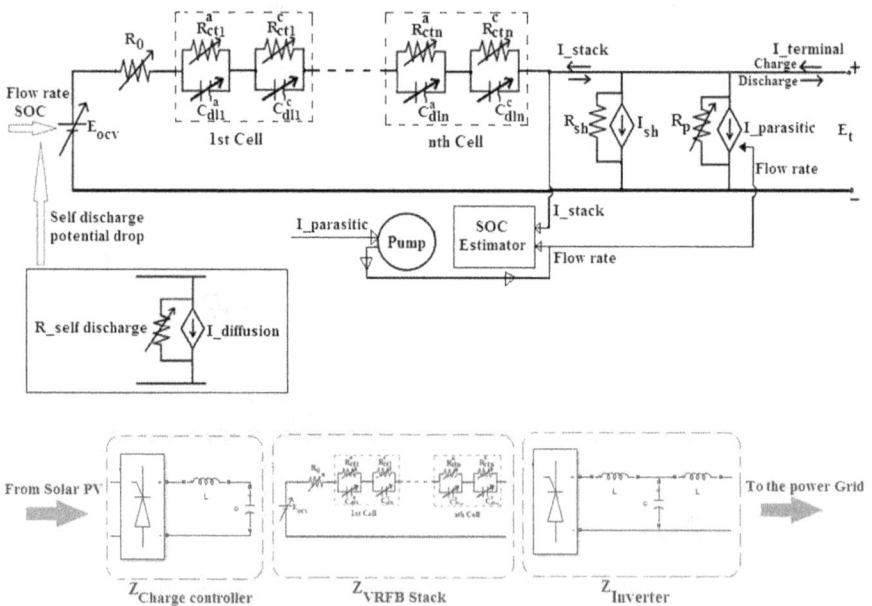

FIGURE 3.2 (A) Electrical equivalent circuit of the VRFB system. (Reproduced from Bhattacharjee et al. 2018b. Copyright 2018, with permission from Elsevier.) (B) Overall impedance matching of the system. (Reproduced from Bhattacharjee et al. 2018b. Copyright 2018, with permission from Elsevier.)

TABLE 3.1

Specifications of the Proposed VRFB System

Parameters	Quantity	Unit
Total count of cells in a VRFB stack	20	–
Quantity of electrolyte in a tank	180	L
Maximum power capacity of a VRFB	1	kW
Maximum energy capacity of a VRFB	6	kWh
Range of voltage allowed in a VRFB stack	20–32	V
VRFB equilibrium potential	28	V
Maximum capacity of "C"	187	Ah
Maximum pump power rating (×2)	85(×2)	W
Maximum pump voltage rating (AC)	220	V
Maximum pump power rating	0.5	A
Solar PV array power capacity	2	kW_P

Source: Reproduced from Bhattacharjee et al. (2018b). Copyright 2018, with permission from Elsevier.

Considering the effect of self-discharge loss in the Nernst potential equation (Bhattacharjee et al. 2018a), the VRFB stack open circuit voltage (OCV) is determined by Eq. (3.1)

$$E_{stack(OCV)}(t) = n \times \left\{ E_{Cell\_eq(at\ 50\%\ SOC)} + \frac{2RT}{F} \ln\left(\frac{SOC}{1-SOC} \right) - I_d R_{self\_discharge}(t) \right\} \quad (3.1)$$

where, E_{stack} and E_{cell} represent the stack and cell voltage of a VRFB in (V), SOC represents the VRFB state of charge in (%), R represents the universal gas constant and it is given by 8.3144 J K^{-1}mol^{-1}, T represents the ambient temperature in (K), and F represents Faraday's constant given by 96,485 C mol^{-1}. Diffusion current of a VRFB is represented by I_d. The self-discharge of a VRFB is represented by $R_{self-discharge}$.

The diffusion current is calculated by Eq. (3.2)

$$I_d = C_i D_m \frac{dC_i}{dx} \quad (3.2)$$

where, C represents the vanadium species concentration in (mol L^{-1}) and x represents the thickness of the membrane in (μm).

Considering the self-discharge loss inside the VRFB stack, the OCV is determined by the basic Nernst potential equation, taking into account the diffusion concentration of V$^+$ (ions)

$$E^+(t) = E_+^0 + \frac{RT}{F} \ln\left(\frac{C_O^+(t)}{C_R^+(t)} \right) \quad (3.3)$$

$$E^-(t) = E_-^0 + \frac{RT}{F} \ln\left(\frac{C_O^-(t)}{C_R^-(t)}\right) \tag{3.4}$$

where, term C_o^+ represents the concentration of oxidized vanadium species in the positive electrolyte side in (mol L^{-1}) and term C_R^- represents the concentration of reduced vanadium species in the negative electrolyte side in (mol L^{-1}).

Then, the net potential during self-discharge becomes

$$E_{self\ discharge}(t) = E^+(t) - E^-(t) \tag{3.5}$$

$$R_{self\_discharge}(t) = \frac{E_{self\_discharge}(t)}{I_d(t)} \tag{3.6}$$

The VRFB stack terminal voltage E_t is estimated by Eq. (3.7)

$$E_t(t) = E_{stack(OCV)}(t) \pm I_{Stack} R_{int}(Q, I) \tag{3.7}$$

where, R_{int} represents the internal resistance of the VRFB stack in (Ω) and Q represents the electrolyte flow rate in (ml/sec). In Eq. (3.7), the "+" sign means the charging and "–" sign connotes discharging condition.

3.3.2 CHARGE CONTROLLER TOPOLOGIES

Three different topologies of solar PV-based charging algorithms for the VRFB system are discussed in this section. The charging efficiency is calculated by Eq. (3.8)

$$\eta_{charging} = \frac{P_{VRFB}}{P_{PV}} \tag{3.8}$$

3.3.2.1 Maximum Power Points Tracking (MPPT) Technique

Maximum Power Point Tracking (MPPT) techniques are utilized in solar PV systems to obtain the maximum power from the solar modules by tracking the continuously varying Maximum Power Point (MPP). MPP is the operating point at which the solar modules produce maximum power. MPP depends on factors such as solar irradiance and module temperature. Among the several available MPPT techniques, the perturb and observe (P&O) MPPT algorithm had been selected in this system for its ease of implementation. Figure 3.3A shows the MPP attained for the P-V curve and I-V curve at different irradiance values and two different temperatures. The operating criteria for the P&O algorithm is such that, in the other scenario, if the power drawn from the solar module decreases, this means that the operating point has moved away from the MPP. Now, the direction of the operating voltage perturbation is reversed to attain MPP (Femia et al. 2005).

The VRFB system efficiency is evaluated for the different values of SOC. The magnitude of flow rate, at which the efficiency of a VRFB is of maximum value, is considered as optimal flow rate (Q_{opt}) as shown in Figure 3.3B. The overpotential

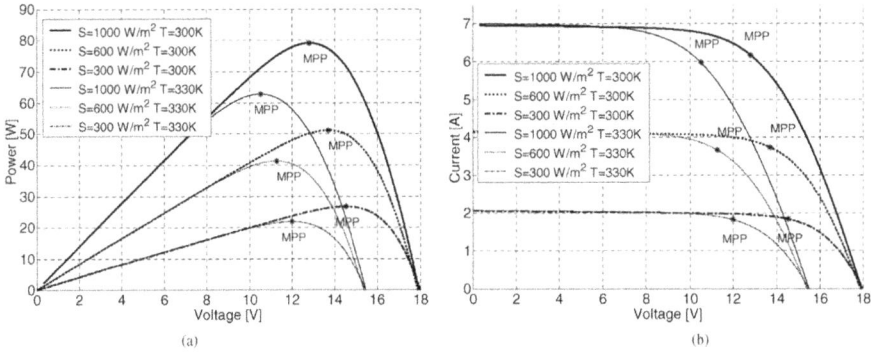

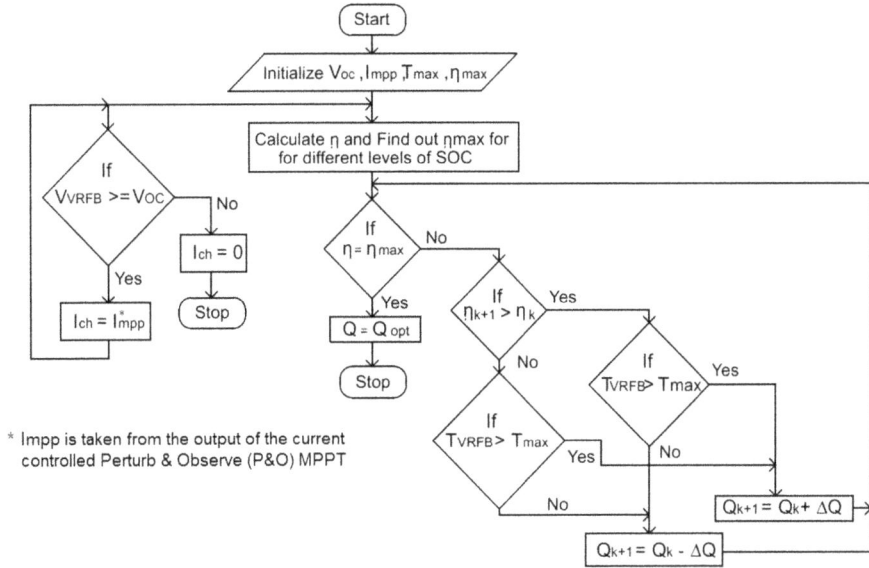

FIGURE 3.3 (A) (a) Power-voltage curve and (b) current-voltage curve for three different irradiance values and two different module temperatures. (B) Work flow diagram for the MPPT-based charging of a VRFB integrated with real-time flow control. (Reproduced from Bhattacharjee et al. 2018b. Copyright 2018, with permission from Elsevier.)

problem that is bound to occur in a VRFB is addressed by the continuous monitoring of the E_t magnitude. When the overcharge limit is attained by the E_t magnitude, the MPPT charging is instantly turned off.

3.3.2.2 Work Flow of Three Levels CC-CV Charging of a VRFB

In addition to the overpotential problem, there are other issues such as thermal instabilities inside a VRFB stack. In order to attain complete charging of a VRFB and address thermal issues, three levels charging of a VRFB is enforced. The three levels of VRFB charging include trickle charging level, bulk charging level, and float charging level. Initially, VRFB charging is commenced with the trickle charging

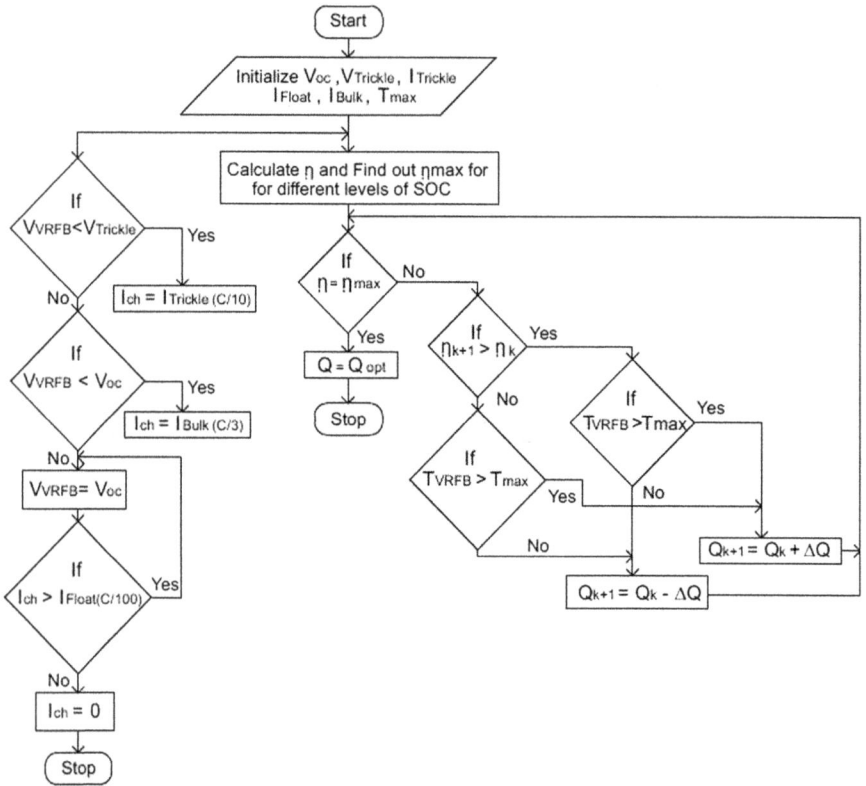

FIGURE 3.4 Work flow diagram for the three levels CC-CV solar PV charging of a VRFB. (Reproduced from Bhattacharjee et al. 2018b. Copyright 2018, with permission from Elsevier.)

level in which the trickle current (C/10) is applied to a VRFB, where "C" is the total ampere-hour (A-h) capacity of a VRFB. Trickle current avoids the extreme heat generation in a VRFB stack.

Followed by the trickle current, bulk charging current (C/3) is applied to the battery. At this instance, the battery voltage is enhanced through continuous charging until the "Voc" limit is attained. The third level of charging is called as float charging level that ensures the complete charging of a VRFB. In this level of charging, constant voltage is maintained across the E_t magnitude and the magnitude of charging current is reduced significantly to float current (C/100), until it reaches zero magnitude as shown in Figure 3.4.

3.3.2.3 Modified Three Levels for Solar PV Charging of a VRFB

RES are dynamically varying in nature. Modified three levels of charging are utilized in order to enhance the charging efficiency of a VRFB when interfaced with solar energy. I_{MPP} is considered as the reference magnitude of charging current at bulk charging level as shown in Figure 3.5. The rest of the operation is similar to

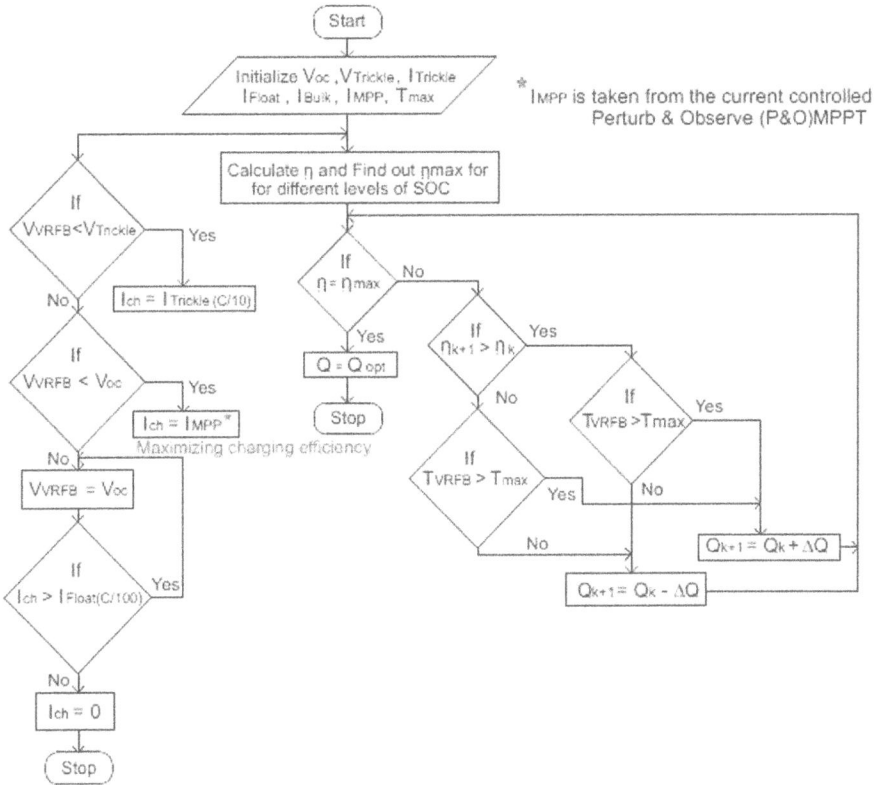

FIGURE 3.5 Work flow diagram of the modified three levels for the solar PV charging of a VRFB. (Reproduced from Bhattacharjee et al. 2018b. Copyright 2018, with permission from Elsevier.)

three levels charging. The modified three levels charging is found to be the most suitable charging topology in practically varying conditions.

3.3.3 DC Stage Conversion for Optimized Electrical Interface of a VRFB

In order to ensure the maximum power transfer from a solar PV to VRFB storage during a suitable charging phase, the optimized electrical interface of the VRFB system is required to be developed. In this work, a DC-DC buck converter is adopted as a charging interface for VRFB storage. When solar PV input voltage is higher than the threshold magnitude for VRFB charging, a DC-DC buck converter operates by limiting the input voltage magnitude to the range suitable for VRFB charging. MPPT also makes use of the DC-DC buck converter for regulating the operating voltage at MPP and to provide load matching for the maximum power transfer. To provide high current for fast charging and low voltage for a VRFB stack, the buck converter is operated in continuous conduction mode (CCM) with the magnitude of inductor

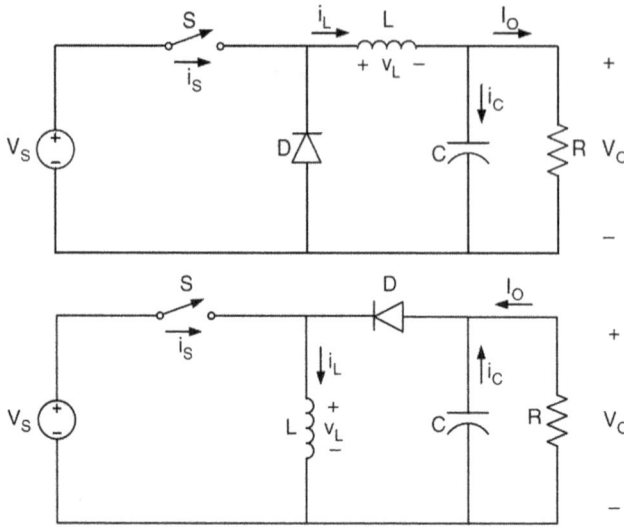

FIGURE 3.6 (A) Schematic of a DC-DC buck converter. (B) Schematic of a DC-DC buck-boost converter.

current (i_L) always greater than zero (Rashid 2001). As shown in Figure 3.6A, when the switch "S" is turned ON, the Diode "D" is reverse biased, thus blocking the voltage and conducting the current. Switch "S" will be triggered by a pulse width modulation (PWM) signal received from the charge controller based on the signals taken from the solar PV side and VRFB side. In a DC-DC buck converter, the voltage gain is always less than 1, meaning the output voltage (V_o) is always lower than the supply voltage (V_s).

A buck converter is utilized in the proposed system for operational convenience. However, future researchers can investigate using the theoretically more efficient buck-boost converter. Using the buck-boost converter, the supply voltage (V_s) can be converted either to a higher or lower magnitude at the output terminals. As shown in Figure 3.6B, when the switch "S" is turned ON, inductor "L" operates and inductor current i_L increases, while diode "D" is kept up at the OFF condition. When the switch "S" is turned OFF, diode "D" provides the path for inductor current i_L. The polarity of the diode reverses by drawing the current from the output terminals (Rashid 2001). In a steady-state condition, voltage transfer function of the buck-boost converter is given by Eq. (3.9)

$$\frac{Vo}{Vs} = -\frac{D}{1-D} \tag{3.9}$$

where, the "–" sign implies that the output voltage is negative with respect to the ground. When the magnitude of duty ratio D is 0.5, output voltage is equal to the supply voltage. Depending up on the magnitude of D being greater than or less than 0.5, the magnitude of output voltage will be greater or less than the input voltage. Thus, the buck-boost converter performs both the buck ($V_o < V_s$) as well as boost operation ($V_o > V_s$).

3.3.4 Optimization of Dynamically Varying Flow Rate

As the flow rate (Q) increases, the pump power (P_{pump}) increases, which leads to the increment of overall VRFB power consumption. In this chapter, the dynamic optimal flow rate is recognized for the VRFB system under various charging condition to optimize the pump power loss and to improve the VRFB system efficiency.

$$P_{Charging} = \left(P_{stack} + P_{int\_loss} + P_{pump}(Q) \right) \tag{3.10}$$

$$P_{Discharging} = \left(P_{stack} - P_{int\_loss} - P_{pump}(Q) \right) \tag{3.11}$$

$$P_{stack} = V_t \times I_{stack} \tag{3.12}$$

$$P_{int\_loss} = I_{stack}^2 \times R_{int} \tag{3.13}$$

The power consumed by flow pumps corresponding to the dynamic flow rates is shown in Figure 3.7.

The overall system efficiency (η) of the VRFB system is formulated by Eq. (3.14)

$$\eta = \frac{\int_0^{t_d} P_{Discharge}(t)dt}{\int_0^{t_c} P_{Charge}(t)dt} = \frac{\int_0^{t_d} \left(P_{stack} - P_{pump} - P_{int\_loss} \right)dt}{\int_0^{t_c} \left(P_{stack} + P_{pump} + P_{int\_loss} \right)dt} \tag{3.14}$$

The variation of optimal flow rate with respect to the VRFB SOC is shown in Figure 3.8. The average value of the overall VRFB system efficiency displayed in Table 3.2 is around 88% and it is higher than the values reported in the literature.

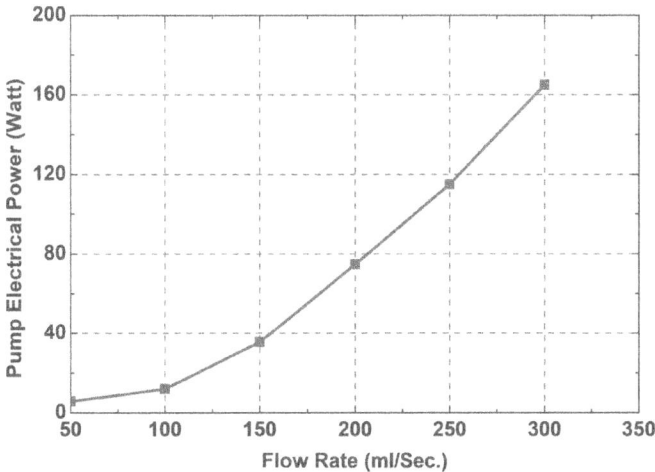

FIGURE 3.7 Variation of pump power consumption based on dynamic flow rates. (Reproduced from Bhattacharjee et al. 2018b. Copyright 2018, with permission from Elsevier.)

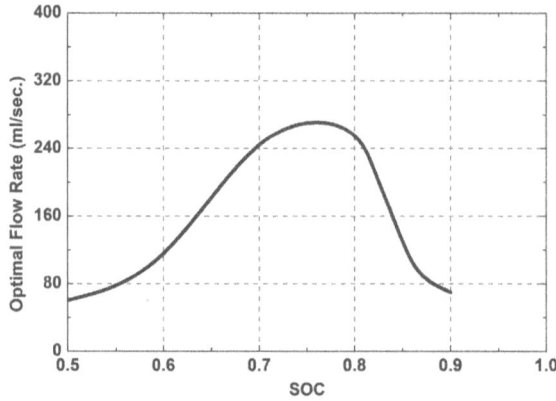

FIGURE 3.8 Variation of optimal flow rate with respect to the VRFB SOC. (Reproduced from Bhattacharjee et al. 2018b. Copyright 2018, with permission from Elsevier.)

TABLE 3.2

Estimation of Maximum VRFB System Efficiency

SOC (%)	Dynamically Varying Optimal Flow Rate (ml/sec.)	P_{min_charge} (Watt)	$P_{max\_discharge}$ (Watt)	η_{max} (%)
50	60	942.7	842	89.31
60	120	967	858	88.72
70	240	1018	889	87.33
80	250	1059	922	87.06
90	70	930	825	88.70

Source: Reproduced from Bhattacharjee et al. (2018b). Copyright 2018, with permission from Elsevier.

3.4 DEVELOPMENT OF THE OPTIMIZED VRFB CHARGE CONTROLLER

The significant components of the charge controller include a 2 kW$_P$ solar PV array, 1 kW 6 h VRFB system, and a charge controller module as shown in Figure 3.9. The DC-DC buck converter is designed in the proposed system to efficiently interface the 1 kW, 6 h VRFB system with 2 kW$_P$ solar PV array. A MATLAB®/Simulink software package is utilized for the design of the charge converter module. For controlling the speed of two pumps of the VRFB system, a DC-AC inverter is designed in the system. The VRFB system is charged by the solar PV source. The comparison of the simulation and experimental results of a VRFB OCV over different SOC is shown in Figure 3.10.

3.5 PERFORMANCE VALIDATION

To validate the performance of the model discussed in Section 3.3, a dsPIC processor is used to implement the charging topologies.

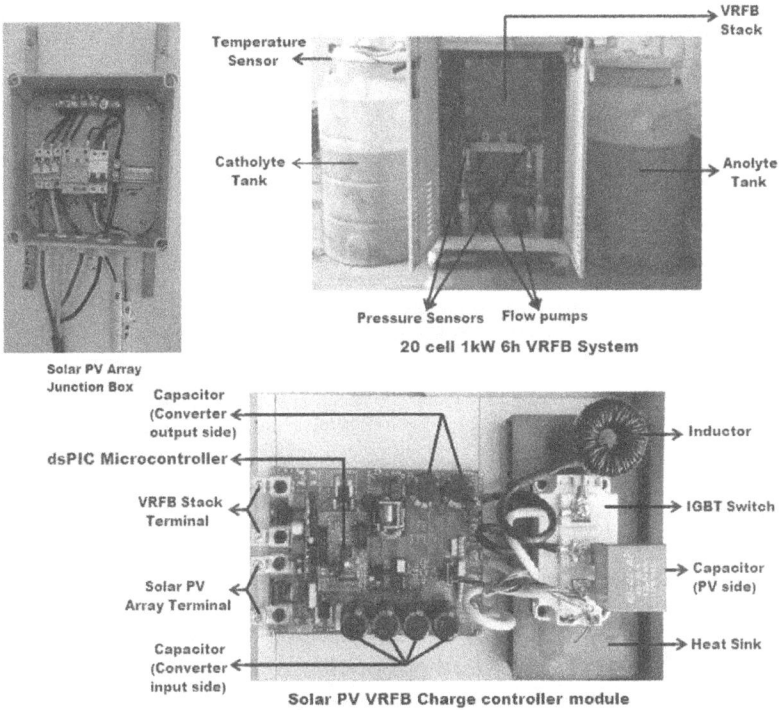

FIGURE 3.9   Pictorial representation of a solar PV-based MPPT charge controller module. (Reproduced from Bhattacharjee et al. 2018b. Copyright 2018, with permission from Elsevier.)

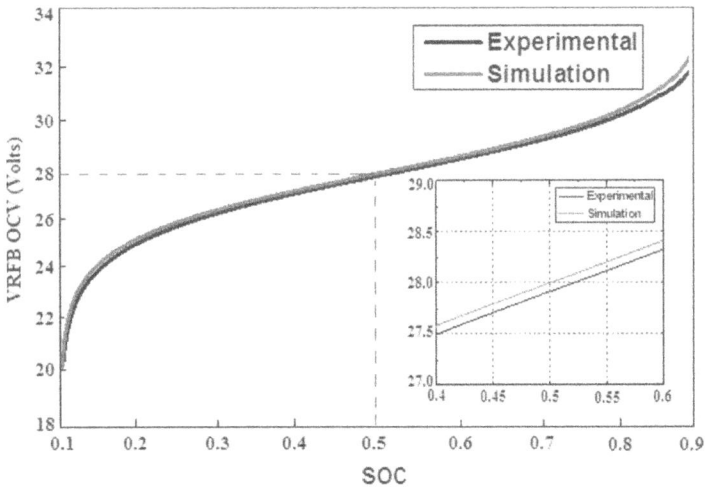

FIGURE 3.10   Comparison of the simulation and experimental results of a VRFB OCV over different SOC. (Reproduced from Bhattacharjee et al. 2018b. Copyright 2018, with permission from Elsevier.)

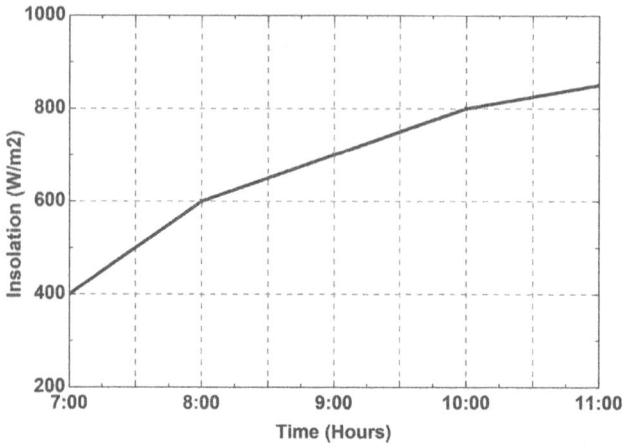

FIGURE 3.11 Representation of the solar insolation variation during a sunny day. (Reproduced from Bhattacharjee et al. 2018b. Copyright 2018, with permission from Elsevier.)

3.5.1 VALIDATION OF THE PROPOSED CHARGE CONTROLLER UNDER PRACTICAL CONDITIONS

3.5.1.1 Condition 1: Sunny Weather Conditions

Among the three charging topologies discussed in this chapter, the MPPT-based VRFB charging topology discussed in Section 3.3.2.1 is not suitable for renewable energy sources applications such as solar PV power systems because of incomplete charging of the VRFB until 75% SOC only due to the overpotential restriction at 32 V as shown in Figures 3.11–3.15. This leads to battery life degradation as a result

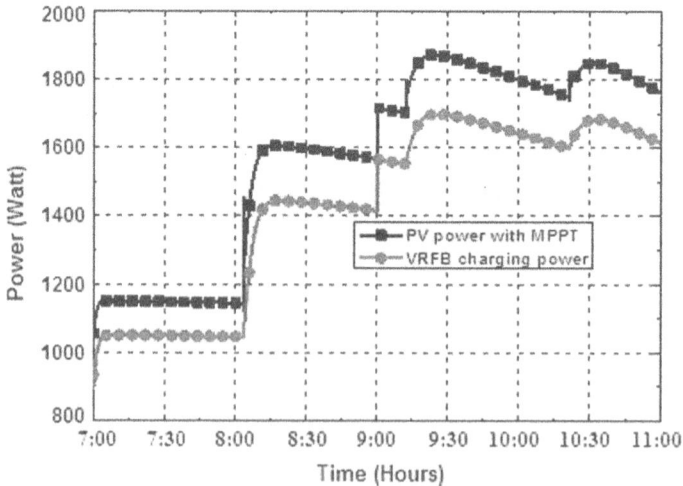

FIGURE 3.12 Representation of a VRFB charging power and MPPT-based PV power on a sunny day. (Reproduced from Bhattacharjee et al. 2018b. Copyright 2018, with permission from Elsevier.)

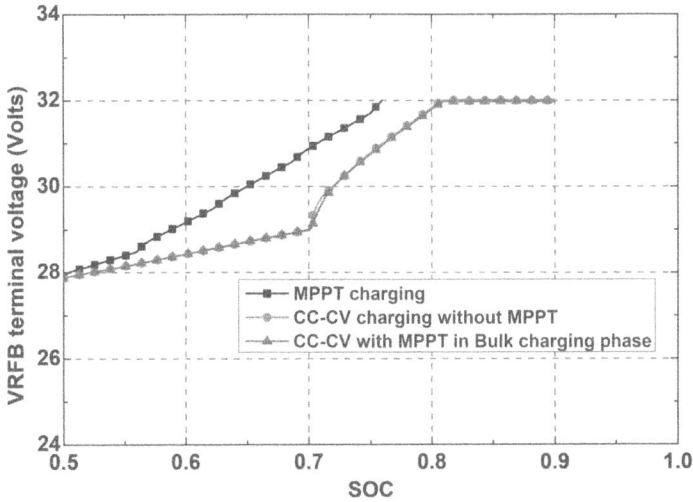

FIGURE 3.13 Comparison of a VRFB terminal voltage during a sunny day. (Reproduced from Bhattacharjee et al. 2018b. Copyright 2018, with permission from Elsevier.)

of repetitive incomplete charging. Thus, a higher range of charging cycle is obtained with the help of modified three-stage charging topology.

3.5.1.2 Condition 2: Cloudy Weather Conditions

The solar PV-based VRFB charge controller performance is studied under unexpected appearance of cloud. It is evident that the issue of the incomplete charging is also continued in the overcast weather condition. It is seen from Figures 3.16–3.20 that the VRFB SOC reaches close to a magnitude of 90% for the three levels as well

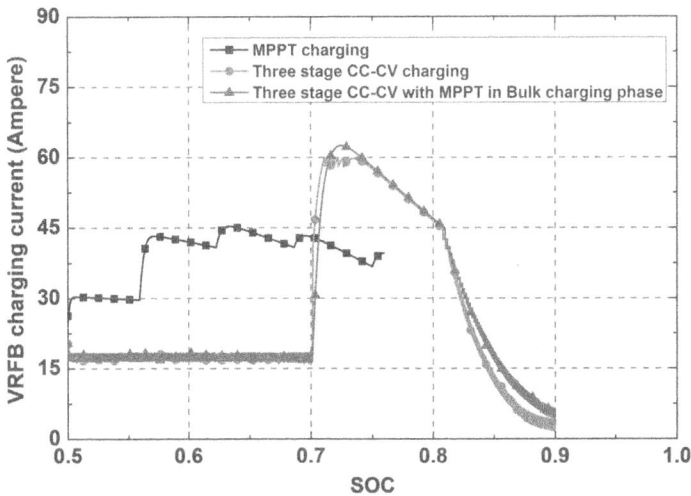

FIGURE 3.14 Comparison of a VRFB charging current during a sunny day. (Reproduced from Bhattacharjee et al. 2018b. Copyright 2018, with permission from Elsevier.)

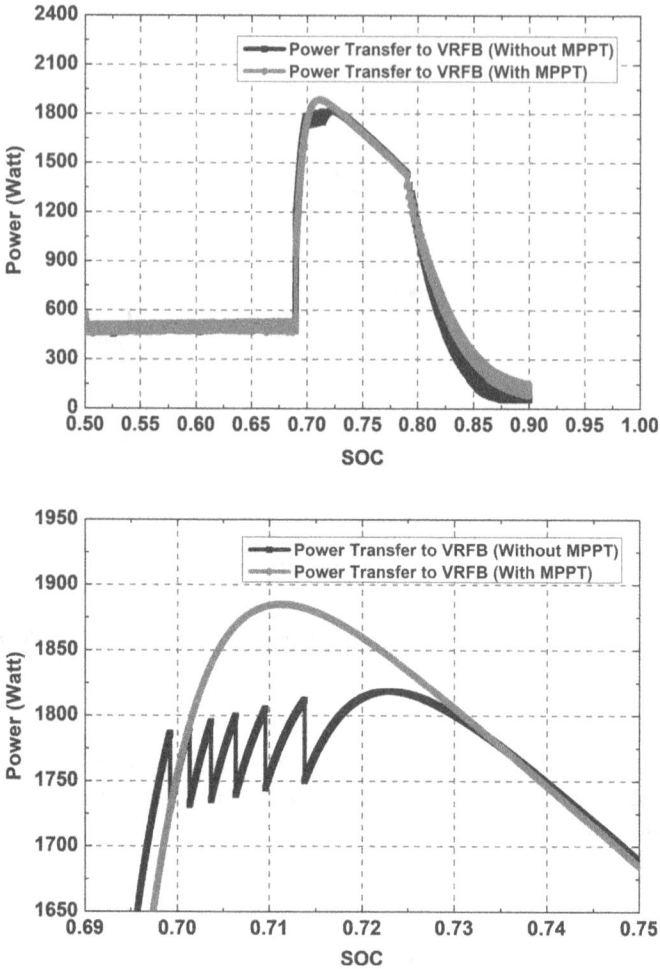

FIGURE 3.15   (A) Representation of variation between power transfer to a VRFB during sunny weather conditions, comparing with and without usage of an MPPT charge controller. (Reproduced from Bhattacharjee et al. 2018b. Copyright 2018, with permission from Elsevier.) (B) Zoomed in version of the variation between power transfer to a VRFB during sunny weather conditions, comparing with and without usage of an MPPT charge controller. (Reproduced from Bhattacharjee et al. 2018b. Copyright 2018, with permission from Elsevier.)

as the modified three levels charging topologies discussed in Sections 3.3.2.2 and 3.3.2.3, respectively. Thus, the incomplete charging is avoided by utilizing the above-mentioned topologies. This results in the improved efficiency of the charge controller with the additional average power transfer of 45 W in modified three levels charging as shown in Figures 3.20A and 3.20B. Therefore, the modified three phases charging topology is the most suitable algorithm for the efficient electrical interface of a solar PV source with the VRFB system.

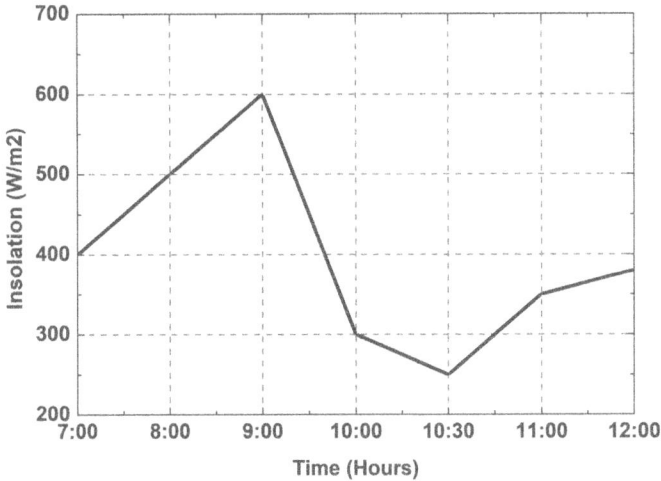

FIGURE 3.16 Representation of the solar insolation variation during a cloudy day. (Reproduced from Bhattacharjee et al. 2018b. Copyright 2018, with permission from Elsevier.)

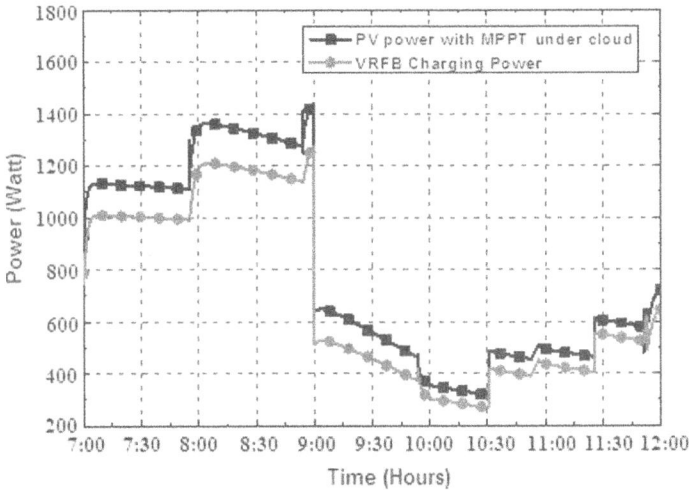

FIGURE 3.17 Representation of VRFB charging power and MPPT-based PV power on a cloudy day. (Reproduced from Bhattacharjee et al. 2018b. Copyright 2018, with permission from Elsevier.)

3.6 CONCLUSION

This chapter focused on the optimized interfacing of a VRFB with RES through a solar PV-based charge controller. The performance has been demonstrated under practical dynamic irradiance profiles (Bhattacharjee et al. 2018b). P&O MPPT charging topology, which is generally utilized for conventional batteries such as lead acid and Li-ion batteries is further modified by the inclusion of dynamic electrolyte

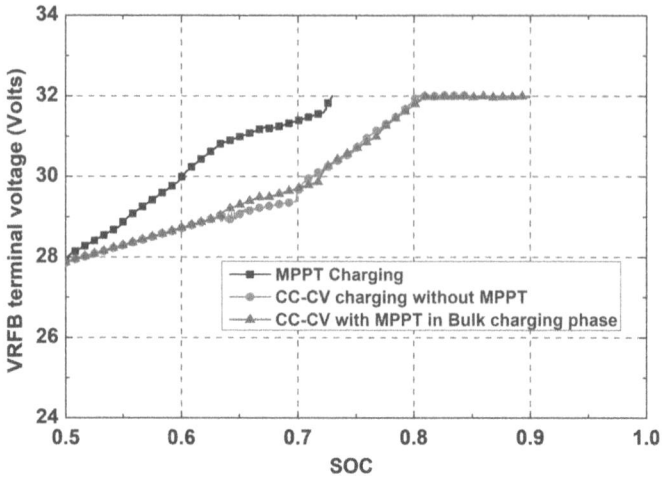

FIGURE 3.18 Comparison of a VRFB terminal voltage during a cloudy day. (Reproduced from Bhattacharjee et al. 2018b. Copyright 2018, with permission from Elsevier.)

flow rate control of VRFB storage. The performance validation shows a significant improvement in the charging efficiency from 83% to 94.5%. The dynamic flow rate optimization of a VRFB discussed in this chapter improvises the thermal management capabilities of the VRFB system (Bhattacharjee and Saha 2018). The premature thermal shutdown of the controller is prevented by maintaining the stack temperature within the safe limits. The optimized electrical interface for the VRFB storage system discussed in this chapter is a generalized solution and it will be useful

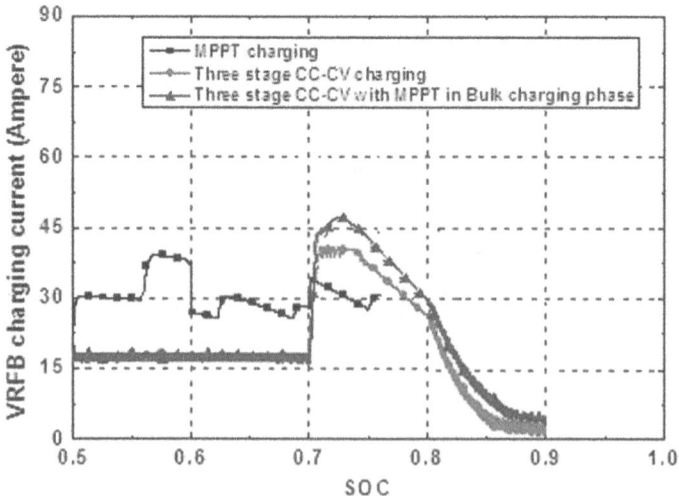

FIGURE 3.19 Comparison of a VRFB charging current during a sunny day. (Reproduced from Bhattacharjee et al. 2018b. Copyright 2018, with permission from Elsevier.)

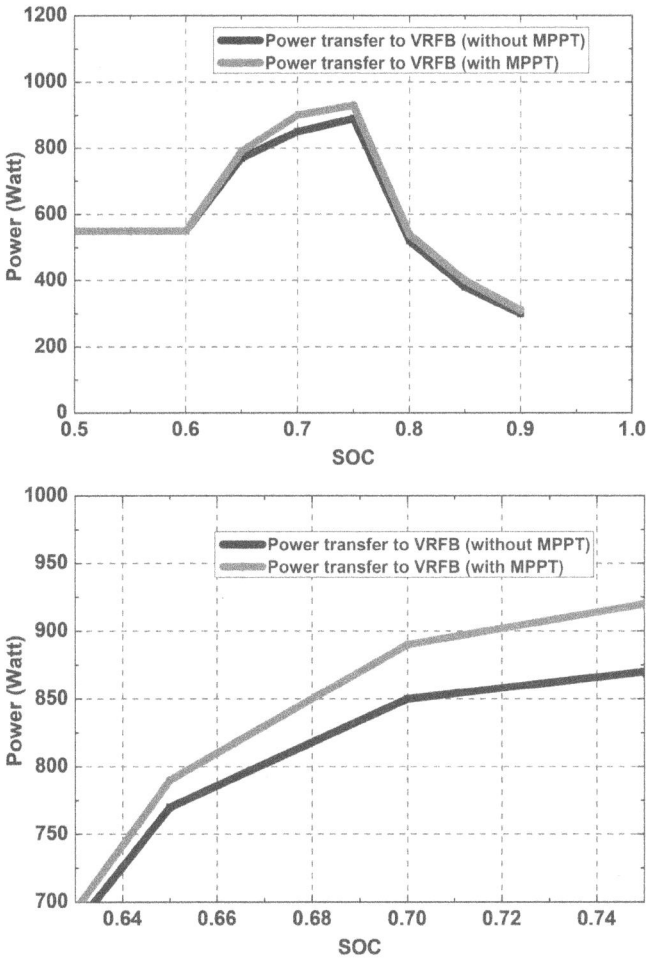

FIGURE 3.20   (A) Representation of variation between power transfer to a VRFB during cloudy weather conditions, comparing with and without usage of an MPPT charge controller. (Reproduced from Bhattacharjee et al. 2018b. Copyright 2018, with permission from Elsevier.) (B) Zoomed in version of the variation between power transfer to a VRFB during cloudy weather conditions, comparing with and without usage of an MPPT charge controller. (Reproduced from Bhattacharjee et al. 2018b. Copyright 2018, with permission from Elsevier.)

for future researchers to explore and integrate a scalable VRFB energy storage system with RES.

REFERENCES

Arenas, L. F., C. Ponce de León, and F. C. Walsh. 2017. "Engineering aspects of the design, construction and performance of modular redox flow batteries for energy storage." *Journal of Energy Storage* 11: 119–53. doi:10.1016/j.est.2017.02.007.

Bhattacharjee, Ankur, and Hiranmay Saha. 2018. "Development of an efficient thermal management system for vanadium redox flow battery under different charge-discharge conditions." *Applied Energy* 230 (August): 1182–92. doi:10.1016/j.apenergy.2018.09.056.

Bhattacharjee, Ankur, Anirban Roy, Nipak Banerjee, Snehangshu Patra, and Hiranmay Saha. 2018a. "Precision dynamic equivalent circuit model of a vanadium redox flow battery and determination of circuit parameters for its optimal performance in renewable energy applications." *Journal of Power Sources* 396 (June): 506–18. doi:10.1016/j.jpowsour.2018.06.017.

Bhattacharjee, Ankur, Hiranmay Samanta, Nipak Banerjee, and Hiranmay Saha. 2018b. "Development and validation of a real time flow control integrated MPPT charger for solar PV applications of vanadium redox flow battery." *Energy Conversion and Management* 171 (June): 1449–62. doi:10.1016/j.enconman.2018.06.088.

Femia, Nicola, Giovanni Petrone, Giovanni Spagnuolo, and Massimo Vitelli. 2005. "Optimization of perturb and observe maximum power point tracking method." *IEEE Transactions on Power Electronics* 20 (4): 963–73. doi:10.1109/TPEL.2005.850975.

Franco, A. A., and C. Frayret. 2015. "Modeling the design of batteries for medium- and large-scale energy storage." *Advances in Batteries for Medium and Large-Scale Energy Storage: Types and Applications.* Elsevier Ltd. doi:10.1016/B978-1-78242-013-2.00015-7.

Guarnieri, Massimo, Paolo Mattavelli, Giovanni Petrone, and Giovanni Spagnuolo. 2016. "Vanadium redox flow batteries: potentials and challenges of an emerging storage technology." *IEEE Industrial Electronics Magazine* 10 (4): 20–31. doi:10.1109/MIE.2016.2611760.

Lourenssen, Kyle, James Williams, Faraz Ahmadpour, Ryan Clemmer, and Syeda Tasnim. 2019. "Vanadium redox flow batteries: A comprehensive review." *Journal of Energy Storage* 25 (July). doi:10.1016/j.est.2019.100844.

Parasuraman, Aishwarya, Tuti Mariana Lim, Chris Menictas, and Maria Skyllas-Kazacos. 2013. "Review of material research and development for vanadium redox flow battery applications." *Electrochimica Acta* 101: 27–40. doi:10.1016/j.electacta.2012.09.067.

Qiu, Xin, Tu A. Nguyen, Joe D. Guggenberger, M. L. Crow, and A. C. Elmore. 2014. "A field validated model of a vanadium redox flow battery for microgrids." *IEEE Transactions on Smart Grid* 5 (4): 1592–1601. doi:10.1109/TSG.2014.2310212.

Rashid, M.H. 2001. *The Power Electronics Handbook: Industrial Electronics Series.* San Diego, CA: Elsevier Academic Press.

Tang, Ao, Jie Bao, and Maria Skyllas-Kazacos. 2012. "Thermal modelling of battery configuration and self-discharge reactions in vanadium redox flow battery." *Journal of Power Sources* 216: 489–501. doi:10.1016/j.jpowsour.2012.06.052.

Yan, Yitao, Maria Skyllas-Kazacos, and Jie Bao. 2017. "Effects of battery design, environmental temperature and electrolyte flowrate on thermal behaviour of a vanadium redox flow battery in different applications." *Journal of Energy Storage* 11: 104–18. doi:10.1016/j.est.2017.01.007.

Zhang, Yu, Jiyun Zhao, Peng Wang, Maria Skyllas-Kazacos, Binyu Xiong, and Rajagopalan Badrinarayanan. 2015. "A comprehensive equivalent circuit model of all-vanadium redox flow battery for power system analysis." *Journal of Power Sources* 290: 14–24. doi:10.1016/j.jpowsour.2015.04.169.

4 A Review on Recent Advancements in Chamber-Based Microfluidic PCR Devices

Madhusudan B Kulkarni and Sanket Goel

CONTENTS

4.1 Introduction 49
4.2 Design and Development of Various PCR Microfluidic Devices 52
    4.2.1 Material and Apparatus 53
    4.2.2 Fabrication Techniques and Bonding Process 53
4.3 Single Chamber Based PCR Devices 56
4.4 Multi-Chamber Array PCR Devices 58
4.5 Droplet-Based Virtual Chamber-Based PCR Microfluidic Devices 61
4.6 Isothermal Amplification Based Microfluidic PCR Devices 62
    4.6.1 Loop-Mediated Isothermal Amplification 62
    4.6.2 Recombinase Polymerase Amplification 63
    4.6.3 Nucleic Acid Amplification Test 63
4.7 Conclusion and Future Prospects 65
References 65

4.1 INTRODUCTION

In recent times, microfluidics has emerged in both scientific research and innovative technology for many applications including biological and biochemical (Ohno, Tachikawa, and Manz 2008; Melin and Quake 2007). It is a multidisciplinary domain entailing areas such as biochemistry, nanotechnology, engineering, and biotechnology. Since the initiation of microfluidic technology by Ahrberg, Manz, and Chung (2016), it has gained immense recognition (Wang, Chen, and Sun 2017). The microfluidics era began after the initial venture in developing a microfluidic system in the early 1950s; subsequently, numerous mechanisms were investigated and used to drive and regulate the fluid stream in microfluidic chips. Thereafter, during the 1990s, the strategy for research and development in the microfluidics area accelerated and started showing diverse potential to revolutionize the existing laboratory traditions, especially in biological sciences (Knoška et al. 2020), analytical sciences (Kugimiya et al. 2020), and biochemical studies (Chen, Shao, and Xianyu 2020). Hence, the microfluidic devices were also called micro-total-analysis-systems

49

(µTAS), or lab-on-a-chip (LOC), which can accomplish basic laboratory functions on a tiny chip to accelerate various studies and analyses. Further, it can decrease the volume of reaction mixture consumption and reaction time, thus delivering resourceful and ingenious solutions for pharmaceutical, chemical, and food industries and healthcare fields (Kudr et al. 2017; Cui and Wang 2019; Kuswandi et al. 2007; Kant et al. 2018; Das and Srivastava 2016). Usually, microfluidic deals with mixing, transferring, separating, and processing the fluids. Subsequently, the microfluidics emerged as one of the leading tools in the development of deoxynucleic acid (DNA) chips, inkjet printheads, micropropulsion, microbial fuel cell, acoustic droplet ejection, and microthermal technologies. The simplest present microfluidic system comprises of microchannels fabricated with a polymer that is bonded to a plane surface. It relates to the accurate control, conduction, and management of solutions that are geometrically restricted to a microrange at which capillary flow regulates the mass transport. Microfluidic polymerase chain reaction (PCR) devices can be manufactured on a commercial scale using a state-of-the-art approach. Figure 4.1 shows the evolution phase of PCR innovation. Also, the figure depicts the simple appearance of a thermocycler and microfluidic PCR device.

A majority of the earlier works show that the microfluidic platforms were mainly focused on the evolution of microfluidic biochip-based electrophoresis, demonstrating widespread advantages in various fields such as biology (Ren, Chen, and Wu 2014), chemistry (Wixforth 2006), environment (Jokerst, Emory, and Henry 2012), healthcare (Yeo, Kenry, and Lim 2016), and medicine (Rivet et al. 2011). This made it possible for analysis of miniaturization and separation of biochemical substances. Figure 4.2 illustrates the various benefits of microfluidic PCR devices.

PCR has turned out to be an inimitable and powerful tool for the extremely sympathetic molecular studies of target genetic code sequencing ever since its creation by Kary Mullis in 1984 (Mullis et al. 1992). The PCR technique is mainly used to amplify the genetic code templates for the functional analysis and scrutiny of genes in the detection of diseases, bacteria, viruses, and pathogens. During the amplification process, several million copies of DNA can be produced from a single molecule by executing only a few thermal loops of the PCR technique. The first microfluidic

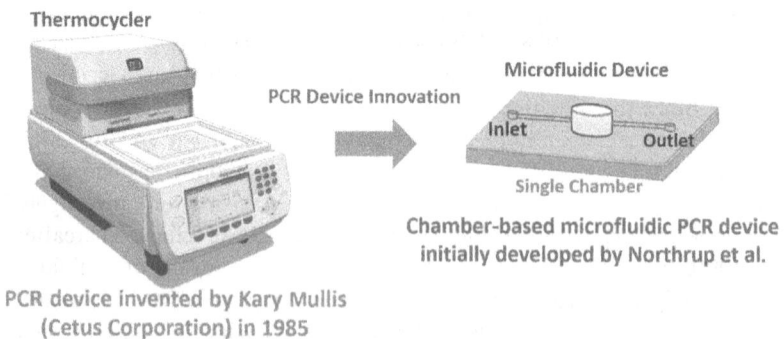

FIGURE 4.1   Evolution phase of a PCR device innovation.

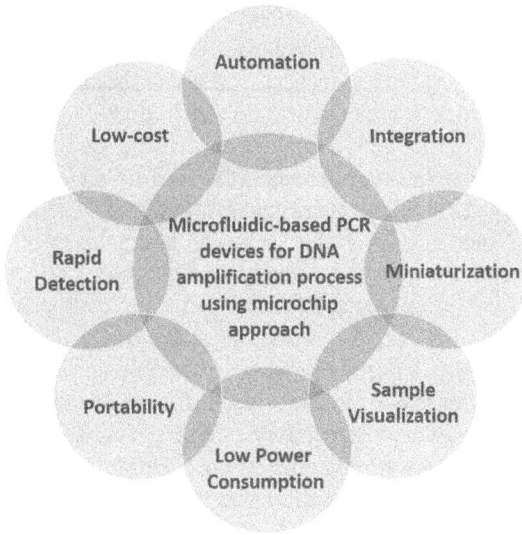

FIGURE 4.2 A simple representation depicting the benefits of microfluidic PCR devices.

PCR platform was described in 1993 (Northrup et al. 1995), whereby three distinct thermal loops, involved in the PCR method, were performed in a chamber-based approach with a reservoir for sample storage. The chamber-based PCR is also called a time-domain microfluidic PCR, wherein the sample stays motionless with the change in temperature.

Figure 4.3 represents the three discrete temperature stages of the PCR process, such as melting, annealing, and elongation, involved during the DNA amplification.

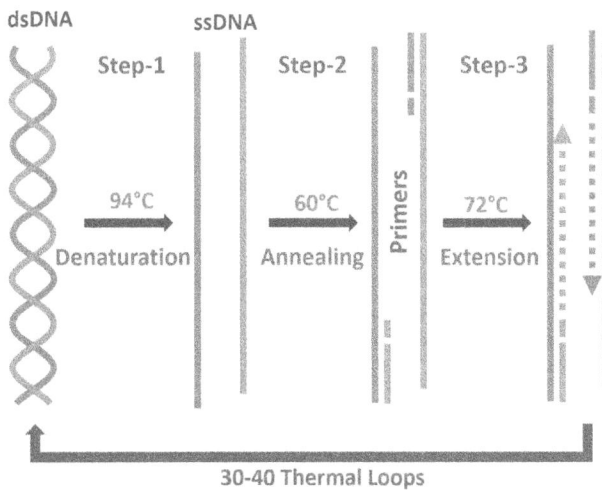

FIGURE 4.3 Three distinct thermal steps of PCR involved during the DNA amplification process: denaturation, annealing, and extension zones.

A PCR cycle normally comprises of trio temperature regions. In the first region, denaturation (94°C), the double-stranded (dsDNA) gets separated into two separate single strands. Thereafter, the thermal scale is reduced for the annealing stage (56°C) where small primers of complementary templates to that of target genes can temper the single-stranded (ssDNA) template. Further, primers were attached to the complete strand and the temperature is increased for the third region, i.e., extension stage at 72°C. The activity of the primer increases, synthesizing an additional complementary nucleic acid strand from nucleotides in reaction samples (Welch 2012). Initially, the ds DNA is passed through three distinct stages, where the denaturation separates the double strand into single at 94°C and annealing adds short primers to these single strands at 58°C. Further, during the elongation stage at 72°C, the primers get elongated to form dsDNA. This process goes on multiplying to form exponential factor 2^N. Consequently, this exponential increase forms double-stranded concentration nucleic acid by repeating the aforementioned thermal stages.

In this chapter, we discuss chamber-based microfluidic PCR systems, along with their merits and demerits. Specific focus is given on recent progress made in chamber-based on-chip microfluidic PCR devices to comprehend real-time portable microfluidic PCR devices with automation, fabrication, integration, and miniaturization applicable for point-of-care (POC) analyzing, and diagnosing genetical and bacterial diseases. This review focuses on the recent advancements in temporal-based PCR devices. In this work, we discuss the design and development of recent trends in chamber-based miniaturized microfluidic PCR devices. In chamber-based microfluidic PCR devices, there are few shortcomings such as execution time is longer for thermal cycles, chances of sample loss due to evaporation, and high thermal inertia. Further, this can be resolved by varying the microchannel dimensions and implying different heaters. In this chapter, we focus on various parameters such as the design of the microchip, materials, and methods, and finally, validation of results. Lastly, the potential challenges and future prospects associated with integrated time domain based microfluidic PCR chips are discussed.

4.2 DESIGN AND DEVELOPMENT OF VARIOUS PCR MICROFLUIDIC DEVICES

The microchannel-based PCR platform has endured comprehensive growth in the biological and biochemical research area. Nowadays, these have become an indispensable platform in miniaturization technology with many advantages and in diverse applications. In the last few years, micro-electro-mechanical system (MEMS) evolution (Grafton et al. 2011) has been believed as a potential mechanism to develop a miniaturized device, because the tremendous transformation has been noticed in the development of microfluidic devices due to its small volume, better performance with low power, large surface-to-volume ratio, reduced cost of reagents, high throughput, and can be collectively used with supplementary devices as an LoC platform (Erickson and Li 2004). Microfluidic chips have previously been used in biological and chemical analysis (Vyawahare, Griffiths, and Merten 2010), clinical diagnostics (Lee et al. 2010), environmental monitoring (Pol et al. 2017), cell study and analysis (Dmytryshyn 2011), and drug delivery (Kleinstreuer, Li, and Koo 2008).

Among all other various microfluidic chips, the PCR chip has grown as a highly indispensable device in current biomedical and biological research areas, and also within related areas. The PCR microdevices have been extensively used as a genetic biological mechanism to amplify or exponentially increase the DNA templates by three distinct steps involved in the thermocycling process. After the initialization and invention of the PCR chip by Northrup et al. (1995), several research labs, academia, and bio-companies have begun to study the working principle on a miniaturized platform to overcome the existing conventional method. This led to the huge growth toward the development of PCR microfluidic devices with different methods, techniques, and concepts. During the progress of microchannel-based PCR devices, various approaches have been depicted and determined by several researchers. Presently, herein, the PCR microfluidic devices can be categorized into a pair of kinds: chamber-based PCR chips and flow-through PCR chips. Table 4.1 summarizes an overview of different chamber-based microfluidic PCR devices with parameters such as design, method, and approaches utilized in the development of the microfluidic PCR chips including commercial PCR devices.

4.2.1 Material and Apparatus

The microfluidic chip utilizes different substrate and materials for microchip fabrication, which include glass, silicon, metals, quartz, alloys, and organic polymers such as flexible plastic, epoxy (Koerner et al. 2005), polydimethylsiloxane (Zhang et al. 2010), polyethylene terephthalate (Hu and Chen 2018), polymethyl methacrylate (Kulkarni et al. 2020), polyimide (Metz, Holzer, and Renaud 2001), and polycarbonate (Su et al. 2019). Amongst all, silicon (Martinoia et al. 1999) and glass (Roeraade and Stjernström 1998) have good thermal constancy and good chemical inertness. Comparatively, glass has a higher mechanical strength than silicon material, thus it is a superior material for microfluidic PCR chips. In organic polymers, polydimethylsiloxane (PDMS) material shows great biocompatibility and dominant softness, and its flat layer has a very fine attraction among several materials, which makes it comfortable for reversible encapsulation, and can be utilized for biochemical analytic systems (Yamada and Seki 2004). However, there are few drawbacks of PDMS for fabricating PCR chips such as poor heat dissipation and low thermal conductivity. These shortcomings can be overcome by developing hybrid technology (Lee et al. 2011) chips that use PDMS-glass or PDMS-PMMA (poly(methyl methacrylate)) materials, where the heat dissipation and optical properties are enhanced concurrently (Xue, Khan, and Yang 2009).

4.2.2 Fabrication Techniques and Bonding Process

A suitable and sustainable fabrication process plays an instrumental role in the development and advancement of microfluidic devices that can be exactly processed by relatively new technologies such as soft lithography (Kim et al. 2008), computer numerical control (CNC) milling (Guckenberger et al. 2015), CO_2 laser ablation

TABLE 4.1

A Summary Table Showing an Overview of Various Designs and Approaches of Chamber-Based Microfluidic PCR Devices

Technique	Design and Method	Approach	Fabrication Material	Target Pathogen	Detection	Cycles/Time	Reaction Volume	Year	Ref
Well-based	Single chamber	Real-time fluorescence signal identification in a single chamber. The single-use microchip that provides double-sided heating of the PCR chamber for rapid thermal cycling.	Cellulose-based filter paper	Genomic 305 bp of DNA *Bacillus cereus*	Fluorescence signal and 1.5% agarose gel electrophoresis	40 cycles	25 µl	2014	Qiu and Mauk (2015)
	Multi-chamber array	The dimensions are 6 × 3 mm with 920 µm, 12 × 3 mm with 460 µm, and 18 × 4 mm with 230 µm thin-film PCR multi-chips. A POCT system to detect foodborne pathogens through genetic analysis.	PET, PVC adhesive film	*Bacillus cereus* gene	Gel electrophoresis	30 cycles	20 µl	2018	Bae et al. (2018)
	Droplet-based virtual-chamber	A multivolume multilevel vertical dividing microchannel dPCR microchip. The dPCR chip attained a dynamic range of over 10^4 for DNA amplification with 1,792 chambers. The PDMS micro-array layer comprises two sorts of cylindrical chambers with different volumes and quantities.	PDMS and glass sandwiched substrate	KRAS plasmid DNA fragment	Fluorescence intensity	45 cycles; 80 min	1 µl typical volume for a single array	2020	Si et al. (2020)
		The gradient is measured over encapsulated water-based drops of a thermal-dependent dye inside mineral oil. The emission spectrum of fluorescent droplets through two-step PCR amplification.	Glass slide	DNA segment of Avian virus	2% agarose gel electrophoresis and Fluorescence signal	30 min	25 µl	2017	Li, Wu, and Manz (2017)

(Continued)

Technique	Design and Method	Approach	Fabrication Material	Target Pathogen	Detection	Cycles/Time	Reaction Volume	Year	Ref
		To generate droplets in a 4-fold tapered Teflon tube of even size using only two 34 g needles without mixing any surfactant to the mineral oil.	PDMS and Teflon tube	Clinical serum samples of HBV	CMOS-based fluorescence video analysis	40 cycles; 30 to 40 mins	50 µl	2020	Li et al. (2020)
		Solar-based power operation, no need for external power supply for portable single digital PCR device. Absolute quantitative of the hepatitis B virus for the detection in human serum. Two temperature-based PCR devices.	CNC-processed copper plates wrapped Teflon tube	HBV	Fluorescence droplet images for analysis	40 cycles	18 µl	2020	Jiang, Manz, and Wu (2020)
Isothermal	LAMP	An isothermal temperature at a constant 65°C. Rapid and inexpensive detection for different food matrices.	Polypropylene tube	*Salmonella spp.*	2% agarose gel and EtBr for visualization	60 min	10 µl	2012	Ziros et al. (2012)
	RPA	An electrochemical-based stem-loop (S-L) probe DNA detection technique. An RPA-based isothermal temperature at a constant 38°C.	Acetate tube	*Staphylococcus epidermidis*	Gel electrophoresis	30 min	50 µl	2020	Khaliliazar et al. (2020)
		A portable uMED device was used as a POC diagnostic testing module to detect infectious diseases. The thermoregulation was incorporated and is accomplished with an accuracy of +/-0.1°C	Cellulose membrane (paper-based)	*Mycobacterium smegmatis* for *M. tuberculosis*	Electrochemical detection	20 min	25 µl	2018	Tsaloglou et al. (2018)
	NAAT	The reaction sample is mixed before injecting it into the chamber and infused through a pipette. A heater was placed at the bottom of the microchip at 65°C constantly.	Clear plastic microchamber	Proteinase K, Lysosome	Fluorescence-based CCD camera	65°C, 20 min	25 µl	2018	Mauk et al. (2018)

(Chen, Li, and Shen 2016), and 3D printing (Puneeth and Goel 2019). Similarly, the available traditional methods, such as injection molding (Attia, Marson, and Alcock 2009), micro-machining (Ziaie et al. 2004), laser ablation (Isiksacan et al. 2016), and hot embossing (Becker and Dietz 1998), have also been functional in fabricating the microfluidic devices. Further, during microchannel encapsulation, strong and perpetual bonding can be accomplished by instant adhesive strength after using oxygen plasma treatment or ultraviolet (UV) irradiation (Sun et al. 2012). In microfluidic devices, the bonding of the microchannels is very essential to provide smooth movement of the fluid within the microchannel, without any leakages and blockages in the channel. There are different ways of bonding process microfluidic devices such as oxygen plasma surface treatment (Xu Li et al. 2008), chemical bonding (Sivakumar and Lee 2020), thermal bonding (Sun, Kwok, and Nguyen 2006), and solvent bonding (Chen and Duong 2016).

4.3 SINGLE CHAMBER BASED PCR DEVICES

In a chamber-based PCR device, the PCR reagents are inserted into the reservoir and then the complete microchip, along with the reaction sample, is heated from the bottom of the chamber and cooled with a fan fixed on top of the microchannel actively via predefined thermal loops associated with temperature zones. The sample remains stationary and the temperature varies with time. This approach is a time-domain process because the sample temperature is reliant on time. In this, the temperature distribution is not reliant on microchannel designing aspects. Further, any variations to PCR thermal loops can be completed by an amendment to the heating and cooling procedures. The chamber-based PCR is also known as temporal domain PCR, static PCR chips, or microchamber PCR. The chamber-based PCR chips are the miniaturized edition of a conventional thermocycler and they comprise four types: single chamber (Cheong et al. 2008), multi-chamber array (Frey et al. 2007), droplet virtual chamber PCR chip (Wang and Burns 2009), and isothermal amplification process (Zanoli and Spoto 2013).

In 1993, a single-chamber PCR microdevice was designed by Northrup et al. (1995) for the fabrication of the single chamber PCR chip that had a simple structure with silicon glass used as a substrate. The silicon wafer was engraved aeolotropically, and the top portion of the PCR chamber was bonded anodically to a borosilicate glass slide. A β-actin gene sample volume of 50 µl reaction was amplified using this well-based PCR device. The well-based PCR approach was the first attempt reported in the advancement technical apparatus with a small volume of sample and simplified experimental setup of the chamber-based chip. However, the fabrication of the chip and bonding of PCR required exorbitant tools and procedures, which require sophisticated microfabrication amenities. Giordano et al. (2001) developed a simple chamber-based single microchannel chip, and polyimide was used as a fabrication material with infrared (IR) heaters as a heating source. The authors amplified the fragment of 500 base pairs (bp) successfully, which were detected in 4 minutes.

Qiu and Mauk (2015) reported an integrated single chamber PCR chip, whereby a cellulose-based filter membrane was utilized to extract the DNA sample (Figure 4.4A). In this work, a single disposable microfluidic chip with 25 µl of sample volume, with a portable instrument, was used. The devices offered dual side heating

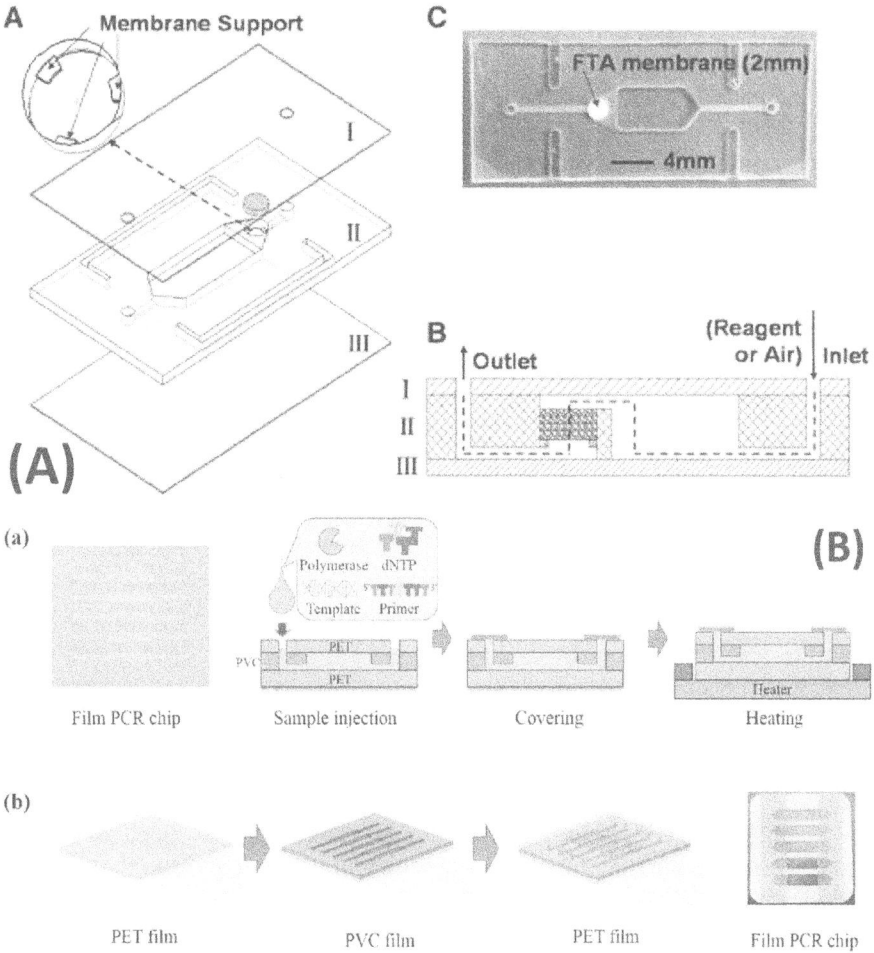

FIGURE 4.4 (A) Single chamber based PCR system: (a) exploded outlook; (b) cross section; (c) a fabricated plastic microchip with polycarbonate (PC) substrate. (Images reprinted with permission of copyright from respective journals: Qiu and Mauk 2015.) (B) A multi-chamber-based microfluidic thin-film PCR microchip: (a) a schematic representation of DNA amplification by using the thin film PCR chip that could efficiently amplify the genetic pathogen by smearing to the minimized heating system; (b) the developed multiplex film-based PCR microchip, which contains polyethylene terephthalate (PET) and PVC was fabricated by stacking each with equipped film layer-by-layer; (c) describes a photograph of the dPCR archetype device with red dye included in it. (Images reprinted with permission of copyright from respective journals: Bae et al. 2018.)

for a single chamber PCR chip exhibiting quick thermal cycling. Real-time fluorescence detection was incorporated with a commercially available fluorescence reader. A cellulose-based filter paper (Whatmann FTA) acted as a medium for sample placement where a multiple pass sample was loaded onto the membrane. A nucleic acid segment of *Bacillus cereus* was successfully amplified and executed a detection limit of 10^3 target cells.

4.4   MULTI-CHAMBER ARRAY PCR DEVICES

According to Zou et al. (2005), the chamber-based PCR can be miniaturized with multi-chamber assembly for high PCR throughput by fabricating microchambers where the silicon was etched aeolotropically. The silicon-based microchamber of multiple array was bonded with a printed circuit board (PCB) through the novel flip-chip bonding process. Four microchambers were fabricated on the PCR chip, which was used for DNA quantification of the β-glucuronidase gene. Here, a template of the target DNA sample was amplified successfully for 30 thermal loops with an initial sample volume of 3 µl in 30 minutes. In this study, the authors performed the DNA amplification of four reactions simultaneously by integrated the heater and sensor on a single metal plate. Here, for real-time precise control and measurement of temperature, the heating element and sensing source was positioned onto the bottom layer of the silicon-based microchambers for proper heat transfer by depositing metals such as platinum and titanium by lift-off technique. However, this deposition of metal needed special equipment for deposition and ensuing chemical measures, which make the platform not flexible to users. Taylor et al. (1997) described a PCR microdevice with a total of 48 microwells in a 2 mm² sized silicon medium where 0.5 µl of the sample was filled in each chamber. A β-actin of fragment 297 bp DNA template was quantified and detected through the fluorescence method successfully in real time by using TaqMan polymerase. Zou et al. (2002) demonstrated that the device can be fabricated on a thermoplastic sheet, known as polyethylene terephthalate (PET) substrate, and a silicon heating block was used as a heat source with a temperature controller module. The device with a microchamber thermocycler was competent in multiplexing up to 16 simultaneous reactions. A plasmid DNA sample was successfully amplified with sample volume up to 20 µl within 8 minutes for 30 thermal cycles. Further, Nagai et al. (2001) carried out the amplification for an 85 pico liter (pl) sample of a pGFP gene segment. In this work, the fabrication was done on a silicon wafer through aeolotropically engraving the multi-chamber microarray. During the DNA amplification process, the fluorescence detection signal was generated for real-time PCR. It was detected using a dye (YOYO-1), thus capable of real-time detection of the fluorescence signal. The authors also reported in this work that simultaneous performance of both quantification and identification for parallel reaction mixture without withdrawing the amplified samples outside for later identification of the DNA fragments.

Bae et al. (2018), designed a multi-chamber film-based disposable PCR detection device (Figure 4.4B). Here, a simplified fabrication technology was described where two substrates were used such as polyethylene terephthalate (PET) and a double-sided adhesive thin layer material (polyvinyl chloride (PVC)) was used to have perfect bonding with no leakages. After the thermal bonding on the stereotyped device, the *Bacillus cereus* DNA segment was amplified successfully on a sole film PCR microchip. The demonstrated device was used to identify the foodborne virulent through gene analysis. A total of five individual thin-film microfluidic devices were fabricated on a single substrate. The stacking method was used to place three polymeric film layers on the same platform.

Si et al. (2020), described a multivolume branch of microchannel setup with an inexpensive and easy-to-use digital PCR (dPCR) device (Figure 4.5A) The microchip

A

(A)

Inlet

Branch channel

Main channel

Big chamber

Small chamber

B

Cover glass

PDMS micro-array layer

Substrate glass

C

(B)

Camera

Lens+emission filter

LED

Computer

Excitation filter

Droplet array

Cover glass

Silicon wafer

Copper plate

HEAT SINK

Pt100 TEC1 TEC2 Pt100

TEC control module

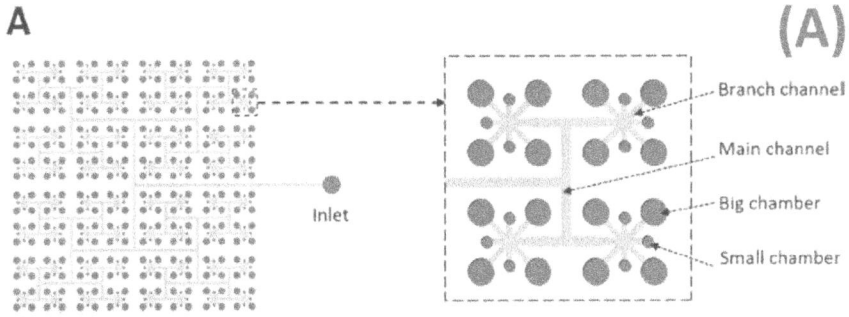

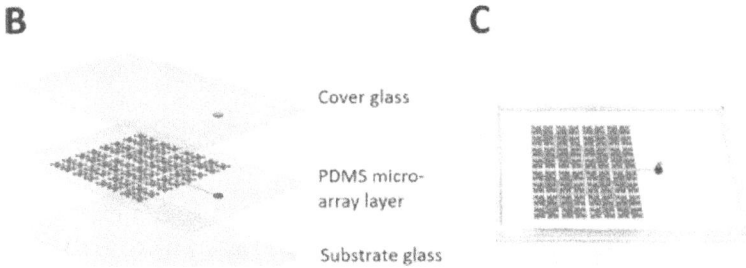

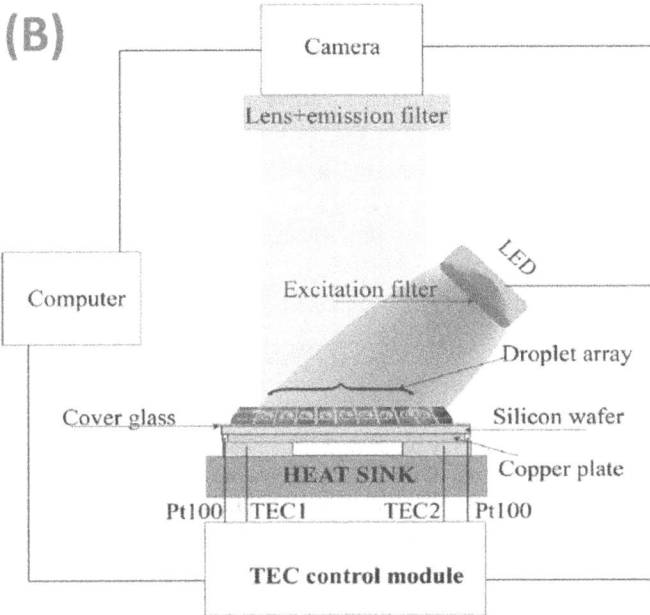

FIGURE 4.5 Continued

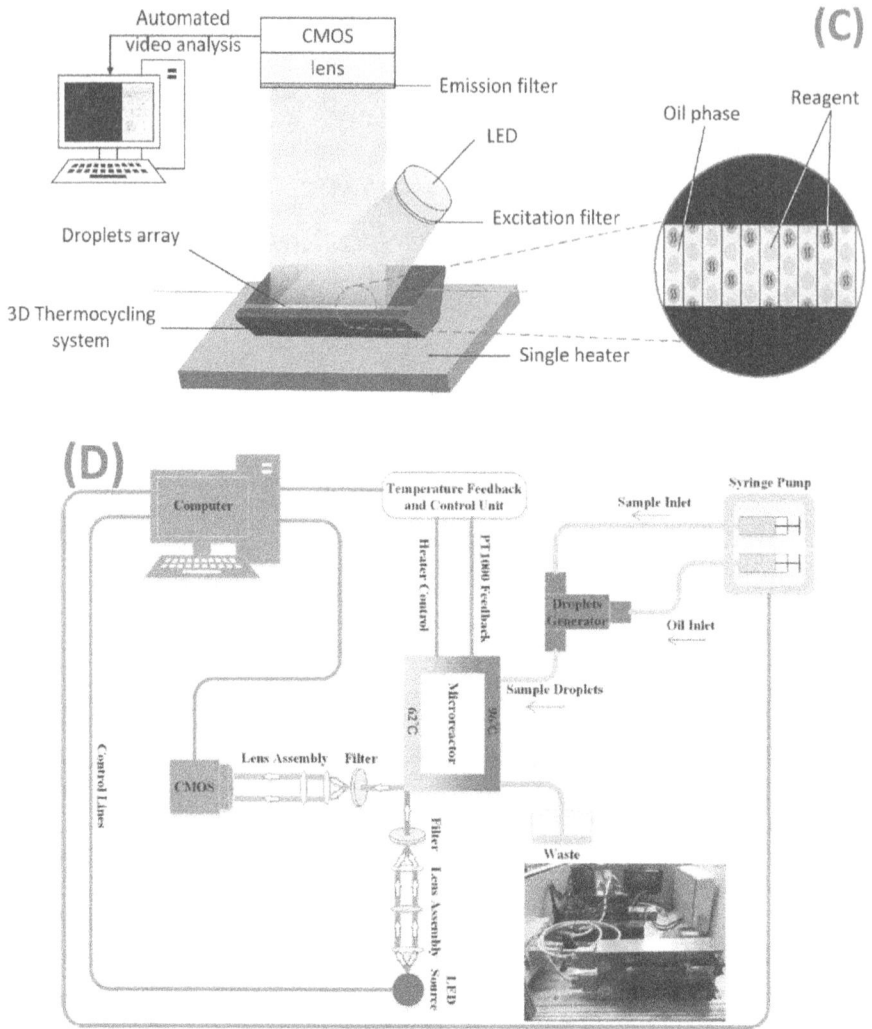

FIGURE 4.5   (A) Droplet-based virtual chamber and integrated microreactor PCR devices:
(a) a representation of a microarray-based multi-chamber platform for nucleic acid amplification; (b) a representation of the dPCR chip structural design; (c) the microchip comprises
two categories of microchambers with diverse volumes and quantities. (Image reprinted with
permission of copyright from respective journals: Si et al. 2020.) (B) A simple illustration
of a droplet-based microfluidic PCR chip with peltier element (TEC) as a heating source
and LED-based fluorescence signal with a feedback camera connected to a computer for
analysis and detection. (Image reprinted with permission of copyright from respective journals: Li, Wu, and Manz 2017.) (C) A unique CMOS-based automated detection unit for the
microfluidic droplet array is executed on a single heater. (Image reprinted with permission
of copyright from respective journals: Li et al. 2020.) (D) A fully automated and integrated
microfluidic module for a microfluidic PCR microreactor for DNA amplification using an
automated sample injection system with a droplet generator. (Image reprinted with permission of copyright from respective journals: Jiang, Manz, and Wu 2020.)

with a multivolume strategy can be attained with a microfluidic platform over 10^4 for DNA quantification with 1,792 microchambers, allowing for parallel processing. It includes two kinds of microchambers with discrete volume and quantity that can be effectively used for the operation utilizing the entire space on the microdevice. In this analysis, this method realizes 100% bifurcation and compartmentalization of reaction composition without mixing and with parallel amplification. A glass-PDMS-based hybrid chip was fabricated and sandwiched accordingly. Additionally, the microchip was driven by negative density offered by degassed PDMS. The dPCR microdevice was appraised by analyzing a 10-fold sequential dilution using a plasmid template. Besides, it included many features such as wide active range, low cost, no reaction loss, and simple detection process. The microchannel was separated by a vertical lane with several subdivisions process-by-process. The green line in Figure 4.5A signified the microchannels, and the red circles represent the microchambers. Also, the individual end of the microchannel was associated with four big and three small microchambers via constricted subdivision microchannels. The large microchamber dimensions of the diameter and height were 225 µm and 100 µm, respectively, and the small microchamber dimension was 100 µm for both height and diameter. The width of the core channel was around 80 µm and that of the branch microchannels was 50 µm and the height was 30 µm.

4.5 DROPLET-BASED VIRTUAL CHAMBER-BASED PCR MICROFLUIDIC DEVICES

Neuzil, Pipper, and Hsieh (2006), demonstrated a droplet-based virtual reaction chamber using mineral oil to generate droplets and to reduce sample evaporation effectively. The designed microfluidic chip could easily perform faster heating and cooling operation and the PCR product purity was validated by the capillary electrophoresis and melting curve analysis approach. Besides, the multi-chamber array was created by proliferating the high-throughput of a sole chamber-based PCR microchannel.

Li, Wu, and Manz (2017), developed an open-source device with a fluorometric visualization and temperature gradient for the three temperature zones of the PCR. Here, a unique approach was designed where two thermoelectric devices (Peltier elements) were utilized as a heating element. The thermal ramp was measured via covered water-based beads of the thermal-dependent dye placed inside mineral oil, thus generating the droplet array-based virtual reaction chamber. A typical sample volume of 0.7 µl was used as droplets (Figure 4.5B). Temperature-sensitive inert dye was utilized as intrinsic fluorescence for the detection of intensity level at a known unvarying thermal value with the calculated calibration of unknown temperature. This approach of temperature ramping function was conveniently utilized for the optimization of the enzymatic reaction concerning the PCR reaction.

Li et al. (2020) demonstrated a miniaturized digital PCR system for the amplification of genes that significantly reduced the need for micro-droplet reaction and generation kit. The PDMS was united with a Teflon tube to form 3D microfluidics. The main benefit of this methodology is the requirement of a single heater for heating the microchip (Figure 4.5C). Two 34 g needles were used to generate and transmit micro-droplets in a tapered Teflon tube, because this is the simplest way to generate the digital PCR droplets. A complementary metal-oxide-semiconductor (CMOS) based optical-device was used for the identification unit to automatically analyze fluorescence video. Biomedical serum samples of hepatitis B virus (HBV) were realized in this technique. The 3D thermocycling system was incorporated with on the single heater. The total reaction time was about 30 to 40 minutes.

Jiang, Manz, and Wu (2020) demonstrated a fully automated and integrated microfluidic PCR system for DNA quantification. The platform included a computer, sample pumping system, micro-droplet generation, thermocycling, signal analysis unit, and fluorescence data acquisition system (Figure 4.5D). The device was able to parallelly perform multistep operations and had a potential for multiplexing. Absolute DNA quantification was monitored from the HBV in serum samples, which were further detected on this system. The reaction mixture was injected through the droplet generator module and it was fed to the microreactor for processing. Later, the fluorescence tool with an optical module was used to analyze the result. Resistance temperature-dependent based temperature sensors (PT1000) were used for monitoring the heater and the relay module was used as a driver circuit to drive the heater in terms of the current load. Labview software was used as a data acquisition system (DAQ) for controlling and monitoring the entire device.

4.6   ISOTHERMAL AMPLIFICATION BASED MICROFLUIDIC PCR DEVICES

Isothermal amplification is an approach used for the amplification of DNA and RNA samples that are extensively used in biological tests. It allows simplification in temperature control because the temperature remains constant. Besides, its mechanism uses a relatively low temperature for nucleic acid amplification in less reaction time. The PCR reaction mixture sample remains static in the chamber during the quantification process and can be functional in the lab-on-a-disc (LOD) form. Thus, the approach can be contemplated as an expanded edition of a chamber-based PCR device. When using this LOD system, it works on a compact disc-based chip for analyzing several DNA reaction samples and centrifugal processes for the insertion of a sample. In this method, the integration with the sample extraction and purification unit at the upstream of the PCR chip, and target detection unit toward the downstream, can be easily designed and realized with any complication on-chip. There are many different types of isothermal *Bacillus cereus* processes available. Among these methods, the widely used isothermal amplification techniques are loop-mediated isothermal amplification (LAMP), recombinase polymerase amplification (RPA), and the nucleic acid amplification test (NAAT).

4.6.1   LOOP-MEDIATED ISOTHERMAL AMPLIFICATION

In the LAMP process for nucleic acid amplification, template DNA can be quantified at a steady thermal range at approximately 58–63°C with a reaction time of

30–60 minutes. Also, this method is highly specific because it uses 2–3 primers for the amplification of genes. Silva, Pardee, and Pena (2019) demonstrated the diagnostics device for the identification of the Zika pathogen (ZIKV). In this work, the authors reported a rapid, handheld, sensitive, and reliable diagnostic tool for the pandemic virus. Ziros et al. (2012) developed a platform for the identification of *Salmonella spp.* strains insulated from foodborne pathogens. The *Salmonella invA* bacteria was amplified at a constant temperature of 65°C for 1 hour. The detection of the LAMP products could be visualized under UV light or ambient daylight. The detection process was faster and suitable for inexpensive detection in food laboratories. Oh et al. (2016) illustrated a fully automated detection unit for food pathogens using the LOD system. The fabrication involves a chamber designed using PMMA and PC that is sealed using dual-sided tape.

4.6.2 RECOMBINASE POLYMERASE AMPLIFICATION

In an RPA based isothermal amplification process, both DNA and RNA can be amplified. Chen et al. (2020) reported isothermal amplification using the RPA method whereby target DNA gene *Thermos thermophilus* was used for the amplification with simplified primer design and pre-denaturation. Fluorescence signals were produced using the FRET sensor known as Proof-Man. The designed device can amplify in the version of 10^2 DNA targets in 30 minutes. Khaliliazar et al. (2020) demonstrated the POC device based on the RPA method for the amplification of the *Salmonella enteritidis* template DNA gene and was electrochemically detected as meagre as 10 copies/µl in 70–75 minutes. Other studies reported that various pathogens such as *Bacillus atrophaeus* (Burke et al. 2004), *Streptococcus agalactiae* (Keefe 1997), *Escherichia coli* (Lenski, Sniegowski, and Gerrish 1997), *and Staphylococcus warneri* (Kamath, Singer, and Isenberg 1992) were detected using the LOD platform. Even though good outcomes were noticed from various examinations in terms of integration of chamber-based PCR devices, these domains focused on a solo restricted platform such as the LOD format and a few shortcomings of these methods are dependent on the thermal gradient and its accuracy.

Tsaloglou et al. (2018) demonstrated a novel approach based on isothermal amplification using electrochemical detection for specific nucleic acid templates used as a POC device (Figure 4.6A). The handy device had four main parts involved in RPA, disposable cellulose membrane strips fabricated using screen printing for the generation of carbon electrodes, a thermal management device with +/-0.1°C temperature precision, and an electrochemical detection method for validation of results using cyclic voltammetry and square-wave voltammetry, and used for the detection of *Mycobacterium smegmatis*.

4.6.3 NUCLEIC ACID AMPLIFICATION TEST

The NAAT is one of the sensitive isothermal amplification methods used to detect low-density infectious diseases, bacteria, and pathogens. It is a molecular assay with a generally more specific and sensitive approach toward nucleic acid amplification and a useful tool in detecting malaria infections and *Chlamydia trachomatis*.

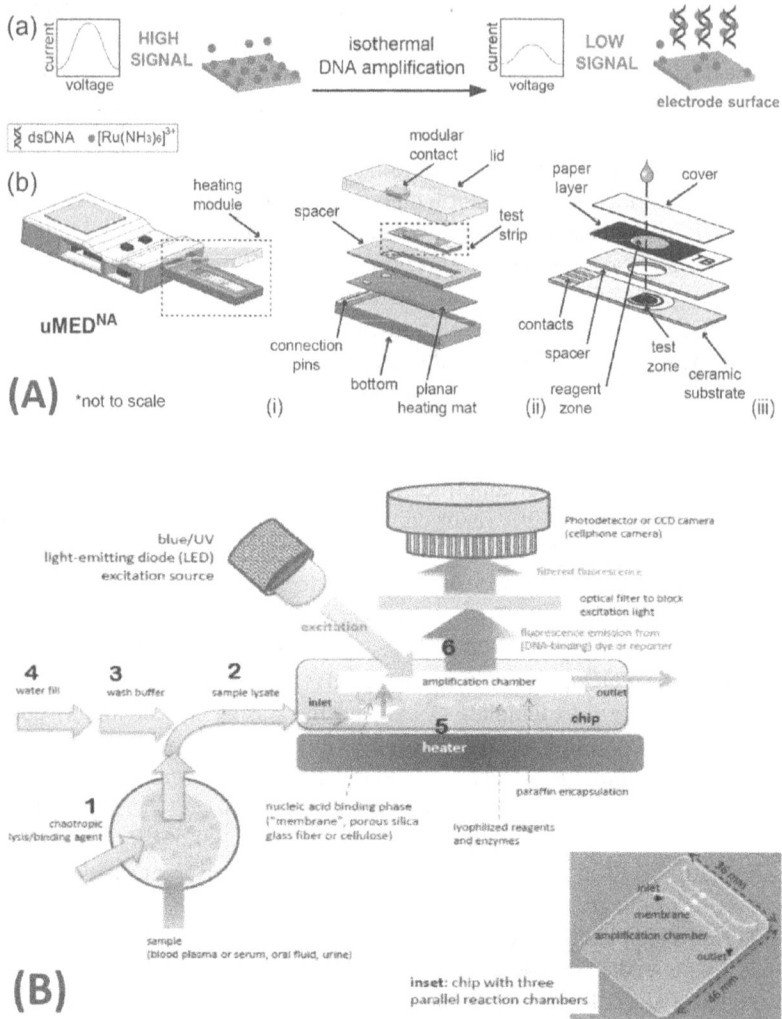

FIGURE 4.6   (A) Isothermal amplification-based PCR devices: (a) illustration of the electrochemical based RPA method; (b) a schematic approach for the portable device for nucleic acid amplification and identification system: (i) uMED module is associated with the heating element, (ii) an exploded outlook of the module, and contacts to test strip, (iii) an exploded outlook of the disposable, paper-based test strip. The dotted line signifies the movement of liquid through the disposable strip. (Images reprinted with permission of copyright from respective journals: Tsaloglou et al. 2018.) (B) Processing steps and schematic cross section of amplification chamber for the NAAT: (1) reaction gets mixed off-chip with primers, buffers, milliQ; (2) reaction sample volume of 100 μl infused into the chip with either syringe or pipette; (3) ethanol and high-salt buffer is inserted into the chip to clean the membrane; (4) microchamber of 25–50 μl is filled and sealed up with tape; (5) microchip is heated at 65°C using a low power electric heater (1 W); (6) the validation using fluorescence signal with UV light or blue LED detection using a CCD camera. (Images reprinted with permission of copyright from respective journals: Mauk et al. 2018.)

Mauk et al. (2018) designed a microfluidic device for genetic diagnostics using the NAAT with minimal instrumentation for POC prospects (Figure 4.6B). Emphasis was given to resource-limited settings during the infectious disease testing. It features an amplification reagents chamber (25 µl) and the plastic chip is fixed on the thermoregulatory stage, which has an emitting light diode, opto-filter, and a photodetector for the validation of optical measurements. A porous plug of glass silica fiber and a paper membrane having a diameter of 1–3 mm and 1 mm of thickness for sample storage is used. The PCR microchip is heated at 60°C and the reaction sample collected from the outlet is further analyzed using the fluorescence-based detection system.

4.7 CONCLUSION AND FUTURE PROSPECTS

In this review, we have discussed various chamber-based microfluidic PCR devices that have drastically undergone ample development and rapid advancement. This has intensely transformed the usage of conventional-based thermocyclers for DNA amplification due to their drawbacks such as bulky instrument, thermal dissipation, laborious technique, expensive devices, more thermal dissipation, requires more power, tedious approach, and manual processing. The chamber-based systems appear with numerous advantages over existing conventional methods such as easy-to-operate, portability, minimum sample volume, automation, integration, miniaturization, low thermal loss, faster reaction time, real-time sample analysis, transparent processing, and minimal power needed to operate. Chamber-based devices were developed to meet challenges in the biological and medicinal fields. Since then, they have gained a wider range of advantages and applications in the detection and quantification of pathogens, bacteria, and viruses. Undoubtedly, chamber-based microfluidic PCR devices have a wide scope of future use with the advantage of inexpensive, miniaturized, and portable PCR microchips that can be widely used in the future in biological and pathological applications for the rapid amplification of nucleic acid and easy real-time detection of various pathogens and viruses. In the future scope of research, microfluidic PCR devices that integrate other features of testing such as sample extraction and preparation would be a major step in a microfluidic environment.

REFERENCES

Ahrberg, Christian D., Andreas Manz, and Bong Geun Chung. 2016. "Polymerase chain reaction in microfluidic devices." *Lab on a Chip* 16 (20): 3866–84. https://doi.org/10.1039/c6lc00984k.
Attia, Usama M., Silvia Marson, and Jeffrey R. Alcock. 2009. "Micro-injection moulding of polymer microfluidic devices." *Microfluidics and Nanofluidics* 7 (1): 1–28. https://doi.org/10.1007/s10404-009-0421-x.
Bae, Nam Ho, Sun Young Lim, Younseong Song, Soon Woo Jeong, Seol Yi Shin, Yong Tae Kim, Tae Jae Lee, et al. 2018. "A disposable and multi-chamber film-based PCR chip for detection of foodborne pathogen." *Sensors (Switzerland)* 18 (9): 1–10. https://doi.org/10.3390/s18093158.
Becker, Holger, and Wolfram Dietz. 1998. "Microfluidic devices for µ-TAS applications fabricated by polymer hot embossing." *Microfluidic Devices and Systems* 3515 (September): 177. https://doi.org/10.1117/12.322081.

Burke, S. A., J. D. Wright, M. K. Robinson, B. V. Bronk, and R. L. Warren. 2004. "Detection of molecular diversity in *Bacillus atrophaeus* by amplified fragment length polymorphism analysis." *Applied and Environmental Microbiology* 70 (5): 2786–90. https://doi.org/10.1128/AEM.70.5.2786-2790.2004.

Chen, Gangyi, Rong Chen, Sheng Ding, Mei Li, Jiayu Wang, Jiawei Zou, Feng Du, et al. 2020. "Recombinase assisted loop-mediated isothermal DNA amplification." *Analyst* 145 (2): 440–44. https://doi.org/10.1039/c9an01701a.

Chen, Pin Chuan, and Lynh Huyen Duong. 2016. "Novel solvent bonding method for thermoplastic microfluidic chips." *Sensors and Actuators, B: Chemical* 237: 556–62. https://doi.org/10.1016/j.snb.2016.06.135.

Chen, Wenwen, Fangchi Shao, and Yunlei Xianyu. 2020. "Microfluidics-implemented biochemical assays: From the perspective of readout." *Small* 16 (9): 1–19. https://doi.org/10.1002/smll.201903388.

Chen, Xueye, T. Li, and J. Shen. 2016. "CO2 laser ablation of microchannel on PMMA substrate for effective fabrication of microfluidic chips." *International Polymer Processing* 31 (2): 233–38. https://doi.org/10.3139/217.3184.

Cheong, Kwang Ho, Dong Kee Yi, Jeong Gun Lee, Jong Myeon Park, Min Jun Kim, Joshua B. Edel, and Christopher Ko. 2008. "Gold nanoparticles for one step DNA extraction and real-time PCR of pathogens in a single chamber." *Lab on a Chip* 8 (5): 810–13. https://doi.org/10.1039/b717382b.

Cui, Ping, and S. Wang. 2019. "Application of microfluidic chip technology in pharmaceutical analysis: A review." *Journal of Pharmaceutical Analysis* 9 (4): 238–47. https://doi.org/10.1016/j.jpha.2018.12.001.

Das, Susmita, and Vimal Chandra Srivastava. 2016. "Microfluidic-based photocatalytic microreactor for environmental application: A review of fabrication substrates and techniques, and operating parameters." *Photochemical and Photobiological Sciences* 15 (6): 714–30. https://doi.org/10.1039/c5pp00469a.

Dmytryshyn, Bogdan. 2011. "Microfluidic cell culture systems and cellular analysis." *2011 Proceedings of 7th International Conference on Perspective Technologies and Methods in MEMS Design, MEMSTECH 2011* 1 (1): 193–96.

Erickson, David, and Dongqing Li. 2004. "Integrated microfluidic devices." *Analytica Chimica Acta* 507 (1): 11–26. https://doi.org/10.1016/j.aca.2003.09.019.

Frey, Olivier, Sonja Bonneick, Andreas Hierlemann, and Jan Lichtenberg. 2007. "Autonomous microfluidic multi-channel chip for real-time PCR with integrated liquid handling." *Biomedical Microdevices* 9 (5): 711–18. https://doi.org/10.1007/s10544-007-9080-4.

Giordano, B. C., J. Ferrance, S. Swedberg, A. F.R. Hühmer, and J. P. Landers. 2001. "Polymerase chain reaction in polymeric microchips: DNA amplification in less than 240 seconds." *Analytical Biochemistry* 291 (1): 124–32. https://doi.org/10.1006/abio.2000.4974.

Grafton, Meggie M. G., Teimour Maleki, Michael D. Zordan, Lisa M. Reece, Ron Byrnes, Alan Jones, Paul Todd, and James F. Leary. 2011. "Microfluidic MEMS hand-held flow cytometer." *Microfluidics, BioMEMS, and Medical Microsystems IX* 7929: 79290C. https://doi.org/10.1117/12.874299.

Guckenberger, David J., Theodorus E. De Groot, Alwin M.D. Wan, David J. Beebe, and Edmond W.K. Young. 2015. "Micromilling: A method for ultra-rapid prototyping of plastic microfluidic devices." *Lab on a Chip* 15 (11): 2364–78. https://doi.org/10.1039/c5lc00234f.

Hu, Z. L., and Xue Ye Chen. 2018. "Fabrication of polyethylene terephthalate microfluidic chip using Co2 laser system." *International Polymer Processing* 33 (1): 106–9. https://doi.org/10.3139/217.3447.

Isiksacan, Ziya, M. Tahsin Guler, Berkan Aydogdu, Ismail Bilican, and Caglar Elbuken. 2016. "Rapid fabrication of microfluidic PDMS devices from reusable PDMS molds using laser ablation." *Journal of Micromechanics and Microengineering* 26 (3). https:// doi.org/10.1088/0960-1317/26/3/035008.

Jiang, Yangyang, Andreas Manz, and Wenming Wu. 2020. "Fully automatic integrated continuous-flow digital PCR device for absolute DNA quantification." *Analytica Chimica Acta* 1125: 50–56. https://doi.org/10.1016/j.aca.2020.05.044.

Jokerst, Jana C., Jason M. Emory, and Charles S. Henry. 2012. "Advances in microfluidics for environmental analysis." *Analyst* 137 (1): 24–34. https://doi.org/10.1039/c1an15368d.

Kamath, U., C. Singer, and H. D. Isenberg. 1992. "Clinical significance of Staphylococcus warneri bacteremia." *Journal of Clinical Microbiology* 30 (2): 261–64. https://doi. org/10.1128/jcm.30.2.261-264.1992.

Kant, Krishna, Mohammad Ali Shahbazi, Vivek Priy Dave, Tien Anh Ngo, Vinayaka Aaydha Chidambara, Linh Quyen Than, Dang Duong Bang, and Anders Wolff. 2018. "Microfluidic devices for sample preparation and rapid detection of foodborne pathogens." *Biotechnology Advances* 36 (4): 1003–24. https://doi.org/10.1016/j. biotechadv.2018.03.002.

Keefe, Gregory P. 1997. "Streptococcus agalactiae mastitis: A review." *Canadian Veterinary Journal* 38 (7): 429–37.

Khaliliazar, Shirin, Liangqi Ouyang, Andrew Piper, Georgios Chondrogiannis, Martin Hanze, Anna Herland, and Mahiar Max Hamedi. 2020. "Electrochemical detection of genomic DNA utilizing recombinase polymerase amplification and stem-loop probe." *ACS Omega* 5 (21): 12103–9. https://doi.org/10.1021/acsomega.0c00341.

Kim, Pilnam, Keon Woo Kwon, Min Cheol Park, Sung Hoon Lee, Sun Min Kim, and Kahp Yang Suh. 2008. "Soft lithography for microfluidics: A review." *Biochip Journal* 2 (1): 1–11.

Kleinstreuer, Clement, Jie Li, and Junemo Koo. 2008. "Microfluidics of nano-drug delivery." *International Journal of Heat and Mass Transfer* 51 (23–24): 5590–97. https://doi. org/10.1016/j.ijheatmasstransfer.2008.04.043.

Knoška, Juraj, Luigi Adriano, Salah Awel, Kenneth R. Beyerlein, Oleksandr Yefanov, Dominik Oberthuer, Gisel E. Peña Murillo, et al. 2020. "Ultracompact 3D microfluidics for time-resolved structural biology." *Nature Communications* 11 (1): 1–12. https:// doi.org/10.1038/s41467-020-14434-6.

Koerner, Terry, Laurie Brown, Ruixi Xie, and Richard D. Oleschuk. 2005. "Epoxy resins as stamps for hot embossing of microstructures and microfluidic channels." *Sensors and Actuators, B: Chemical* 107 (2): 632–39. https://doi.org/10.1016/j.snb.2004.11.035.

Kudr, Jiri, Ondrej Zitka, Martin Klimanek, Radimir Vrba, and Vojtech Adam. 2017. "Microfluidic electrochemical devices for pollution analysis—A review." *Sensors and Actuators, B: Chemical* 246: 578–90. https://doi.org/10.1016/j.snb.2017.02.052.

Kugimiya, Akimitsu, Akane Fujikawa, Xiao Jiang, Z. Hugh Fan, Toshikazu Nishida, Jiro Kohda, Yasuhisa Nakano, and Yu Takano. 2020. "Microfluidic paper-based analytical device for histidine determination." *Applied Biochemistry and Biotechnology*, 3–4. https://doi.org/10.1007/s12010-020-03365-z.

Kulkarni, Madhusudan B., Yashas, Prasanth K. Enaganti, Khairunnisa Amreen, and Sanket Goel. 2020. "IoT enabled portable thermal management system with microfluidic platform to synthesize MnO2 nanoparticles for electrochemical sensing." *IOP Nanotechnology*: 1–8. https://doi.org/10.1088/1361-6528/ab9ed8.

Kuswandi, Bambang, Nuriman, Jurriaan Huskens, and Willem Verboom. 2007. "Optical sensing systems for microfluidic devices: A review." *Analytica Chimica Acta* 601 (2): 141–55. https://doi.org/10.1016/j.aca.2007.08.046.

Lee, Kang Wook, Akihiro Noriki, Kouji Kiyoyama, Takafumi Fukushima, Tetsu Tanaka, and Mitsumasa Koyanagi. 2011. "Three-dimensional hybrid integration technology of CMOS, MEMS, and photonics circuits for optoelectronic heterogeneous integrated systems." *IEEE Transactions on Electron Devices* 58 (3): 748–57. https://doi.org/10.1109/ TED.2010.2099870.

Lee, Won Gu, Yun Gon Kim, Bong Geun Chung, Utkan Demirci, and Ali Khademhosseini. 2010. "Nano/microfluidics for diagnosis of infectious diseases in developing countries." *Advanced Drug Delivery Reviews* 62 (4–5): 449–57. https://doi.org/10.1016/j. addr.2009.11.016.

Lenski, Richard E, P. D. Sniegowski, and P. J. Gerrish. 1997. "Evolution of high mutation rates in experimental populations of E. coli." *Nature* 387 (June): 703–5.

Li, Xiangping, Wenming Wu, and Andreas Manz. 2017. "Thermal gradient for fluorometric optimization of droplet PCR in virtual reaction chambers." *Microchimica Acta* 184 (9): 3433–39. https://doi.org/10.1007/s00604-017-2353-6.

Li, Xu, Junfei Tian, Thanh Nguyen, and Wei Shen. 2008. "Paper-based microfluidic devices by plasma." *Lab on a Chip* 80 (2008903553): 9131–34. https://doi.org/10.1039/ b811135a.10.1021/ac801729t.

Li Bin, Yuanming Li, Andreas Manz, and Wenming Wu. 2020. "Miniaturized continuous-flow digital PCR for clinical-level serum sample based on the 3D microfluidics and CMOS imaging device." *Sensors (Basel)* 20 (9). https://doi.org/10.3390/s20092492.

Martinoia, Sergio, Marco Bove, Mariateresa Tedesco, Benno Margesin, and Massimo Grattarola. 1999. "A simple microfluidic system for patterning populations of neurons on silicon micromachined substrates." *Journal of Neuroscience Methods* 87 (1): 35–44. https://doi.org/10.1016/S0165-0270(98)00154-X.

Mauk, Michael G., Jinzhao Song, Changchun Liu, and Haim H. Bau. 2018. "Simple approaches to minimally-instrumented, microfluidic-based point-of-care nucleic acid amplification tests." *Biosensors* 8 (1). https://doi.org/10.3390/bios8010017.

Melin, Jessica, and Stephen R. Quake. 2007. "Microfluidic large-scale integration: The evolution of design rules for biological automation." *Annual Review of Biophysics and Biomolecular Structure* 36: 213–31. https://doi.org/10.1146/annurev.biophys.36. 040306.132646.

Metz, Stefan, Raphael Holzer, and Philippe Renaud. 2001. "Polyimide-based microfluidic devices." *Lab on a Chip—Minituarization for Chemistry and Biology* 1 (1): 29–34. https://doi.org/10.1039/b103896f.

Mullis, K., F. Faloona, S. Scharf, R. Saiki, G. Horn, and H. Erlich. 1992. "Specific enzymatic amplification of DNA in vitro: The polymerase chain reaction. 1986." *Biotechnology (Reading, Mass.)* 24 (Table 1): 17–27.

Nagai, H., Y. Murakami, Y. Morita, K. Yokoyama, and E. Tamiya. 2001. "Development of a microchamber array for picoliter PCR." *Analytical Chemistry* 73 (5): 1043–47. https:// doi.org/10.1021/ac000648u.

Neuzil, Pavel, Juergen Pipper, and Tseng Ming Hsieh. 2006. "Disposable real-time microPCR device: Lab-on-a-chip at a low cost." *Molecular BioSystems* 2 (6): 292–98. https://doi. org/10.1039/b605957k.

Northrup, M. A., C. Gonzalez, D. Hadley, R. F. Hills, P. Landre, S. Lehew, R. Saiki, J. J. Sninsky, R. Watson, and R. Watson. 1995. "MEMS-based miniature DNA analysis system." *International Conference on Solid-State Sensors and Actuators, and Eurosensors IX, Proceedings* 1: 764–67.

Northrup, M Allen, Carlos Gonzalez, Stacy Lehew, and Rob Hills. 1995. "Development of a PCR-microreactor," In *Micro Total Analysis Systems*. Springer, Dordrecht. doi. org/10.1007/978-94-011-0161-5_13.

Oh, Seung Jun, Byung Hyun Park, Goro Choi, Ji Hyun Seo, Jae Hwan Jung, Jong Seob Choi, Do Hyun Kim, and Tae Seok Seo. 2016. "Fully automated and colorimetric foodborne pathogen detection on an integrated centrifugal microfluidic device." *Lab on a Chip* 16 (10): 1917–26. https://doi.org/10.1039/c6lc00326e.

Ohno, Ken Ichi, Kaoru Tachikawa, and Andreas Manz. 2008. "Microfluidics: Applications for analytical purposes in chemistry and biochemistry." *Electrophoresis* 29 (22): 4443–53. https://doi.org/10.1002/elps.200800121.

Pol, Roberto, Francisco Céspedes, David Gabriel, and Mireia Baeza. 2017. "Microfluidic lab-on-a-chip platforms for environmental monitoring." *TrAC - Trends in Analytical Chemistry* 95: 62–68. https://doi.org/10.1016/j.trac.2017.08.001.

Puneeth, S. B., and Sanket Goel. 2019. "Novel 3D printed microfluidic paper-based analytical device with integrated screen-printed electrodes for automated viscosity measurements." *IEEE Transactions on Electron Devices* 66 (7): 3196–3201. https://doi.org/10.1109/TED.2019.2913851.

Qiu, Xianbo, and Michael G. Mauk. 2015. "An integrated, cellulose membrane-based PCR chamber." *Microsystem Technologies* 21 (4): 841–50. https://doi.org/10.1007/s00542-014-2123-x.

Ren, Kangning, Yin Chen, and Hongkai Wu. 2014. "New materials for microfluidics in biology." *Current Opinion in Biotechnology* 25: 78–85. https://doi.org/10.1016/j.copbio.2013.09.004.

Rivet, Catherine, Hyewon Lee, Alison Hirsch, Sharon Hamilton, and Hang Lu. 2011. "Microfluidics for medical diagnostics and biosensors." *Chemical Engineering Science* 66 (7): 1490–1507. https://doi.org/10.1016/j.ces.2010.08.015.

Roeraade, Johan, and Marten Stjernström. 1998. "Method for fabrication of microfluidic systems in glass." *Journal of Micromechanics and Microengineering* 8: 33–38.

Si, Huaqing, Gangwei Xu, Fengxiang Jing, Peng Sun, Dan Zhao, and Dongping Wu. 2020. "A multi-volume microfluidic device with no reagent loss for low-cost digital PCR application." *Sensors and Actuators, B: Chemical* 318 (April). https://doi.org/10.1016/j.snb.2020.128197.

Silva, Severino Jefferson Ribeiro Da, Keith Pardee, and Lindomar Pena. 2019. "Loop-mediated isothermal amplification (LAMP) for the diagnosis of Zika virus: A review." *Viruses* 12 (1): 1–20. https://doi.org/10.3390/v12010019.

Sivakumar, Rajamanickam, and Nae Yoon Lee. 2020. "Microfluidic device fabrication mediated by surface chemical bonding." *The Analyst* 145 (12): 4096–4110. https://doi.org/10.1039/d0an00614a.

Su, Shisheng, Gaoshan Jing, Miaoqi Zhang, Baoxia Liu, Xiurui Zhu, Bo Wang, Mingzhu Fu, Lingxiang Zhu, Jing Cheng, and Yong Guo. 2019. "One-step bonding and hydrophobic surface modification method for rapid fabrication of polycarbonate-based droplet microfluidic chips." *Sensors and Actuators, B: Chemical* 282: 60–68. https://doi.org/10.1016/j.snb.2018.11.035.

Sun, Yi, Yien Chian Kwok, and Nam Trung Nguyen. 2006. "Low-pressure, high-temperature thermal bonding of polymeric microfluidic devices and their applications for electrophoretic separation." *Journal of Micromechanics and Microengineering* 16 (8): 1681–88. https://doi.org/10.1088/0960-1317/16/8/033.

Sun, Yi, Ivan Perch-Nielsen, Martin Dufva, David Sabourin, Dang Duong Bang, Jonas Høgberg, and Anders Wolff. 2012. "Direct immobilization of DNA probes on non-modified plastics by UV irradiation and integration in microfluidic devices for rapid bioassay." *Analytical and Bioanalytical Chemistry* 402 (2): 741–48. https://doi.org/10.1007/s00216-011-5459-4.

Taylor, Theresa B., Emily S. Winn-Deen, Enrico Picozza, Timothy M. Woudenberg, and Michael Albin. 1997. "Optimization of the performance of the polymerase chain reaction in silicon-based microstructures." *Nucleic Acids Research* 25 (15): 3164–68. https://doi.org/10.1093/nar/25.15.3164.

Tsaloglou, Maria Nefeli, Alex Nemiroski, Gulden Camci-Unal, Dionysios C. Christodouleas, Lara P. Murray, John T. Connelly, and George M. Whitesides. 2018. "Handheld isothermal amplification and electrochemical detection of DNA in resource-limited settings." *Analytical Biochemistry* 543: 116–21. https://doi.org/10.1016/j.ab.2017.11.025.

Vyawahare, Saurabh, Andrew D. Griffiths, and Christoph A. Merten. 2010. "Miniaturization and parallelization of biological and chemical assays in microfluidic devices." *Chemistry and Biology* 17 (10): 1052–65. https://doi.org/10.1016/j.chembiol.2010.09.007.

Wang, Fang, and Mark A. Burns. 2009. "Performance of nanoliter-sized droplet-based microfluidic PCR." *Biomedical Microdevices* 11 (5): 1071–80. https://doi.org/10.1007/s10544-009-9324-6.

Wang, He, Liguo Chen, and Lining Sun. 2017. "Digital microfluidics: A promising technique for biochemical applications." *Frontiers of Mechanical Engineering* 12 (4): 510–25. https://doi.org/10.1007/s11465-017-0460-z.

Welch, Hazel M. 2012. "The polymerase chain reaction." *Methods in Molecular Biology* 878: 71–88. https://doi.org/10.1007/978-1-61779-854-2_5.

Wixforth, Achim. 2006. "Acoustically driven programmable microfluidics for biological and chemical applications." *Journal of Laboratory Automation* 11 (6): 399–405. https://doi.org/10.1016/j.jala.2006.08.001.

Xue, Chang Ying, Saif A. Khan, and Kun Lin Yang. 2009. "Exploring optical properties of liquid crystals for developing label-free and high-throughput microfluidic immunoassays." *Advanced Materials* 21 (2): 198–202. https://doi.org/10.1002/adma.200801803.

Yamada, Masumi, and Minoru Seki. 2004. "Nanoliter-sized liquid dispenser array for multiple biochemical analysis in microfluidic devices." *Analytical Chemistry* 76 (4): 895–99. https://doi.org/10.1021/ac0350007.

Yeo, Joo Chuan, Kenry, and Chwee Teck Lim. 2016. "Emergence of microfluidic wearable technologies." *Lab on a Chip* 16 (21): 4082–90. https://doi.org/10.1039/c6lc00926c.

Zanoli, Laura Maria, and Giuseppe Spoto. 2013. "Isothermal amplification methods for the detection of nucleic acids in microfluidic devices." *Biosensors* 3 (1): 18–43. https://doi.org/10.3390/bios3010018.

Zhang, Mengying, Jinbo Wu, Limu Wang, Kang Xiao, and Weijia Wen. 2010. "A simple method for fabricating multi-layer PDMS structures for 3D microfluidic chips." *Lab on a Chip* 10 (9): 1199–1203. https://doi.org/10.1039/b923101c.

Ziaie, Babak, Antonio Baldi, Ming Lei, Yuandong Gu, and Ronald A. Siegel. 2004. "Hard and soft micromachining for BioMEMS: Review of techniques and examples of applications in microfluidics and drug delivery." *Advanced Drug Delivery Reviews* 56 (2): 145–72. https://doi.org/10.1016/j.addr.2003.09.001.

Ziros, Panos G., Kostas Papanotas, Apostolos Vantarakis and Petros Kokkinos. 2012. "Loop-mediated isothermal amplification (lamp) for the detection of Salmonella spp. isolated from different food types." *Journal of Microbiology, Biotechnology and Food Sciences* 2 (1): 152–61.

Zou, Quanbo, Yubo Miao, Yu Chen, Uppili Sridhar, Chong Ser Chong, Taichong Chai, Yan Tie, Christina Hui Leng Teh, Tit Meng Lim, and Chew Kiat Heng. 2002. "Micro-assembled multi-chamber thermal cycler for low-cost reaction chip thermal multiplexing." *Sensors and Actuators, A: Physical* 102 (1–2): 114–21. https://doi.org/10.1016/S0924-4247(02)00384-9.

Zou, Zhi Qing, Xiang Chen, Qing Hui Jin, Meng Su Yang, and Jian Long Zhao. 2005. "A novel miniaturized PCR multi-reactor array fabricated using flip-chip bonding techniques." *Journal of Micromechanics and Microengineering* 15 (8): 1476–81. https://doi.org/10.1088/0960-1317/15/8/014.

5 A Classification and Evaluation of Approximate Multipliers

U. Anil Kumar and Syed Ershad Ahmed

CONTENTS

5.1 Introduction 71
5.2 A General Multiplier Architecture 72
5.3 Approximation in Multipliers 73
    5.3.1 Approximation at Partial Product Generation Stage 73
    5.3.2 Approximation at Partial Product Reduction Stage 73
    5.3.3 Approximate Compressor in Partial Product Reduction Tree 76
5.4 Experimental Results of Various Multiplier Architectures 77
    5.4.1 Error Characteristics 77
    5.4.2 Circuit Characteristics 78
5.5 Multipliers in Image Sharpening Application 78
5.6 Conclusion 84
References 84

5.1 INTRODUCTION

Modern-day applications in areas such as deep learning and multimedia processing have become increasingly complex and tend to consume more energy. Traditional methods involving technology scaling have played an important role over the last few decades in the improvement of energy consumption. However, due to the limitations and diminishing returns of technology scaling, alternate methods to mitigate the same has to be explored. It has been found that most multimedia applications can tolerate errors up to a certain extent in computations [1]. For instance, it is a well-established fact that in certain image processing applications, a range of image resolutions is acceptable. In these types of applications, savings in power can be obtained by truncating the arithmetic modules [2], such as a multiplier. Truncation tends to introduce errors in computing and hence poses a challenge between the amount of error that can be tolerated and the benefits achieved in terms of area, speed, and power. In such applications, approximate computing has become an efficient technique to achieve an improvement in circuit parameters [3]. Approximations can be introduced at various levels of abstractions namely algorithmic, arithmetic, and Boolean levels. In this chapter, the focus will be on approximation at the arithmetic level, which is achieved by removing a portion of logic.

Multiplication is a ubiquitous and frequently executed arithmetic operation in most digital signal applications. Significant research work has been carried out to improve the performance of multipliers by truncating them. The criteria used to measure their performance include die-area, speed of operation, and power consumed. Consequently, many techniques were proposed at the design/architecture level, which helps to achieve improvement in the area, latency, and power.

This chapter presents various design methods to implement approximate multiplier circuits. Several multiplier designs have been reported in the literature and found to be a potential solution for efficient implementation. Among these techniques, most of them are either array or recursive-based [4]. Section 5.2 presents an introduction to approximate multipliers while state-of-the-art approximate multiplier circuits are discussed in Section 5.3. Error and synthesis results are provided in Section 5.4, and implementation on image processing applications are presented in Section 5.5. Finally, conclusions are drawn in Section 5.6.

5.2 A GENERAL MULTIPLIER ARCHITECTURE

A typical multiplier architecture depicted in Figure 5.1 consists of partial product generation (PPG), partial product reduction (PPR), and final carry propagate addition stages.

Partial product generation is accomplished using AND gates while the reduction is carried out by partial product reduction structure. Two popular PPR tree structures

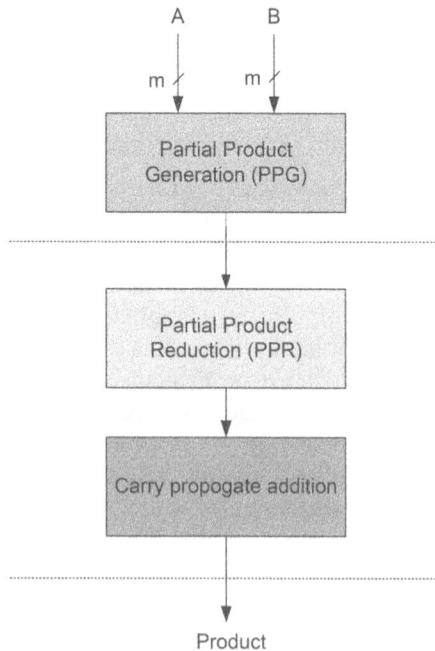

FIGURE 5.1   A typical multiplier architecture.

suggested by Dadda and Wallace [5] are used to compress the partial products. The output from the PPR structure is reduced to the final product using a ripple carry or a two-operand fast adder [6].

5.3 APPROXIMATION IN MULTIPLIERS

Approximation can be introduced in multipliers at partial product generation [7] or at partial product reduction [8–14] stages. Approximation at PPG level reduces the number of partial products generated, thus simplifying the generation hardware.

The other method of reducing the multiplier complexity is at the reduction stage. Various techniques have been proposed in the literature to minimize the complexity at the PPR stage. One scheme is to truncate the lower significant partial products because they contribute least to the final result. However, as a result, error creeps into the final result, which can be minimized by replacing the truncated portion with a constant or variable. Another approach is to prune the compressor modules used to reduce the partial products.

It has been found that approximation at PPR stage results in better savings compared to the PPG stage. Hence, most of the works [8–14] focus on reducing the partial products at the PPR stage because it results in improvement in area, latency, and power. Section 3.1 discusses the design that approximates the partial products at the PPG stage, simplification at the PPR stage is presented in Section 3.2, while Section 3.3 presents the approximation methods adopted in the compressor design.

5.3.1 APPROXIMATION AT PARTIAL PRODUCT GENERATION STAGE

Recursive multipliers based on the Karatsuba-Ofman Algorithm (KOA) [15] are found to have a hierarchical architecture comprising of several submultipliers. Recursive multipliers have been proven to be efficient as opposed to the array multipliers in terms of area and power consumption.

Work in [7] presents an underdesigned recursive multiplier (UDM) consisting of the multiplier (B) partitioned into B_H and B_L while multiplicand (A) is divided into A_H and A_L as shown in Figure 5.2.

The recursive multiplication is carried out by splitting the operands A and B into four submultipliers, namely $A_L B_L$, $A_H B_L$, $A_L B_H$, and $A_H B_H$. Each submultiplier is optimized by altering an entry in the K-map and is shown in Table 5.1.

5.3.2 APPROXIMATION AT PARTIAL PRODUCT REDUCTION STAGE

A recursive multiplier namely AWTM based on the Wallace tree was proposed by Bharadwaj et al. [8] and is shown in Figure 5.3. The most significant portion in this multiplier is calculated using an exact circuit while the LSB portion was computed using approximate logic.

Revanth et al. [9] proposed a recursive multiplier (EAM) of the form shown in Figure 5.4 (a). In this multiplier, $A_H B_L$, $A_L B_H$, $A_L B_L$ are approximate while $A_H B_H$ is accurate. A portion of approximate submultiplier highlighted in Figure 5.4 (b)

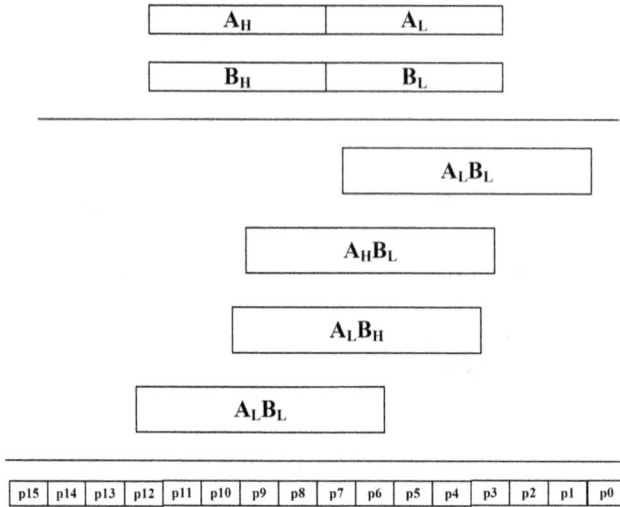

FIGURE 5.2 A UDM scheme based on recursive multiplier architecture.

TABLE 5.1

K-Map of a 2 × 2 UDM Submultiplier

A_1	A_0	B_1	B_0	$Y_2Y_1Y_0$
0	0	0	0	000
0	0	0	1	000
0	0	1	0	000
0	0	1	1	000
0	1	0	0	000
0	1	0	1	001
0	1	1	0	010
0	1	1	1	011
1	0	0	0	000
1	0	0	1	010
1	0	1	0	100
1	0	1	1	110
1	1	0	0	000
1	1	0	1	011
1	1	1	0	110
1	1	1	1	111

Source: As shown in Table 5.1, the output combination "1001" is approximated to "111," thereby saving one output bit. Consequently, the
resulting submultiplier is much simpler and consumes less area.

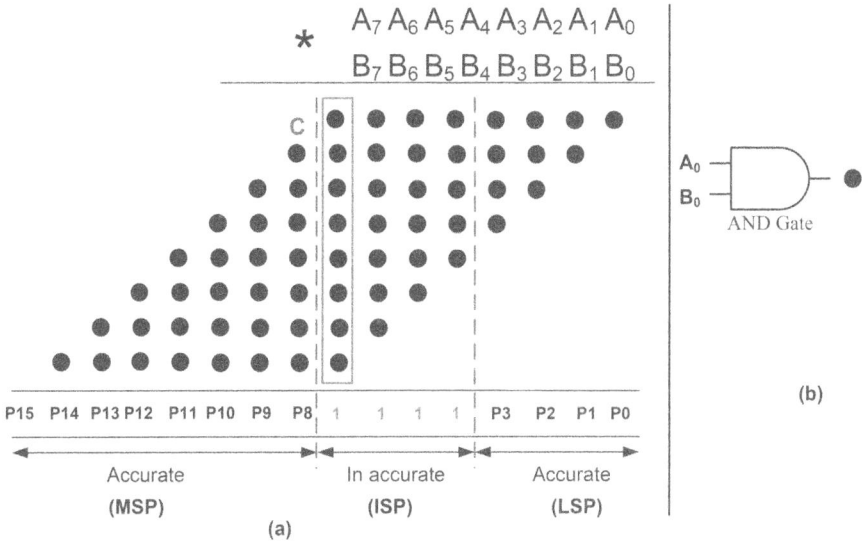

FIGURE 5.3   (a) An 8×8 approximate Wallace tree based multiplier (AWTM) based on the Dadda scheme; (b) partial product generation using the AND gate.

corresponding to $A_L B_L$ is simply replaced with "1100." The constant in the $A_L B_L$ is computed with intent to make the average error zero. This constant was computed by taking the average error between estimated outcome and actual outcome obtained from all possible input combinations in $A_L B_L$. In a similar manner, a constant ("110") was obtained for $A_L B_H$ and $A_H B_L$. The truncation of approximate submultipliers and the replacement of a correction term that compensates the error significantly reduce the area, delay, and power.

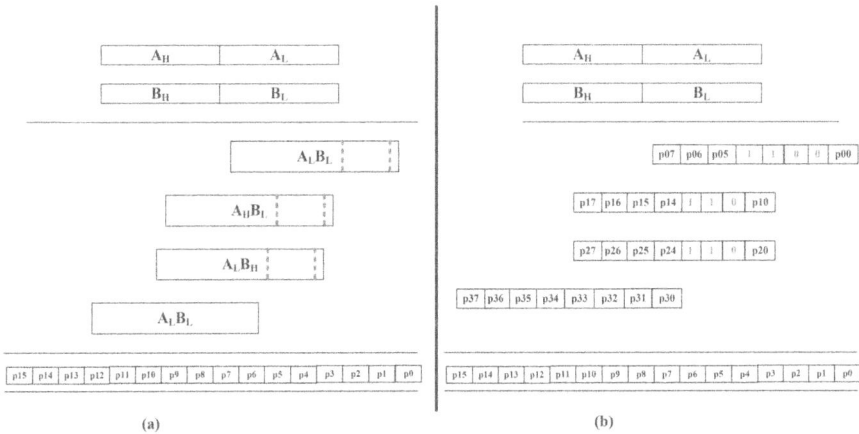

FIGURE 5.4   (a) Recursive multiplier with highlighted portions of $A_L B_L$, $A_H B_L$, and $A_L B_H$ replaced with constant; (b) error compensating constant used in $A_L B_L$, $A_H B_L$, and $A_L B_H$.

5.3.3 Approximate Compressor in Partial Product Reduction Tree

The fundamental computing module in the PPR stage is a compressor. The area, critical path delay, and power metrics in a multiplier can be improved by using approximate compressors in the place of conventional accurate designs [10–14]. This section presents the approximate compressor-based multiplier designs.

Momeni et al. [10] proposed two approximate (4:2) compressor designs with the intent to minimize the complexity in a multiplier. The area and delay optimized compressor designs were achieved by manipulating the truth table entries. These compressor designs have been used in Dadda-based multiplier architectures as depicted in Figure 5.5. The partial products in region 1 and region 2 are computed using an approximate compressor whereas partial products in region 3 are computed using an exact circuit.

By altering the K-map entry in an exact compressor circuit [11], Yang et al. improved Momeni et al. compressors and proposed three new approximate compressor designs. These designs were then used at the PPR stage in a Dadda multiplier architecture. This multiplier is divided into three regions: least significant portion (LSP), intermediate significant portion (ISP), and most significant portion (MSP) (as shown in Figure 5.5).

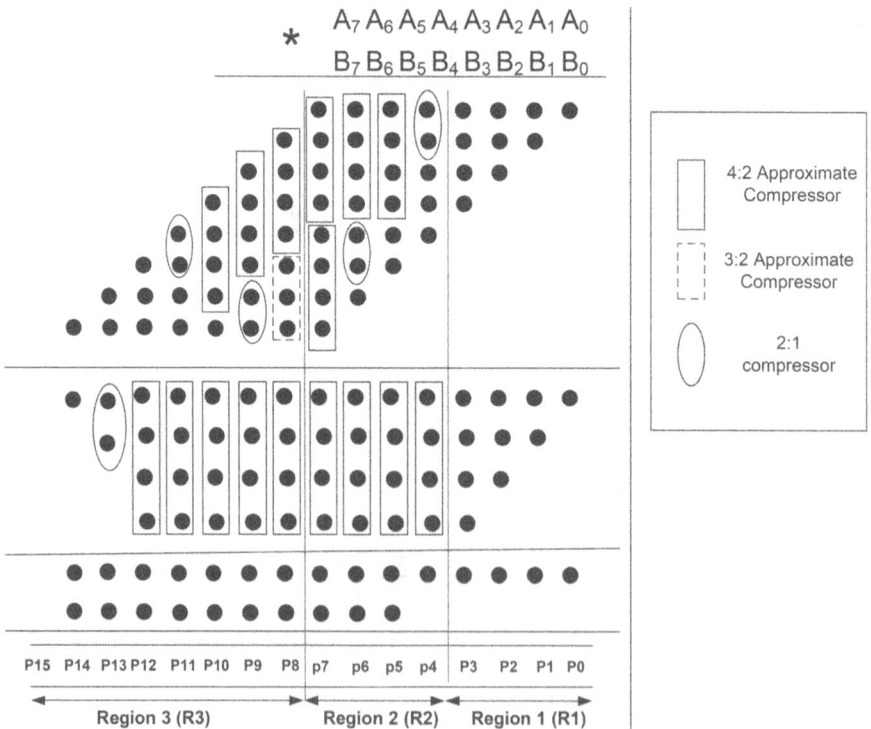

FIGURE 5.5   Dadda multiplier with approximate compressor designs.

The LSP region is truncated; an approximate compressor was used in the ISP portion while exact compressors were used in the MSP region. Due to truncation and approximate compressors, substantial decrease in area, power, and delay compared to Momeni et al. have been achieved [10].

Ha et al. [12] improved the 4:2 compressor design of Yang et al., and additionally introduced an error compensating circuit to improve the accuracy. The simple error correction circuit comprising of basic gates derives the input from the MSP of ISP and generates a carry to the MSP. Consequently, it improves the area, delay, power, and accuracy compared to other approximate compressor-based designs [10, 11].

Work in [13] encoded the partial products using OR and AND gates, and then the PPR was simplified by deploying improved approximated compressor circuits. Two multiplier architectures of the form as shown in Figure 5.5 were presented. In the first design, all three regions of the PPR stage are approximated, while in the second, region 1 and region 2 form LSP, whereas region 3 forms the MSP. The LSP region is approximated using approximate compressors, approximate half adders, and approximate full adders, while exact compressors are used in the MSP region.

Work in [14] simplified the PPR by deploying approximate compressor circuits. This compressor design has been used in Dadda-based multiplier architecture as shown in Figure 5.5. The partial products in region 1 and region 2 are computed using the approximate compressor whereas partial products in region 3 are computed using an exact circuit.

5.4 EXPERIMENTAL RESULTS OF VARIOUS MULTIPLIER ARCHITECTURES

5.4.1 ERROR CHARACTERISTICS

A comprehensive error analysis was performed on various state-of-the-art multiplier architectures using MATLAB® for all input combinations. To evaluate the accuracy, the quality metrics [16] error rate (ER), normalized mean error distance (NMED), and mean relative error distance (MRED) are used.

The error distance (ED) is given as ED = $|Q - Q`|$. Where $Q$ is the exact result and $Q`$ is the approximate result. The mean of all EDs is the mean error distance (MED), whereas NMED is calculated by dividing MED with the maximum output of the exact design. Metric RED is the ratio of ED to Q, while MRED is the mean of all relative error distances (REDs). The metrics ER, NMED, and MRED of various multiplier architectures [7–14] are shown in Figures 5.6 and 5.7.

From Figure 5.6, it can be observed that in most of the designs, ER is over 80%; however, due to approximation done at the LSP, Li et al. [14] achieve a lower error rate of 66.24%.

It is evident from Figure 5.7 (a) and (b) that Ha et al. and Yang et al. have lower NMED and MRED compared to other designs owing to efficient compressors. Li et al. [14], Momeni et al. [10], and Multiplier2 [13] have moderate NMED while AWTM has high MRED compared to other designs and is shown in Figure 5.7 (b).

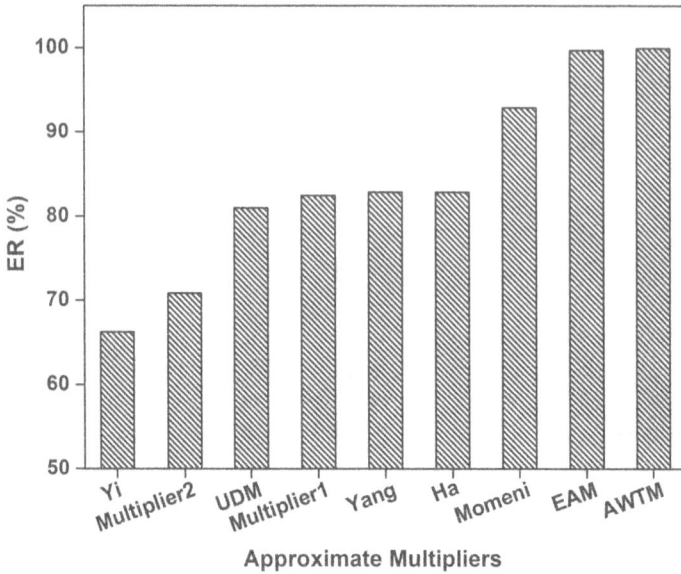

FIGURE 5.6   Error rate (ER) comparison of various approximate multipliers.

5.4.2   CIRCUIT CHARACTERISTICS

For the fair analysis, all 8-bit multiplication schemes are created in Verilog Hardware Description Language. Exhaustive functional simulation was performed using Cadence Inclusive Unified Simulator while hardware synthesis was performed using Cadence RTL Compiler v7.1 at TSMC 180 nm process node. Figure 5.8 (a–c) shows the area, power, and delay comparison of various multipliers. The power-delay product (PDP) and area-delay product (ADP) of various designs are shown in Figure 5.9.

It can be observed that EAM [9] has least die-area whereas UDM [7] consumes more area compared to other designs. The designs EAM and multiplier1 [13] tend to consume less power while multiplier2 [13] consumes more power. Compared to all the other designs, EAM is the fastest scheme whereas UDM has the highest critical path delay.

It is obvious that a multiplier with the low area consumption and power dissipation normally has better PDP and ADP. Accordingly, EAM has lower PDP and ADP in comparison with other designs. The Ha et al., Multiplier1, Momeni et al., and Yang et al. [10–13] schemes have moderate PDP and ADP, while UDM has higher PDP and ADP.

5.5   MULTIPLIERS IN IMAGE SHARPENING APPLICATION

In order to evaluate the performance of various multiplier designs in a real-time application, the image sharpening algorithm is chosen since it involves multiplication operations. Image sharpening improves the visual quality of the image. The

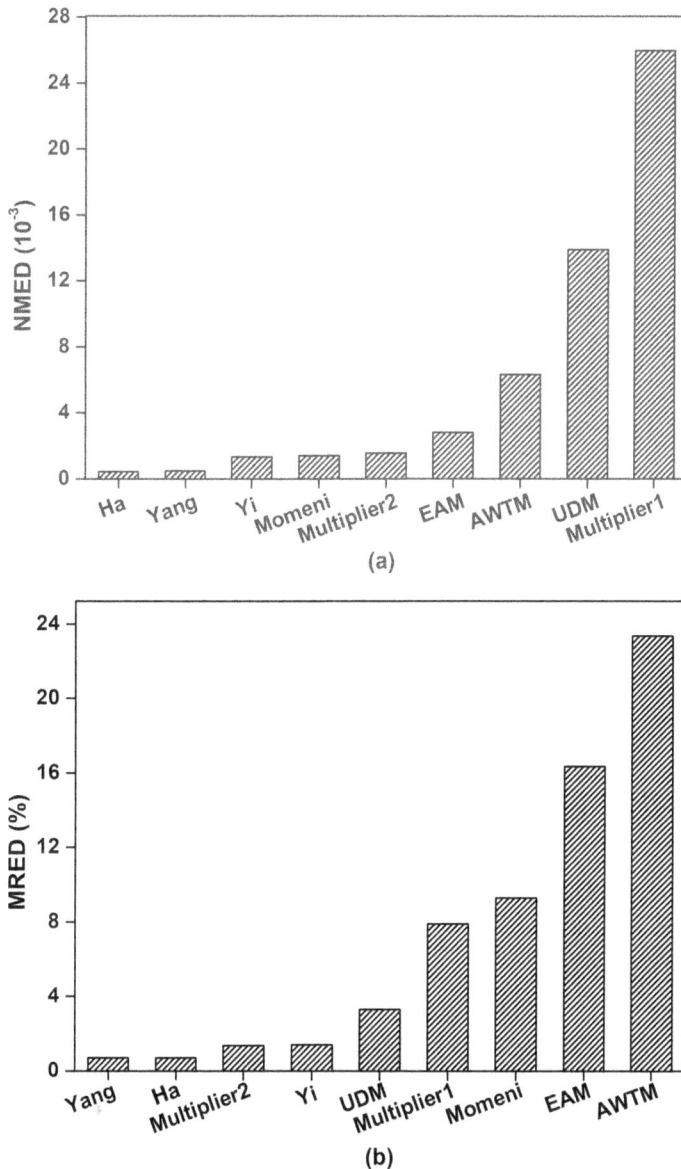

FIGURE 5.7   Evaluation of NMED and MRED of various multipliers: **(a)** designs arranged with increasing NMED; **(b)** designs arranged with increasing MRED.

sharpening algorithm accepts an image (P), processing it using 5×5 kernels to create a high-quality image.

Let Q be the input image, and output image OI can be expressed as

$$OI\,(x,\,y) = 2Q\,(x,\,y) - K \tag{5.1}$$

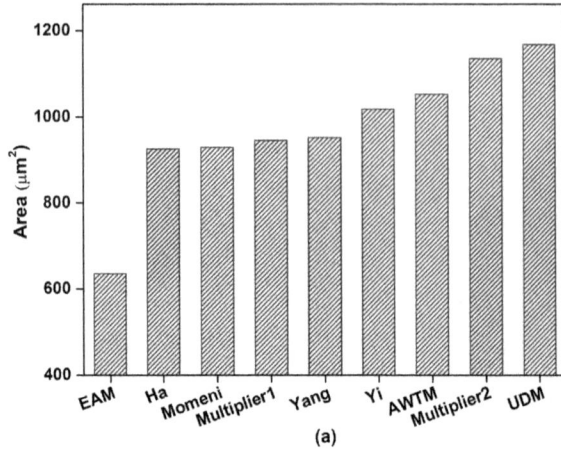

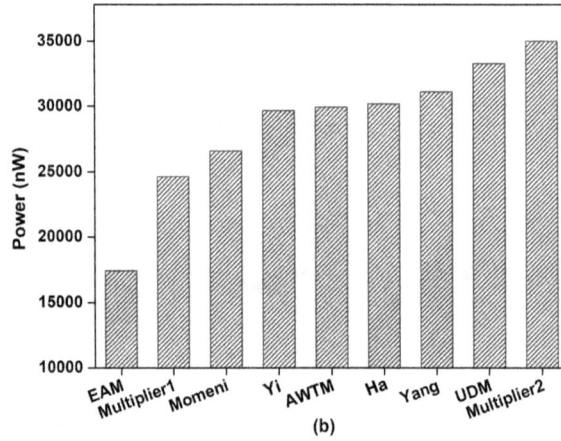

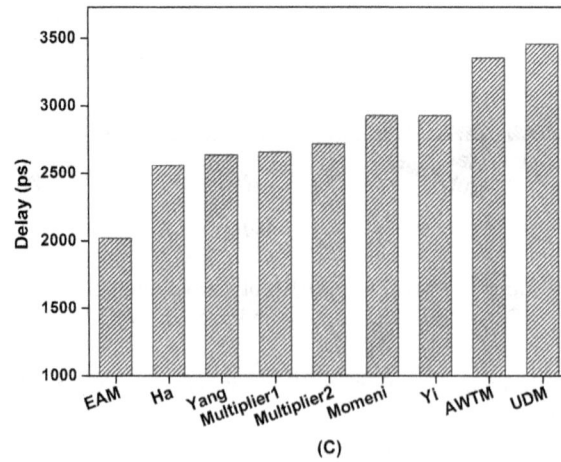

FIGURE 5.8   (a) Area comparison of various multiplier designs, (b) power comparison of various multiplier designs, (c) delay comparison of various multiplier designs.

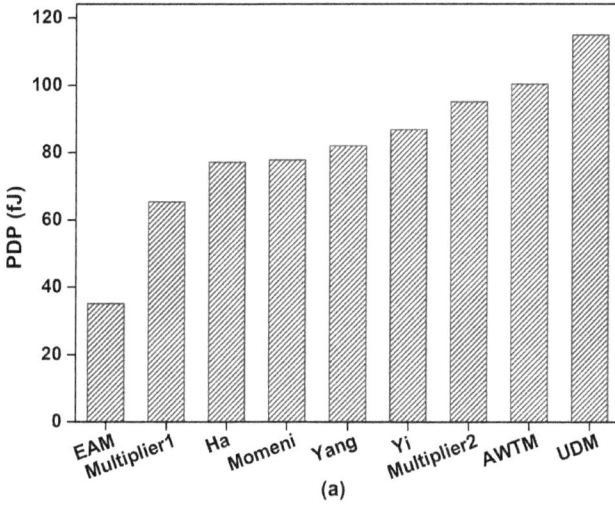

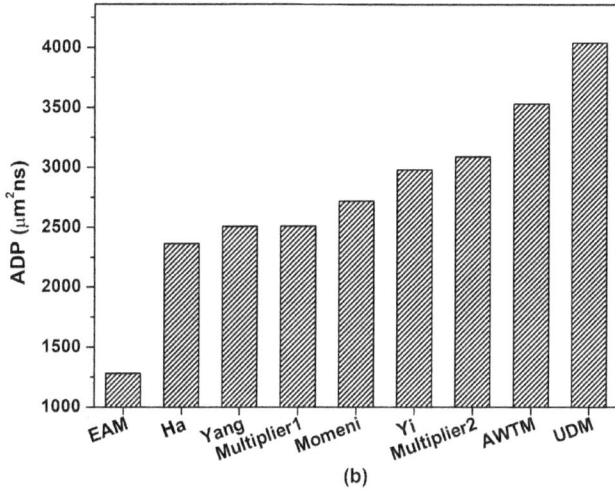

FIGURE 5.9 (a) PDP comparison of various multiplier designs; (b) ADP comparison of various multiplier designs.

where,

$$K = 1/273 \sum_{p=-2}^{2} \sum_{q=-2}^{2} M(p+3, q+3) Q(x-p, y-q) \qquad (5.2)$$

and M is a matrix defined as $M = \begin{bmatrix} 1 & 4 & 7 & 4 & 1 \\ 4 & 16 & 26 & 16 & 4 \\ 7 & 26 & 41 & 26 & 7 \\ 4 & 16 & 26 & 16 & 4 \\ 1 & 4 & 7 & 4 & 1 \end{bmatrix}$

The multiplication operation mentioned in Eq. (5.2) is replaced with the approximate designs while the other operations remain accurate. Figure 5.10 (a–k) shows images obtained by sharpening the input cameraman image using various multiplier designs, including the exact multiplier.

(a) Input Image   (b) Processed using exact multiplier

(c) Processed using Ha [12]   (d) Processed using Yang [11]

(e) Processed using Multiplier1 [13]   (f) Processed using Multiplier2 [13]

FIGURE 5.10   Images sharpened using different approximate multipliers (a–k). (Continued)

(g) Processed using Yi [14]

(h) Processed using Momeni [10]

(i) Processed using AWTM [8]

(j) Processed using UDM [7]

(k) Processed using EAM [9]

FIGURE 5.10   Images sharpened using different approximate multipliers (a–k).

It is evident that images processed using exact and various multiplier designs are almost similar. Ha et al. and Yang et al. got better quality images compared to other designs. Results show that EAM and AWTM achieve visually acceptable image sharpening results.

5.6 CONCLUSION

Applications such as graphics, audio, video, and image processing tend to consume more power due to the computational complexity. Most of these applications inherent a quality of error tolerance. Multipliers are ubiquitous modules in these applications. Approximate computing offers potential benefits in terms of area, delay, and power without significant compromise on accuracy.

In this chapter, various multiplier architectures with approximation at PPG and PPR stages were reviewed. Most of the designs achieve improvement in the area, delay, and power by pruning the LSP; however, with a trade-off in accuracy. The UDM design shows high NMED, moderate ER, and MRED. Ha et al. and Yang et al. schemes perform well in terms of NMED and MRED compared to all the other designs. The EAM design achieves the best circuit characteristics (area, delay, and power) than other designs. Finally, all the designs were implemented in image sharpening algorithms, and the quality of the obtained images proves that the application can tolerate errors up to a certain limit.

REFERENCES

1. Mittal, Sparsh. "A survey of techniques for approximate computing," *ACM Computing Surveys (CSUR)* 48, no. 4 (2016): 1–33.
2. Reda, Sherief, and Muhammad Shafique. *Approximate circuits*. Cham, Switzerland: Springer, 2019.
3. Han, Jie, and Michael Orshansky. "Approximate computing: An emerging paradigm for energy-efficient design," in *2013 18th IEEE European Test Symposium (ETS)*, pp. 1–6. IEEE, 2013.
4. Danysh, Albert N., and Earl E. Swartzlander. "A recursive fast multiplier," in *Conference Record of Thirty-Second Asilomar Conference on Signals, Systems and Computers (Cat. No. 98CH36284)*, vol. 1, pp. 197–201. IEEE, 1998.
5. Townsend, Whitney J., Earl E. Swartzlander Jr, and Jacob A. Abraham. "A comparison of Dadda and Wallace multiplier delays," in *Advanced signal processing algorithms, architectures, and implementations XIII*, vol. 5205, pp. 552–60. International Society for Optics and Photonics, 2003.
6. Parhami, Behrooz. *Computer arithmetic*. Vol. 20, no. 00, Oxford University Press, 1999.
7. Kulkarni, Parag, Puneet Gupta, and Milos Ercegovac. "Trading accuracy for power with an underdesigned multiplier architecture," In *2011 24th International Conference on VLSI Design*, pp. 346–51. IEEE, 2011.
8. Bhardwaj, Kartikeya, Pravin S. Mane, and Jörg Henkel. "Power-and area-efficient approximate Wallace tree multiplier for error-resilient systems," In *Fifteenth International Symposium on Quality Electronic Design*, pp. 263–69. IEEE, 2014.
9. Reddy, C. Sai Revanth, U. Anil Kumar, and Syed Ershad Ahmed. "Design of efficient approximate multiplier for image processing applications," In *International Conference on Modelling, Simulation and Intelligent Computing*, pp. 511–18. Springer, Singapore, 2020.

10. Momeni, Amir, Jie Han, Paolo Montuschi, and Fabrizio Lombardi. "Design and analysis of approximate compressors for multiplication," *IEEE Transactions on Computers* 64, no. 4 (2014): 984–94.
11. Yang, Zhixi, Jie Han, and Fabrizio Lombardi. "Approximate compressors for error-resilient multiplier design," In *2015 IEEE International Symposium on Defect and Fault Tolerance in VLSI and Nanotechnology Systems (DFTS)*, pp. 183–86. IEEE, 2015.
12. Ha, Minho, and Sunggu Lee. "Multipliers with approximate 4–2 compressors and error recovery modules," *IEEE Embedded Systems Letters* 10, no. 1 (2017): 6–9.
13. Venkatachalam, Suganthi, and Seok-Bum Ko. "Design of power and area efficient approximate multipliers," *IEEE Transactions on Very Large Scale Integration (VLSI) Systems* 25, no. 5 (2017): 1782–86.
14. Yi, Xilin, Haoran Pei, Ziji Zhang, Hang Zhou, and Yajuan He. "Design of an energy-efficient approximate compressor for error-resilient multiplications," In *2019 IEEE International Symposium on Circuits and Systems (ISCAS)*, pp. 1–5. IEEE, 2019.
15. Fan, Haining, Jiaguang Sun, Ming Gu, and K-Y. Lam. "Overlap-free Karatsuba–Ofman polynomial multiplication algorithms," *IET Information Security* 4, no. 1 (2010): 8–14.
16. Liang, Jinghang, Jie Han, and Fabrizio Lombardi. "New metrics for the reliability of approximate and probabilistic adders," *IEEE Transactions on Computers* 62, no. 9 (2012): 1760–71

6 Optical MEMS Accelerometers

A Review

Balasubramanian Malayappan and
Prasant Kumar Pattnaik

CONTENTS

6.1   Introduction 87
6.2   Micromachined or MEMS Accelerometers..88
6.3   Accelerometer Performance Metrics 89
6.4   Accelerometer Sensing Techniques 92
6.5   Optical Sensing Principles and Detection Methods for Acceleration
      Measurement..94
      6.5.1   Optical Sensing Techniques..95
      6.5.2   Common Optical Interrogation Methods 100
6.6   Conclusion 103
References.. 103

6.1   INTRODUCTION

This chapter presents a brief review of the sensing techniques and interrogation methods used in optical micro-electro-mechanical-system (MEMS) and nano-photonic accelerometers. In Section 6.2, brief historical development of accelerometers and particularly the MEMS accelerometer highlighting its varied application domains is presented. Most micromachined accelerometers work on the same basic principle. The proof mass attached to a compliant suspension experiences inertial force in response to acceleration of the body/frame. The deflection of the proof mass or the deformation of the suspension are transduced and form the metric for acceleration measurement. The description of the performance metrics of accelerometers is given in Section 6.3. Section 6.4 gives a brief overview of sensing techniques used in MEMS accelerometers. Section 6.5 contains a brief description of both various optical sensing techniques and the commonly used interrogation methods that enable the application of optical MEMS accelerometers for various application domains. The conclusions of the chapter are presented in Section 6.6.

6.2 MICROMACHINED OR MEMS ACCELEROMETERS

Seeking to find directions is a behavior that has been observed among humans since ancient times. Earlier methods were crude such as direction of a star or length of shadows. In today's world, driven by rapidly evolving technology, there are far more advanced techniques to find directions. Special purpose satellites orbiting Earth enable determination of any object with a high degree of precision. In addition, accelerometers and gyroscopes are the inertial sensors that can provide us with this sense of direction even in absence of any external reference like the satellite. Invention and evolution of MEMS technology has resulted in miniaturization of these position tracking devices such that they are hand-held and can be carried around.

In the modern world, tracking the position or movement of an object is an engineering problem in various application domains. Thus, micromachined inertial sensors have been a subject of extensive research since the mid-twentieth century. The requirements of performance metrics of the accelerometers vary significantly with the application domain. Figure 6.1 shows typical range-bandwidth requirements for different accelerometers. Originally, inertial sensors were restricted to applications such as military and aerospace systems where the cost was not a major concern. However, advent of micromachining opened the doors to applying precision inertial sensors for cost-sensitive applications such as automotives and augmented-reality (AR), virtual-reality (VR) domains, structural health monitoring, missile/aircraft guidance, different biomedical parameters, seismometry, machine vibration monitoring, and shock measurement. Acceleration signals are generally categorized as

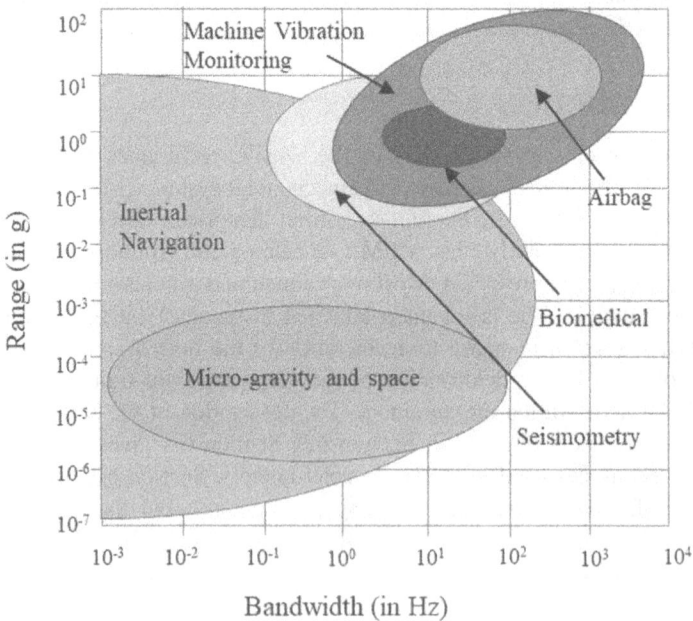

FIGURE 6.1 Accelerometer range-bandwidth performance requirements for different application areas. (Based on Krishnan et al. 2007.)

linear, vibration, or shock accelerations. Linear or quasi-static is the type of acceleration that has large-scale temporal invariance with maximum value in the range of 10 g for most applications (g = acceleration due to gravity). Acceleration being a vector quantity, vibration of a body results in a much larger value of vibrational acceleration than linear accleration, with maximum values in the range of 100 g depending on the amplitude and frequency of the vibration. Shock acceleration are large valued impulse type signals and are typically used as trigger/alarm with maximum value even upto 1,000 g. Based on the application, the type of acceleration and its range vary considerably.

The first MEMS accelerometer was reported in 1979 (Roylance and Angell 1979) in which external acceleration resulted in change in resistance of the suspending beams with sensitivity of 1 mg and a range of 50 g. Since then, various micromachined accelerometers have been designed and demonstrated that can be classified based on fabrication technology, different sensing techniques, and type of control system used (Kraft 1997; Spineanu et al. 1997; Rockstad et al. 1996; Storgaard-Larsen et al. 1996; Lemkin and Boser 1999). Not just in research labs, application-specific commercial (Chu 2012) accelerometers have been developed and are catalogued. The first device from arguably the most popular range of accelerometers from Analog Devices (Devices 1995) was the ADXL05 with a dynamic range of ± 5 g and a resolution of $0.5 \text{ mg}/\sqrt{Hz}$. MEMS accelerometers alongside pressure sensors are now the most voluminous MEMS devices in the market (Ozgen 2010). The following sections describe the different sensing techniques with special attention to advantages and disadvantages of optical sensing techniques.

6.3 ACCELEROMETER PERFORMANCE METRICS

Accelerometers usually consist of a proof mass and a suspension system anchored to a frame. The lumped spring-mass-damper model as shown in Figure 6.2 schematically represents the accelerometer. When an acceleration (a) is experienced by

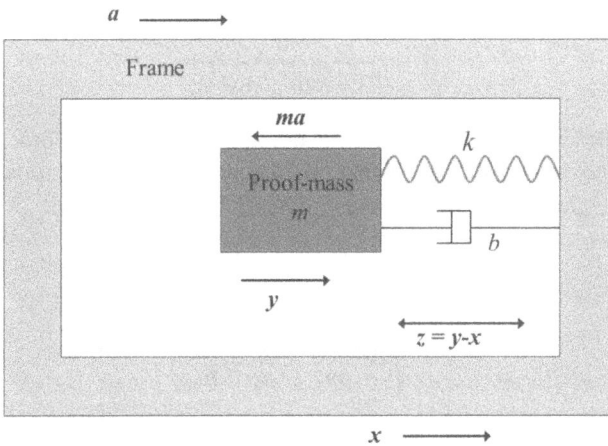

FIGURE 6.2 Lumped model of accelerometer showing basic working principle.

the system/frame, the proof mass (m) experiences a D'Alembert's force equal to mass times applied acceleration ($m \times a$). The deflection of the proof mass for a static acceleration depends on the stiffness of the suspension and the damping in the system. Transduction of the proof mass deflection or change in the spring stiffness into quantities suitable for storage or processing using different techniques like piezo-resistive (Willis and Jimerson 1964; Roylance and Angell 1979), piezo-electric (DeVoe and Pisano 1997), electromagnetic (Abbaspour-Sani et al. 1994), tunneling (Liu and Kenny 2001; Yeh and Najafi 1998; Rockstad et al. 1996), optical (Cooper et al. 2000; Loh et al. 2002), capacitive (Rudolf 1983; Chae et al. 2005), resonant (Pedersen and Seshia 2004; Su and Yang 2001), thermal (Mailly et al. 2003), and others is the method used for acceleration measurement. The usable bandwidth of the system depends on the resonant frequency of the spring-mass-damper system and the damping coefficient.

From fundamentals of mechanics and vibration theory (Senturia 2007), the equations for analysis of mechanical sensitivity of accelerometers are derived as follows. The movement of the frame, denoted by x in response to the acceleration a experienced by the system. The movement of the proof mass is denoted by y. The net extension of the spring and damper is $z = y - x$. By summing the inertial force, spring force, and damping force, the equation of motion for the proof mass is

$$m\ddot{y} + b(\dot{y} - \dot{x}) + k(y - x) = 0 \tag{6.1}$$

where k is the net stiffness of the suspension structures and b is the net damping coefficient. Rewriting the above equation in terms of z, the parameter affecting the spring and damper system, and assuming that the frame's excitation is harmonic, $x = X \sin \omega t$ with amplitude, X, and angular frequency, ω, Eq. (6.1) can be rewritten as

$$m(\ddot{z} + \ddot{x}) + b(\dot{z}) + kz = 0$$
$$\Rightarrow m\ddot{z} + b\dot{z} + kz = -m\ddot{x} = m\omega^2 X \sin(\omega t) \tag{6.2}$$

Denoting amplitude of z as Z, Eq. (6.2) can be solved to obtain the following result:

$$\frac{Z}{X} = \frac{m\omega^2}{\sqrt{(k - m\omega^2)^2 + (b\omega)^2}} \tag{6.3}$$

Noting that the above equation represents the solution of a second-order system, Eq. (6.3) can be rewritten in terms of natural frequency, ω_n, and damping coefficient, x, as

$$\frac{Z}{X} = \frac{\dfrac{\omega^2}{\omega_n^2}}{\sqrt{\left(1 - \dfrac{\omega^2}{\omega_n^2}\right)^2 + \left(2\xi \dfrac{\omega}{\omega_n}\right)^2}} \tag{6.4}$$

where, $\omega_n = \sqrt{k/m}$ and $\xi = b/\sqrt{4km}$. It can be observed that the frequency response depend on the device parameters. From Eq. (6.4), it can be observed that at the system response at frequencies much lower than the natural frequency ($\omega \ll \omega_n$), the denominator approaches unity and we get $Z = \omega^2 X / \omega_n^2$. Since the amplitude acceleration to be measured is $a = \omega^2 X$, at low frequencies the relation between the spring, damper system's actuation, and input acceleration is

$$Z = \frac{a}{\omega_n^2} \qquad (6.5)$$

From Eq. (6.5), it is evident that for low frequencies there is a linear relationship between beam actuation and input acceleration, thus making it an apt parameter for transduction to measure the acceleration. Further, it is evident that the mechanical sensitivity is higher for lower natural frequency. However, the trade-off is a reduction in bandwidth, which depends on both the fundamental frequency and the damping factor. For under or over-damped situations, the amplitude ratio remains constant for shorter range of frequencies than in critical damping as seen in Figure 6.3. Further, it is observed that for under-damped condition, there is a sharp resonance with an amplified sensor response for frequencies near the natural frequency. This sharpness of resonance characterized by the mechanical quality factor, Q, is inversely proportional to the damping factor and approaches $Q = 0.5$ for critical damping. However, high Q-factor for an under-damped system results in a ringing response (transients die down only after a long time).

Apart from the mechanical sensitivity and bandwidth, noise, and thus, resolution, linearity, bias stability, and thermal stability are factors that determine the performance of the MEMS accelerometer. The resolution of an accelerometer depends on the noise floor of the system (given in units of $g / \sqrt{Hz}$), including both mechanical and electrical noise. Due to stochastic nature of movement of electrons within the circuit, and random mechanical vibration due to entropy of the system, there is noise

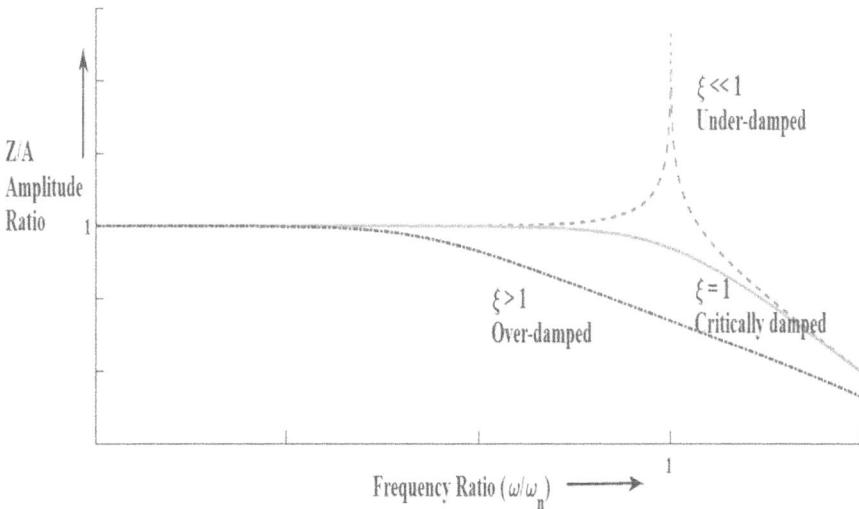

FIGURE 6.3 The frequency response of an accelerometer under various damping conditions.

distribution in every system (for the entire frequency bandwidth). This noise caps the minimum displacement and consequently minimum acceleration that can be measured. From Eq. (6.5), the minimum displacement that can be resolved or sensed by the accelerometer can thus be given as

$$Z_{min} = \frac{a_r}{\omega_n^2} \simeq \frac{a_r}{BW^2} \simeq \frac{a_{nf}\sqrt{BW}}{BW^2} = \frac{a_{nf}}{BW^{3/2}} \tag{6.6}$$

The minimum resolvable acceleration (a_r) for a system is thus a product of the noise floor (a_{nf}) and square-root of system bandwidth (BW). From Eq. (6.6), we can infer that the displacement resolution of an accelerometer depends not only on the noise floor but also on the sensor bandwidth.

6.4   ACCELEROMETER SENSING TECHNIQUES

This section contains a brief description of various techniques for transduction of proof mass deflection into a measurable quantity such as voltage. Of the various techniques available, capactive sensing-based MEMS accelerometers are most popular and commercially available for various application ranges.

1. **Piezo-resistive:** Strain in suspension structure connected to proof mass under acceleration causes change in resistance of a piezo-resistor, which is part of an electronic integrated circuit (Willis and Jimerson 1964; Roylance and Angell 1979). Various electrical circuit configurations such as the Wheatstone bridge are used to convert the change in resistance to an output voltage. Where the suspending structure is a beam, the piezo-resistor is positioned near the foot of the beam to maximize the impact of strain. An accelerometer with this type of sensing is deployed for impact or shock measurement such as air-bag deployment in automobiles. The piezo-resistor, however, has large temperature sensitivity and for decent sensitivity, a large proof mass is required. Further, complex electronics and signal processing is needed to ensure linearity and thermal stability and ultimately good resolution. The sensing range for these accelerometers is from 1 mg to 50 g.

2. **Piezo-electric:** In a piezo-electric accelerometer, a material which produces charge when strained is used as the sensing element (DeVoe and Pisano 1997). Like in a piezo-resistive accelerometer, the position of highest strain is chosen to enhance sensitivity. Still, this sensing technique requires high-voltages to be applied and still has low sensitivity. Since a piezo-electric crystal has to be connected to the MEMS structure, a truly integrated fabrication is not possible, implying low resolution. This type of accelerometer can be used for measuring high acceleration values and to find applications in vibration sensors.

3. **Capacitive:** One of the features of the proof mass is used to form a moving plate of a capacitor whose other end is fixed. The deflection of the proof mass changes the gap between the capacitor plates and thus, the capacitance that is used as a measure of the applied acceleration (Chae et al. 2005;

Rudolf 1983; Jiang et al. 2002). Using comb-like structures, even lateral MEMS accelerometers based on surface micromachining are realized with good sensitivity, linearity, and even multi-axis sensitivity. Use of appropriate electronics to mitigate sources of electronic noise improve the resolution of the capacitive accelerometer. Due to energy storage in the case of capacitors, the power consumption/dissipation is low for these accelerometers. In force-rebalance mode or closed loop operation, the change in capacitance is used to apply a restoring force on the proof mass, thus increasing bandwidth of operation without compromising the sensitivity. The sensing range for these accelerometers are from values as low as 2 mg to 10 g.

4. **Tunneling:** This type of accelerometer is used exclusively in force-rebalance mode as it involves measurement of a tunneling current that is established when two counter-electrodes are brought very close (within a few Å) (Kenny et al. 1991; Rockstad et al. 1996; Kubena et al. 1996). One electrode is designed in the form of a tip similar to the tip used in atomic force microscopy (AFM) and is connected to the proof mass whose deflection changes the gap between electrodes and thus the tunneling current (Liu et al. 1998; Liu and Kenny 2001; Yeh and Najafi 1998; Yeh and Najafi 1997). The dynamic range of this sensor is very small being limited by the range on which tunneling takes place. However, it has very high mechanical sensitivity and excellent resolution. Alongside the challenging electronics for linearization, noise cancellation, and proof mass stabilization, the major challenge experienced in this sensor is the fabrication of the tips sharp to the order of few atoms. The sensing range of this type of accelerometer is from 10 ng to 5 g.

5. **Electromagnetic:** In this type of accelerometer, a coil is placed on the proof mass and another coil on the substrate. Voltage induction in the secondary coil upon excitation of the primary coil depends on the spacing between the coils, which depends on the acceleration experienced (Abbaspour-Sani et al. 1994). On the MEMS scale, electromagnetic effects are weaker than electrostatic ones. Still, an electromagnetic accelerometer provides excellent linearity with simple signal conditioning as a trade-off for low resolution and sensitivity. This type of accelerometer finds application in air-bag deployment and has a sensing range from 0 g to 50g.

6. **Resonant:** Under stress due to deflection caused by applied acceleration, the mechanical properties of the material and thus the structure are modified. Thus, the natural or resonant frequency of the structure also changes. Commonly resonant accelerometers use either vibration-beams or an electrostatic spring. This change in resonant frequency is used as the measure of acceleration (Pedersen and Seshia 2004; Su and Yang 2001; Wang et al. 2018). Resonant accelerometers have high dynamic range, bandwidth, and sensitivity. Yet they are limited by high noise levels and are used as vibration sensors in machine tools. The sensing range of this type of accelerometer is from 5 g to 5,000 g.

7. **Thermal:** In this type of accelerometer, the deflection of proof mass is measured by the movement of hot/cold air within the sensor package. Apart from the MEMS proof mass, the sensor includes a heater and a pair of

thermocouples to measure the difference in temperature caused by movement of hot air due to proof mass deflection. Since hot air has lower density than the cold air, the deflection of the proof mass impacts them differently (Mailly et al. 2003). It has low cost, low noise, less drift, and uses simple signal conditioning circuitry. However, it is limited by a poor resolution and temperature sensitivity. Further, hybrid components and requirement of proper packaging make batch fabrication of this type of sensor difficult. The sensing range of this type of sensor is from 0.5 mg to high g and can be used for applications such as inclination sensing.

8. **Optical:** Optical MEMS accelerometers can be broadly classified as being based on free-space or guided-wave techniques. The deflection of proof mass due to acceleration is picked-up as a change in property of optical signal (intensity, peak wavelength of spectrum, or polarization) (Waters et al. 2002; Storgaard-Larsen et al. 1996; Bhola and Steier 2007; Jaksic et al. 2004; Krishnamoorthy et al. 2008). This type of accelerometer can be used in environments where electromagnetic interference is likely to impact the measurement such as tactical or space-navigation applications. Optical techniques are also attractive solution for biomedical signals measurement and actuation. Further, keeping the signal fully in optical domain enables multiplexing and networking of these sensors. As electronics can be removed from the measurement site, the impact of electronic noise on the sensor can be minimized. Thus, at very low frequencies, only mechanical thermal noise limits the sensor resolution, thus enabling nano-g resolutions. Depending on the sensing technique used, variety of dynamic ranges and sensitivities can be obtained in optical MEMS accelerometers. With advancement of optical MEMS and nanophotonics, there is an increasing maturity in the fabrication and integration of optical components and systems with MEMS structures.

From the above discussion, it is clear that accelerometer performance is quantified by different parameters such as sensitivity, bandwidth, and resolution. Further, it is noted that accelerometer design has now evolved into an application targeted activity with the required specification dictating the optimum design. The following section gives a detailed description of different free-space and guided-wave optical sensing techniques used in acceleration measurement and their relative merits and demerits.

6.5   OPTICAL SENSING PRINCIPLES AND DETECTION METHODS FOR ACCELERATION MEASUREMENT

Optical sensing techniques involve change in properties of light signal due to deflection of proof mass due to inertial force. The properties of light that are used for modulating the acceleration signal are its intensity, phase, polarization, and wavelength. Since light in the infrared (IR) region has high frequency and low wavelength, ideally it has excellent sensitivity and resolution (lower than nano-g) over a large dynamic range (Cooper and Smith 2004). However, practical optical systems are limited by the imperfections in the source and detector circuits. Even in a mode-locked laser, there is an intensity variation at a quantum level thatacts as a noise floor and thus resolution limit for acceleration. Similarly, a detection process using a p-i-n or avalanche photodiode is affected by noise such as relative-intensity noise (RIN). The optical channel itself could have birefringence and mode selectivity for the propagating signal (Jaksic et al. 2004). There is a limit on maximum accuracy in wavelength measurement, which depends on the spread of the signal spectrum. Thus, for realizing the high acceleration resolution with optical sensing techniques, a complex detection circuit is necessitated (Ortega et al. 2003).

6.5.1 OPTICAL SENSING TECHNIQUES

Optical MEMS accelerometers are, however, not limited to micro-gravity or satellite navigation-like applications (Krishnamoorthy et al. 2008; Zandi et al. 2012). They are extensively used in monitoring the health of civil engineering structures and automobiles, vibration measurement, and other biomedical signal measurement (Sreekumar and Asokan 2009; Mita et al. 2000; Stefani et al. 2012). The different optical MEMS accelerometers broadly fit into two categories: free-space light-based sensing and guided light-based sensing. The principles involved for optical transduction are interferometry (Fabry-Pérot, Mach-Zehnder, or Michelson interferometer), use of diffractive gratings or Bragg gratings, change in resonance condition of microring/disk resonators or bandgap in photonic crystal cavities due to proof mass deflection, intensity modulation, and optical gradient force. Table 6.1 presents a summary of various optical sensing techniques for accelerometers. What follows is a brief description of the broad sensing categories.

The optical intensity modulation techniques depend on simple source and detection techniques and can be easily integrated into the sensor to give an electrical output signal depending on the application. These intensity modulating accelerometers present a low-level sophistication compared to the other techniques and are suitable for applications that need resistance to electro-magnetic-interference (EMI) or harsh environments but not very high sensitivity or resolution. In intensity modulation-based accelerometers, proof mass displacement modifies the light coupling between light input and output paths (Guldimann et al. 2001; Plaza et al. 2004; Zhang and Li 2012) or blocks the optical path (Zandi et al. 2011; Hortschitz et al. 2012). Due to low design level sophistication, these sensors are suitable for photonic-integrated-circuit (PIC) integration. However they suffer from poor fundamental resolution limits.

A generic ring resonator consists of an optical waveguide looped back onto itself, where resonance occurs when the optical path length of the resonator is an integer multiple of wavelength (Bogaerts et al. 2012). But a ring resonator is useful only if there is a coupling mechanism to the outside world. Figure 6.4 shows common ring resonator configurations. In the notch or all-pass configuration, there is a notch observed in its transmission spectrum in the transient response to a pulsed input. Add-drop ring resonators act as filter for wavelengths, which are integer multiples of optical path lengths of the closed-loop ring/race-track waveguide. For a ring resonator-based sensor, the acceleration is encoded in the resonance wavelength of the ring. However, as a sensor, the ring resonator has limited dynamic range because the optical coupling takes place only over a small displacement range. Microring

TABLE 6.1

Commonly used Optical Sensing Techniques for Accelerometers

Sensing Principle	Range	Bandwidth	Sensitivity	Features
Intensity modulation (Guldimann et al. 2001; Plaza et al. 2004; Zandi et al. 2011; Hortschitz et al. 2012; Zhang and Li 2012)	Low	Medium	Low	Easy to integrate into PIC; free-space propagation of light; source noise significant
Ring resonators (Mo et al. 2009; Mo et al. 2011; Bramhavar et al. 2019; Wu et al. 2009; Hou et al. 2010)	Low	High	High	Low foot-print; sensor output is nonlinear; guided-wave propagation; acceleration causes wavelength shift
Photonic Crystals (Krause et al. 2012; Sheikhaleh et al. 2017)	Low	High	High	Light in/out coupling is challenging; guided-wave propagation; photonic bandgap changes under acceleration
Diffraction Gratings (Krishnamoorthy et al. 2008)	Low	Low to High	High	Free-space propagation of light; high sensitivity requires complex detection mechanism; complex packaging
Interferometers (Schropfer et al. 1998; Perez and Shkel 2007; Lin et al. 2010; Trigona et al. 2014; Davies et al. 2014; Zhao et al. 2020)	High	Low to High	Low to High	Free-space propagation of light; precision in alignment and positioning needed
Optical gradient force (Flores et al. 2016)	Low	High	High	Pump laser required aside from signal laser; Radio frequency (RF) detuning to be measured needs complex source interrogation setup
Bragg gratings (Berkoff and Kersey 1996a; Storgaard-Larsen et al. 1996; Kersey et al. 1992; Rao 1997; Gagliardi et al. 2008; Lam et al. 2010; Rosenberger et al. 2015a)	High	Low	Medium	Guided-wave propagation; spectrum carries acceleration information; advanced interrogation schemes allow sub-pico strain detection

resonator and racetrack resonator-based integrated accelerometers with a sensitivity of 0.015μm/g suitable for seismic applications have been demonstrated in literature (Mo et al. 2009; Mo et al. 2011). Bramhavar et al. (2019) demonstrated a high-resolution optical microdisk resonator-based integrated resonant accelerometer. The optical microdisk integrated with resonant tether of the accelerometer experiences a change in resonance frequency due to photoelastic effect upon actuation of the accelerometer. By using the highly mechanical sensitive resonant tether rather than

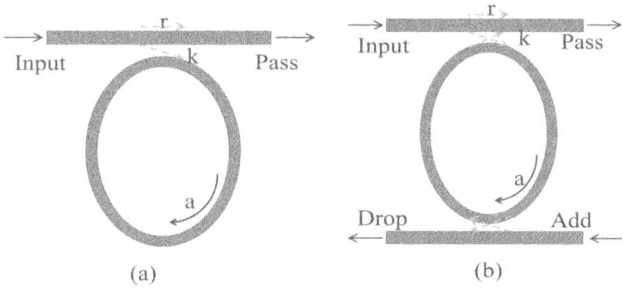

FIGURE 6.4 Common ring resonator configurations: **(a)** all-pass or notch ring; **(b)** add-drop ring.

the suspending beams, the sensitivity of the sensor is increased. By using a narrow linewidth laser locked to an operating wavelength close to resonance peak, the linear range of operation with a DC sensitivity of 6 $\mu g/\sqrt{Hz}$ for device natural frequency of 16.3 kHz was demonstrated. Wu et al. (2009) and Hou et al. (2010) have shown how fiber is drawn into an optical channel and wound into microrings positioned on sensitive regions of MEMS structures. Since the optical channels are drawn on the accelerometer structure, the losses associated with light input and output coupling can be greatly reduced.

Krause et al. (2012) demonstrated a photonic-crystal nanocavity-based high-resolution integrated accelerometer. A resolution in the order of $ng/\sqrt{Hz}$ has been achieved using the evanescent field coupling between precisely positioned tapered fiber held in the vicinity of a photonic-crystal zipper cavity. Krishnamoorthy et al. (2008) had earlier demonstrated a nano-g accelerometer based on subwavelength diffraction gratings. For achieving resolution near the limits of optical sensing, an optimized detection circuit and correspondingly a complex packaging were used.

Interferometry is arguably the most popular technique for realizing optical MEMS accelerometers. The incoming light is passed through a beam-splitter arrangement and part of it is subject to effect of acceleration of the proof mass. The effect of acceleration typically is change of optical path length either by photoelastic effect causing refractive index variation of light guidance medium or strain in the waveguide/fiber. Using a beam-splitter, the light is recombined and a change in power or spatial pattern is observed in the resulting light beam, which is then used for measurement. Figure 6.5 shows the working principles of different optical interferometer configurations. Difference in path lengths of interfering optical signals implies a phase difference between them. Upon interference, the optical waves with phase difference produce an intensity difference. This technique has been widely applied both in free-space and guided-wave modes. Since the information about the acceleration is embedded in the phase of the light signal subject to interference, the range of these accelerometers is limited. Fabry-Pérot (FP) interferometric cavities are a popular choice for accelerometers because the accelerometer proof mass can be prepared as one of the end-faces that form the interferometric cavity. In the FP cavity-based optical accelerometer, the cavity length that determines the output optical signal is

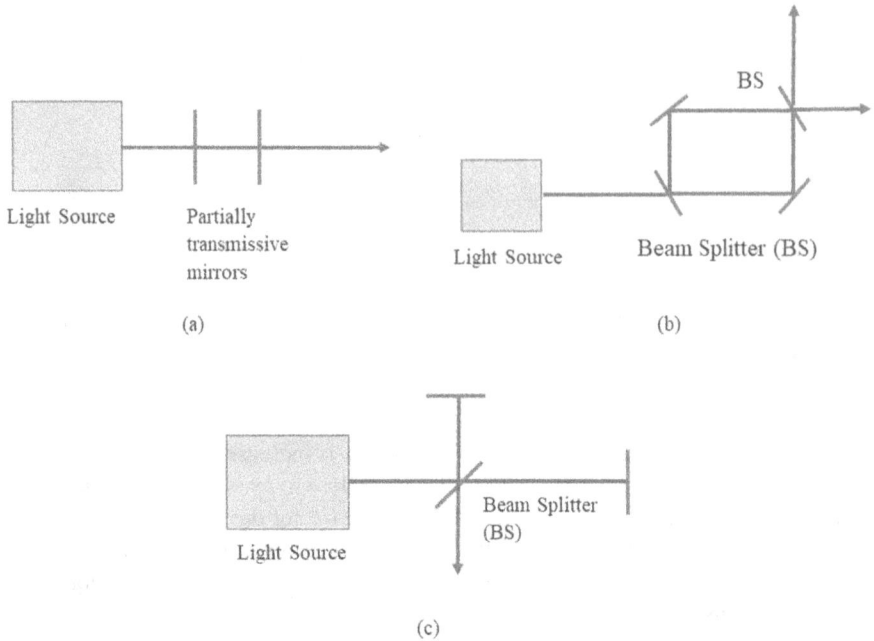

FIGURE 6.5   Working principle of (a) Fabry-Pérot, (b) Mach-Zehnder, and (c) Michelson interferometers.

mechanically modified under external stimulus. This method is based on unguided light propagation or free-space propagation and thus its performance is seriously impacted by quality of alignment and packaging and incurs high optical losses. The simplicity of design has resulted in looking for ways to improve the sensitivity of this class of sensors. Methods such as increasing mechanical sensitivity (Perez and Shkel 2007; Trigona et al. 2014; Davies et al. 2014; Zhao et al. 2020) and moving from intensity-based detection to coherence demodulation (Schropfer et al. 1998) to phase-generated-carrier (PGC) modulation (Lin et al. 2010) have shown great promise in this regard.

Optical fields of sufficient power density exert pressure that can cause mechanical actuation (Van-Thourhout and Roels 2010) in micrometer- to nanometer-sized particles. In combination with nanoelectromechanical systems (NEMS), the optical gradient force acts as a significant actuation mechanism for various applications in communication and photonic integrated circuits (Cai et al. 2013; Luan et al. 2014). From a mechanical perspective, this optical gradient force changes the damping and mechanical resonance of the MEMS/NEMS structure. The shift in oscillation RF under self-induced regenerative oscillations are more than 70-dB intensity peak-to-noise floor. When this nanophotonic structure is coupled to a MEMS spring, proof mass, damper system, under external acceleration, the sustained oscillation mode experience a shift in the peak (Flores et al. 2016). This has been used to realize a high sensitivity of 196 ng/Hz of RF shift along with a high resolution of 730 $ng/\sqrt{Hz}$, but has a very limited range.

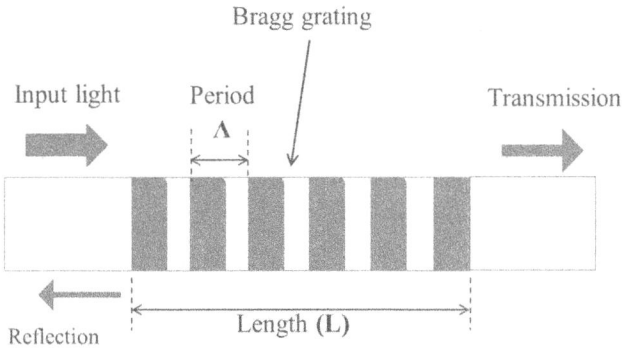

FIGURE 6.6   Working principle of Bragg gratings.

Bragg gratings work on the principle of diffraction occurring for light propagating in guided-wave media with periodic variation of effective refractive index embedded in the waveguide itself as shown in Figure 6.6. Fiber Bragg grating (FBG) based accelerometers are very popular and have been shown great promise for achieving the holy grail of sensing namely high sensitivity with large sensing range (Berkoff and Kersey 1996a; Kersey et al. 1992; Rao 1997). When combined with other resonant structures such as ring resonators (Malara et al. 2014), Fabry-Pérot (FP) cavity (Gagliardi et al. 2007), and P-shifted FBG (PSFBG) (Lam et al. 2010), high resolution has been achieved in FBG-based accelerometers. High acceleration resolution in FBG-based sensors is also achieved using ultrastable lasers (Gagliardi et al. 2006, 2008) and actively locked lasers using RF modulation (Gagliardi et al. 2009; Lam et al. 2010). After the initial excitement for integrated planar waveguide Bragg gratings-based accelerometers using silicon-nitride waveguides (Storgaard-Larsen et al. 1996), integration of planar waveguide Bragg gratings-based accelerometers into photonic integrate circuits have not matured unlike in the case of pressure sensors (Neeharika and Pattnaik 2015; Thondagere et al. 2018). Polymer materials with suitable optical properties are an alternative for integrated optical MEMS sensors because they allow for direct writing onto a substrate. Further planar Bragg gratings can also be written directly onto a substrate. Polymethylmethacrylate (PMMA) and TOPAS-based planar Bragg gratings have been used to demonstrate strain sensors (Rosenberger et al. 2014; Rosenberger et al. 2015a). Apart from the temperature of surroundings, optical properties of polymer Bragg gratings also depend on humidity and thus, robust performance necessitates immunity from humidity (Rosenberger et al. 2015b). For high-resolution Bragg gratings, a narrow Bragg peak is necessary. To realize this, grating length must be high. This in turn means that for realizing integrated MEMS sensors based on Bragg gratings, relatively large MEMS structures with low mechanical bandwith are required. Thus, waveguide Bragg gratings-based accelerometers are suited for high-sensitivity quasi-static acceleration measurements such as in seismometry.

In Malayappan et al. (2020), we proposed the design and method for realizing high-resolution silicon waveguide Bragg gratings-based dual-axis accelerometers as in Figure 6.7. To compensate for the dominant temperature dependence of wavelength

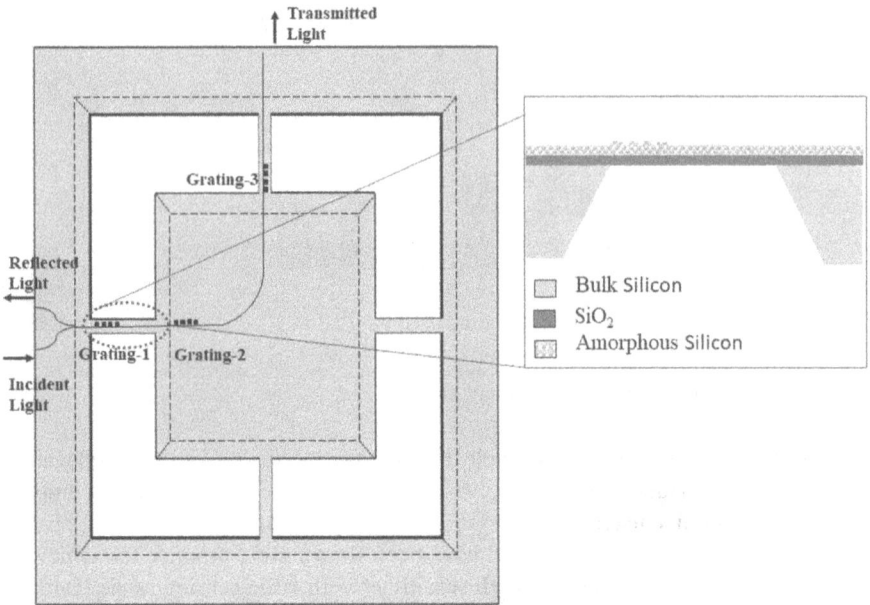

FIGURE 6.7   Top view of schematic of quad-beam accelerometer with integrated wave-guide Bragg gratings. Inset shows cross-sectional view of beam and its material composition.

of Bragg peaks, a differential measurement based on a reference grating is taken. Shift in a reference grating's peak depends only on temperature variation and is made independent of strain by dint of its mechanical position. For the chosen parameters, the designed sensor has a linear response over a large range and a sensitivity of 30 pm/g as shown in Figure 6.8. Low cross-axis sensitivity is achieved by optimizing the design and materials. The proposed design enables a high-resolution well below $1\ \mu g / \sqrt{Hz}$ and is suitable for seismometry and inertial navigation applications.

6.5.2   COMMON OPTICAL INTERROGATION METHODS

For a sensing platform, sensitivity depends on the light source used and the corresponding detection system. For simple intensity-based detection, intensity modulation-direct detection (IM-DD), a noncoherent source and a photo-diode suffice for the interrogation setup as shown in Figure 6.9. As the intensity of light signal reaching the detector is the measurand, there is no stringent requirement for an optically coherent source (Uttamchandani et al. 1992; Hortschitz et al. 2012). An optical isolator or circulator is used in most interrogation setups to avoid the back-scattering light from reaching the source. To improve bandwidth of operation, the lower frequency limit can be improved by reducing 1/f noise (Guldimann et al. 2001). Thus, the intensity modulation testing is done based on a source whose output intensity is driven/modulated by varying the input current. Synchronous demodulation following optical detection complete the setup.

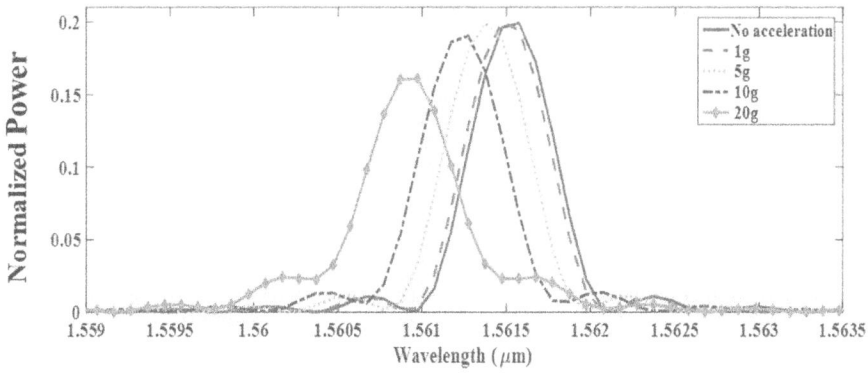

(a)

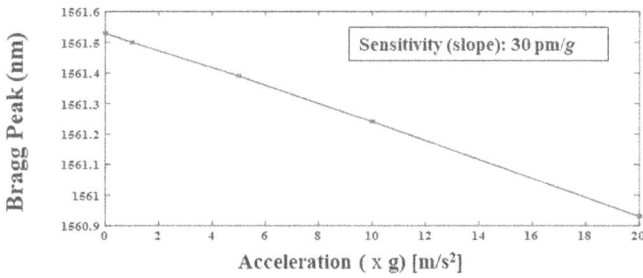

(b)

FIGURE 6.8 Waveguide Bragg gratings dual-axis accelerometer results: (a) reflected spectra for different accelerations; (b) linear variation of Bragg peak with applied acceleration.

However, in sensing techniques such as FP cavities, ring resonators, and Bragg gratings, the spectral response is modified and the position of peak in the spectrum is the measure of experienced acceleration. Since the information is in the frequency, intensity fluctuations due to noise do not adversely impact the sensor performance

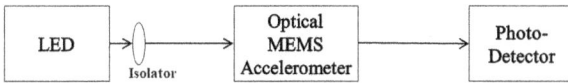

(a)

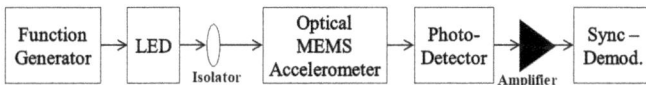

(b)

FIGURE 6.9 Intensity modulation testing setup: (a) basic model; (b) modulated source used for interrogation to reduce impact of noise.

when frequency detection is applied. However, the frequency detection requires cost intensive equipment such as an optical spectrum analyzer or diffraction gratings coupled with a precisely engineered charge-coupled device (CCD) or photo-detector array (Storgaard-Larsen et al. 1996; Carpenter et al. 2013; Wu et al. 2019; Wu et al. 2009). Use of interferometric wavelength discriminators is also a popular method where the wavelength/frequency encoded response of the optical MEMS accelerometer acts as input to the interferometric setup, which can then be tuned using a phase-generated carrier technique (Kersey 1992; Berkoff and Kersey 1996a). The error signal thus generated becomes the measure of strain/acceleration. A Mach-Zehnder unbalanced interferometer-based detector has been used to convert the wavelength shift information into a phase shift (Kersey et al. 1992). This interferometer can also act as a multispectral wavelength shift detector (Berkoff and Kersey 1996b). Use of the above detection methods gives high resolution in the order of nano-strain.

In another detection method, taking advantage of the roll-off of the sensor spectrum, a bias point (wavelength) (Bhola and Steier 2007; Bramhavar et al. 2016; Perez and Shkel 2007) is chosen as shown in Figure 6.10 to maximize linear range of operation. The applied acceleration causes spectral shift corresponding to an intensity change at the bias wavelength. Interrogation using this scheme is based on stable laser source (tunable laser) adjusted to the bias point using appropriate temperature and current compensation (Gagliardi et al. 2008; Sreekumar and Asokan 2009) does not require the spectrometer detector. This method has limited linear range of operation and resolution depends on the intensity and frequency stability of the laser.

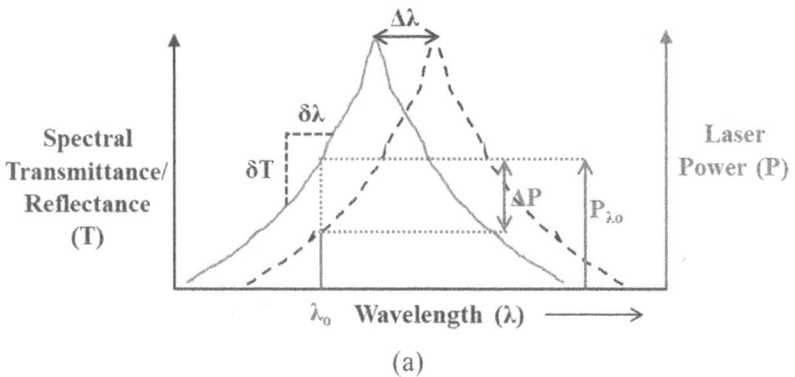

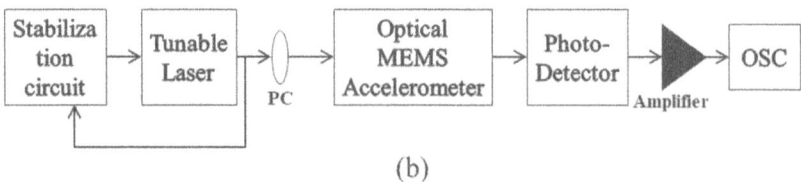

FIGURE 6.10 Intensity modulation due to spectral shift: **(a)** illustration of power modulation at operating wavelength; **(b)** interrogation setup.

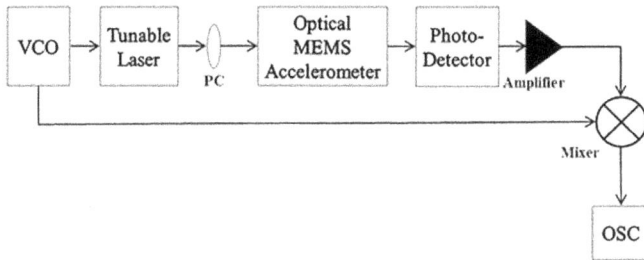

FIGURE 6.11   Interrogation setup based on generation of a Pound-Drever-Hall signal.

Sub-pico strain level high resolution is possible using telecom-grade modulated tunable lasers. The tunable laser is modulated using a voltage controlled oscillator to generate frequency sidebands. The frequency encoded acceleration signal is demodulated using heterodyne detection using an appropriate mixer configuration as shown in Figure 6.11. This demodulation yeilds a highly-dispersive signal with a zero-crossing around the resonant peak (Gagliardi et al. 2010). The frequency shift due to acceleration thus produces a nonzero voltage output that can be used as error signal for Pound-Drever-Hall (Black 2001) frequency locking of the laser and enables continuous tracking of the sensor.

6.6   CONCLUSION

In this chapter, we have presented a summary of different sensing techniques used in MEMS accelerometers with particular focus on optical sensing techniques. Free-space light and guided-wave light-based sensing techniques for optical MEMS accelerometers have been presented with their relative merits and shortcomings. Optical sensing principles not only enable the application of the sensor in harsh environments but can also give high resolution and sensitivity. The commonly used interrogation methods are also discussed to show their impact on realizing high-sensitivity measurement.

REFERENCES

Abbaspour-Sani, E., Huang, R.-S., and Kwok, C. Y. 1994. A linear electromagnetic accelerometer. *Sensors and Actuators A: Physical* 44 (2): 103–09.

Berkoff, T. and Kersey, A. 1996a. Experimental demonstration of a fiber Bragg grating accelerometer. *IEEE Photonics Technology Letters* 8 (12): 1677–79.

Berkoff, T. and Kersey, A. 1996b. Fiber Bragg grating array sensor system using a bandpass wavelength division multiplexer and interferometric detection. *IEEE Photonics Technology Letters* 8 (11): 1522–24.

Bhola, B. and Steier, W. H. 2007. A novel optical microring resonator accelerometer. *IEEE Sensors Journal* 7 (12): 1759–66.

Black, E. D. 2001. An introduction to Pound–Drever–Hall laser frequency stabilization. *American Journal of Physics* 69 (1): 79–87.

Bogaerts, W., De Heyn, P., Van Vaerenbergh, T., De Vos, K., Kumar Selvaraja, S., Claes, T., Dumon, P., Bienstman, P., Van Thourhout, D., and Baets, R. 2012. Silicon microring resonators. *Laser & Photonics Reviews* 6 (1): 47–73.

Bramhavar, S., Kharas, D., and Juodawlkis, P. W. 2016. A photonic integrated resonant accelerometer, in *2016 IEEE Photonics Conference (IPC)*, 1–2. IEEE.

Bramhavar, S., Kharas, D., and Juodawlkis, P. 2019. Integrated photonic inertial sensors. In *Integrated Photonics Research, Silicon and Nanophotonics*, ITh3A–2. Optical Society of America.

Cai, H., Dong, B., Tao, J., Ding, L., Tsai, J., Lo, G.-Q., Liu, A. Q., and Kwong, D. L. 2013. A nanoelectromechanical systems optical switch driven by optical gradient force. *Applied Physics Letters* 102 (2): 023103.

Carpenter, L., Holmes, C., Gates, J., and Smith, P. 2013. MEMS accelerometers utilizing resonant microcantilevers with interrogated single-mode waveguides and Bragg gratings, in *Reliability, Packaging, Testing, and Characterization of MOEMS/MEMS and Nanodevices XII* 8614: 86140K, International Society for Optics and Photonics.

Chae, J., Kulah, H., and Najafi, K. 2005. A monolithic three-axis micro-g micromachined silicon capacitive accelerometer. *Journal of Microelectromechanical Systems* 14 (2): 235–42.

Chu, A. 2012. Accelerometer selection based on applications. *Endevco Technical Paper TP291.*

Cooper, D. J. and Smith, P. W. 2004. Limits in wavelength measurement of optical signals. *JOSA B* 21 (5): 908–13.

Cooper, E., Post, E., Griffith, S., Levitan, J., Manalis, S., Schmidt, M., and Quate, C. 2000. High-resolution micromachined interferometric accelerometer. *Applied Physics Letters* 76 (22): 3316–18.

Davies, E., George, D. S., Gower, M. C., and Holmes, A. S. 2014. MEMS Fabry-Pérot optical accelerometer employing mechanical amplification via a v-beam structure. *Sensors and Actuators A: Physical* 215: 22–9. Special Issue of the Micromechanics Section of Sensors and Actuators based upon contributions revised from the Technical Digest of the 26th IEEE International Conference on Micro Electro Mechanical Systems (MEMS-13; 20–24 January 2013, Taipei, Taiwan).

Devices, A. 1995. 1g to±5g single chip accelerometer with signal conditioning: ADXL05. *Analog Devices, Norwood.* https://scholar.google.com/scholar?hl=en&as_sdt=0%2 C5&q=ADXL05+single+chip+accelerometer+with+signal+conditioning&btnG=#d =gs_cit&u=%2Fscholar%3Fq%3Dinfo%3Ag7mSngvggAYJ%3Ascholar.google.com% 2F%26output%3Dcite%26scirp%3D0%26hl%3Den.

DeVoe, D. L. and Pisano, A. P. 1997. A fully surface-micromachined piezoelectric accelerometer, in *Proceedings of International Solid State Sensors and Actuators Conference (Transducers' 97)* 2: 1205–08. IEEE.

Flores, J. G. F., Huang, Y., Li, Y., Wang, D., Goldberg, N., Zheng, J., Yu, M., Lu, M., Kutzer, M., Rogers, D., et al. 2016. A CMOS-compatible oscillation-mode optomechanical DC accelerometer at 730-ng/hz1/2 resolution, in *2016 IEEE International Symposium on Inertial Sensors and Systems*, 125–27. IEEE.

Gagliardi, G., De Nicola, S., Ferraro, P., and De Natale, P. 2007. Interrogation of fiber Bragg-grating resonators by polarization-spectroscopy laser-frequency locking. *Optics Express* 15 (7): 3715–28.

Gagliardi, G., Salza, M., Ferraro, P., and De Natale, P. 2006. Interrogation of FBG-based strain sensors by means of laser radio-frequency modulation techniques. *Journal of Optics A: Pure and Applied Optics* 8 (7): S507.

Gagliardi, G., Salza, M., Ferraro, P., De Natale, P., Di Maio, A., Carlino, S., De Natale, G., and Boschi, E. 2008. Design and test of a laser-based optical-fiber Bragg-grating accelerometer for seismic applications. *Measurement Science and Technology* 19 (8): 085306.

Gagliardi, G., Salza, M., Lam, T. T.-Y., Chow, J. H., and De Natale, P. 2009. 3-axis accelerometer based on lasers locked to π-shifted fibre Bragg gratings, in *20th International*

Conference on Optical Fibre Sensors 7503: 75033X, International Society for Optics and Photonics.

Gagliardi, G., Salza, M., Ferraro, P., Chehura, E., Tatam, R. P., Gangopadhyay, T. K., Ballard, N., Paz-Soldan, D., Barnes, J. A., Loock, H.-P., et al. 2010. Optical fiber sensing based on reflection laser spectroscopy. *Sensors* 10 (3): 1823–45.

Guldimann, B., Dubois, P., Clerc, P.-A., and de Rooij, N. 2001. Fiber optical MEMS accelerometer with high mass displacement resolution, in *Transducers' 01 Eurosensors XV*, 438–441, Springer.

Hortschitz, W., Steiner, H., Sachse, M., Stifter, M., Kohl, F., Schalko, J., Jachimowicz, A., Keplinger, F., and Sauter, T. 2012. Robust precision position detection with an optical MEMS hybrid device. *IEEE Transactions on Industrial Electronics* 59 (12): 4855–62.

Hou, C.-L., Wu, Y., Zeng, X., Zhao, S., Zhou, Q., and Yang, G. 2010. Novel high sensitivity accelerometer based on a microfiber loop resonator. *Optical Engineering* 49 (1): 014402.

Jaksic, Z., Radulovic, K., and Tanaskovic, D. 2004. MEMS accelerometer with all-optical readout based on twin-defect photonic crystal waveguide, in *2004 24th International Conference on Microelectronics (IEEE Cat. No. 04TH8716)*, 1: 231–4, IEEE.

Jiang, X., Wang, F., Kraft, M., and Boser, B. E. 2002. An integrated surface micromachined capacitive lateral accelerometer with $2\mu g/\sqrt{Hz}$ resolution, in *Proceedings of the Solid-State, Actuator and Microsystems Workshop, Hilton Head Island, South Carolina*, 202–05.

Kenny, T., Waltman, S., Reynolds, J., and Kaiser, W. 1991. Micromachined silicon tunnel sensor for motion detection. *Applied Physics Letters* 58 (1): 100–02.

Kersey, A. 1992. Fiber-grating based strain sensor with phase sensitive detection, in *First European Conference on Smart Structures and Materials* 1777: 17770E, International Society for Optics and Photonics.

Kersey, A., Berkoff, T., and Morey, W. 1992. High-resolution fibre-grating based strain sensor with interferometric wavelength-shift detection. *Electronics Letters* 28 (3): 236–38.

Kraft, M. 1997. *Closed loop accelerometer employing oversampling conversion. Coventry University*. PhD thesis, PhD dissertation.

Krause, A. G., Winger, M., Blasius, T. D., Lin, Q., and Painter, O. 2012. A high-resolution microchip optomechanical accelerometer. *Nature Photonics* 6 (11): 768.

Krishnamoorthy, U., Olsson Iii, R., Bogart, G. R., Baker, M., Carr, D., Swiler, T., and Clews, P. 2008. In-plane MEMS-based nano-g accelerometer with sub-wavelength optical resonant sensor. *Sensors and Actuators A: Physical* 145: 283–90.

Krishnan, G., Kshirsagar, C. U., Ananthasuresh, G., and Bhat, N. 2007. Micromachined high-resolution accelerometers. *Journal of the Indian Institute of Science* 87 (3): 333.

Kubena, R., Atkinson, G., Robinson, W., and Stratton, F. 1996. A new miniaturized surface micromachined tunneling accelerometer. *IEEE Electron Device Letters* 17 (6): 306–8.

Lam, T. T., Gagliardi, G., Salza, M., Chow, J. H., and De Natale, P. 2010. Optical fiber three-axis accelerometer based on lasers locked to π phase-shifted Bragg gratings. *Measurement Science and Technology* 21 (9): 094010.

Lemkin, M. and Boser, B. E. 1999. A three-axis micromachined accelerometer with a CMOS position-sense interface and digital offset-trim electronics. *IEEE Journal of Solid-State Circuits* 34 (4): 456–68.

Lin, Q., Chen, L., Li, S., and Wu, X. 2010. A high-resolution fiber optic accelerometer based on intracavity phase-generated carrier (PGC) modulation. *Measurement Science and Technology* 22 (1): 015303.

Liu, C.-H. and Kenny, T. W. 2001. A high-precision, wide-bandwidth micromachined tunneling accelerometer. *Journal of Microelectromechanical Systems* 10 (3): 425–33.

Liu, C.-H., Barzilai, A. M., Reynolds, J. K., Partridge, A., Kenny, T. W., Grade, J. D., and Rockstad, H. K. 1998. Characterization of a high-sensitivity micromachined tunneling accelerometer with micro-g resolution. *Journal of Microelectromechanical Systems* 7 (2): 235–44.

Loh, N. C., Schmidt, M. A., and Manalis, S. R. 2002. Sub-10 cm/sup 3/interferometric accelerometer with nano-g resolution. *Journal of Microelectromechanical Systems* 11 (3): 182–87.

Luan, X., Huang, Y., Li, Y., McMillan, J. F., Zheng, J., Huang, S.-W., Hsieh, P.-C., Gu, T., Wang, D., Hati, A., et al. 2014. An integrated low phase noise radiation-pressure-driven optomechanical oscillator chipset. *Scientific Reports* 4: 6842.

Mailly, F., Giani, A., Martinez, A., Bonnot, R., Temple-Boyer, P., and Boyer, A. 2003. Micromachined thermal accelerometer. *Sensors and Actuators A: Physical* 103 (3): 359–63.

Malara, P., Campanella, C., Giorgini, A., Avino, S., Zullo, R., Gagliardi, G., and De Natale, P. 2014. Sensitive strain measurements with a fiber Bragg-grating ring resonator, in *23rd International Conference on Optical Fibre Sensors* 9157: 915706, International Society for Optics and Photonics.

Malayappan, B., Krishnaswamy, N., and Pattnaik, P. K. 2020. Novel high-resolution lateral dual-axis quad-beam optical MEMS accelerometer using waveguide Bragg gratings. *Photonics* 7 (3).

Mita, A., Yokoi, I., et al. 2000. Fiber Bragg grating accelerometer for structural health monitoring, in *Fifth International Conference on Motion and Vibration Control (MOVIC 2000), Sydney, Australia.*

Mo, W., Wu, H., Gao, D., and Zhou, Z. 2009. A novel accelerometer based on microring resonator. *Chinese Optics Letters* 7 (9): 798–801.

Mo, W., Zhou, Z., Wu, H., and Gao, D. 2011. Silicon-based stress-coupled optical racetrack resonators for seismic prospecting. *Sensors Journal, IEEE* 11 (4): 1035–39.

Neeharika, V. and Pattnaik, P. K. 2015. Optical MEMS pressure sensors incorporating dual waveguide Bragg gratings on diaphragms. *IEEE Sensors Journal* 16 (3): 681–87.

Ortega, D. R., Magno, W. C., and Cruz, F. 2003. Diode laser stabilization using Pound-Drever-Hall technique. *Annals of Optics* 5: 1–3.

Özgen, C. 2010. *A Tactical grade MEMS accelerometer.* PhD thesis, Middle East Technical University.

Pedersen, C. B. and Seshia, A. A. 2004. On the optimization of compliant force amplifier mechanisms for surface micromachined resonant accelerometers. *Journal of Micromechanics and Microengineering* 14 (10): 1281.

Perez, M. A. and Shkel, A. M. 2007. Design and demonstration of a bulk micromachined Fabry–Pérot μg-resolution accelerometer. *Sensors Journal, IEEE* 7 (12): 1653–62.

Plaza, J. A., Llobera, A., Dominguez, C., Esteve, J., Salinas, I., Garcia, J., and Berganzo, J. 2004. Besoi-based integrated optical silicon accelerometer. *Journal of Microelectromechanical Systems* 13 (2): 355–64.

Rao, Y.-J. 1997. In-fibre Bragg grating sensors. *Measurement Science and Technology* 8 (4): 355.

Rockstad, H. K., Tang, T., Reynolds, J., Kenny, T., Kaiser, W., and Gabrielson, T. B. 1996. A miniature, high-sensitivity, electron tunneling accelerometer. *Sensors and Actuators A: Physical* 53 (1–3): 227–31.

Rosenberger, M., Hessler, S., Belle, S., Schmauss, B., and Hellmann, R. 2014. Compressive and tensile strain sensing using a polymer planar Bragg grating. *Optics Express* 22 (5): 5483–90.

Rosenberger, M., Eisenbeil, W., Schmauss, B., and Hellmann, R. 2015a. Simultaneous 2d strain sensing using polymer planar Bragg gratings. *Sensors* 15 (2): 4264–72.

Rosenberger, M., Hessler, S., Belle, S., Schmauss, B., and Hellmann, R. 2015b. Topas-based humidity insensitive polymer planar Bragg gratings for temperature and multi-axial strain sensing, in *International Seminar on Photonics, Optics, and its Applications (ISPhOA 2014)* 9444: 944408. International Society for Optics and Photonics.

Roylance, L. M. and Angell, J. B. 1979. A batch-fabricated silicon accelerometer. *IEEE Transactions on Electron Devices* 26 (12): 1911–17.

Rudolf, F. 1983. A micromechanical capacitive accelerometer with a two-point inertial-mass suspension. *Sensors and Actuators* 4: 191–98.

Schröpfer, G., Elflein, W., de Labachelerie, M., Porte, H., and Ballandras, S. 1998. Lateral optical accelerometer micromachined in (100) silicon with remote readout based on coherence modulation. *Sensors and Actuators A: Physical* 68 (1–3): 344–49.

Senturia, S. D. 2007. *Microsystem design.* Springer Science & Business Media.

Sheikhaleh, A., Abedi, K., and Jafari, K. 2017. An optical MEMS accelerometer based on a two-dimensional photonic crystal add-drop filter. *Journal of Lightwave Technology* 35 (14): 3029–34.

Spineanu, A., Bénabès, P., and Kielbasa, R. 1997. A digital piezoelectric accelerometer with sigma-delta servo technique. *Sensors and Actuators A: Physical* 60 (1–3): 127–33.

Sreekumar, K. and Asokan, S. 2009. Compact fiber Bragg grating dynamic strain sensor cum broadband thermometer for thermally unstable ambience. *Journal of Optics* 12 (1): 015502.

Stefani, A., Andresen, S., Yuan, W., Herholdt-Rasmussen, N., and Bang, O. 2012. High sensitivity polymer optical fiber-Bragg-grating-based accelerometer. *IEEE Photonics Technology Letters* 24 (9): 763–65.

Storgaard-Larsen, T., Bouwstra, S., and Leistiko, O. 1996. Opto-mechanical accelerometer based on strain sensing by a Bragg grating in a planar waveguide. *Sensors and Actuators A: Physical* 52 (1–3): 25–32.

Su, X.-P. and Yang, H. 2001. Two-stage compliant microleverage mechanism optimization in a resonant accelerometer. *Structural and Multidisciplinary Optimization* 22 (4): 328–34.

Thondagere, C., Kaushalram, A., Srinivas, T., and Hegde, G. 2018. Mathematical modeling of optical MEMS differential pressure sensor using waveguide Bragg gratings embedded in Mach Zehnder interferometer. *Journal of Optics* 20 (8): 085802.

Trigona, C., Andò, B., and Baglio, S. 2014. Design, fabrication, and characterization of Besoi-accelerometer exploiting photonic bandgap materials. *IEEE Transactions on Instrumentation and Measurement* 63 (3): 702–10.

Uttamchandani, D., Liang, D., and Culshaw, B. 1992. A micromachined silicon accelerometer with fibre optic interrogation, in *IEE Colloquium on fibre optics sensor technology*, 4/1–4/4, IET. https://ieeexplore.ieee.org/abstract/document/168467.

Van-Thourhout, D. and Roels, J. 2010. Optomechanical device actuation through the optical gradient force. *Nature Photonics* 4 (4): 211.

Wang, Y., Zhang, J., Yao, Z., Lin, C., Zhou, T., Su, Y., and Zhao, J. 2018. A MEMS resonant accelerometer with high performance of temperature based on electrostatic spring softening and continuous ring-down technique. *IEEE Sensors Journal* 18 (17): 7023–31.

Waters, R. L., Aklufi, M. E., and Jones, T. E. 2002. Electro-optical ultra-sensitive accelerometer, in *2002 IEEE Position Location and Navigation Symposium (IEEE Cat. No.02CH37284)*, 36–43.

Willis, J. and Jimerson, B. 1964. A piezoelectric accelerometer. *Proceedings of the IEEE* 52 (7): 871–72.

Wu, H., Lin, Q., Jiang, Z., Zhang, F., Li, L., and Zhao, L. 2019. A temperature and strain sensor based on a cascade of double fiber Bragg grating. *Measurement Science and Technology* 30 (6): 065104.

Wu, Y., Zeng, X., Rao, Y., Gong, Y., Hou, C., and Yang, G. 2009. MOEMS accelerometer based on microfiber knot resonator. *IEEE Photonics Technology Letters* 21 (20): 1547–49.

Yeh, C. and Najafi, K. 1997. A low-voltage tunneling-based silicon microaccelerometer. *IEEE Transactions on Electron Devices* 44 (11): 1875–82.

Yeh, C. and Najafi, K. 1998. CMOS interface circuitry for a low-voltage micromachined tunneling accelerometer. *Journal of Microelectromechanical Systems* 7 (1): 6–15.

Zandi, K., Zou, J., Wong, B., Kruzelecky, R. V., and Peter, Y.-A. 2011. VOA-based optical mems accelerometer, in *16th International Conference on Optical MEMS and Nanophotonics*, 15–16. IEEE.

Zandi, K., Bélanger, J. A., and Peter, Y.-A. 2012. Design and demonstration of an in-plane silicon-on-insulator optical MEMS Fabry–Pérot-based accelerometer integrated with channel waveguides. *Journal of Microelectromechanical Systems* 21 (6): 1464–70.

Zhang, B. and Li, W. F. 2012. Development of a micro accelerometer based MOEMS, in *2012 International Conference on Manipulation, Manufacturing and Measurement on the Nanoscale (3M-NANO)*, 250–53.

Zhao, M., Jiang, K., Bai, H., Wang, H., and Wei, X. 2020. A MEMS based Fabry–Pérot accelerometer with high resolution. *Microsystem Technologies*, 1–9.

7 A 60-GHz SiGe HBT Receiver Front End for Biomedical Applications

Puneet Singh and Saroj Mondal

CONTENTS

7.1 Introduction 109
7.2 Millimeter Wave Wireless Systems.. 110
7.3 SiGe HBT Receiver Front-End Architecture.. 111
    7.3.1 Receiver Architecture 111
    7.3.2 Low Noise Amplifier 112
        7.3.2.1 Design Considerations for an LNA.................................. 113
        7.3.2.2 Design Methodology for Cascode LNA 114
        7.3.2.3 LNA Simulation Results and Observations 115
    7.3.3 RF Divider 125
    7.3.4 Mixer... 126
        7.3.4.1 Classification of a Mixer.. 127
        7.3.4.2 Single-Balanced Gilbert Cell Mixer Topology 128
        7.3.4.3 Single-Balanced Mixer 129
    7.3.5 Baseband Amplifier 131
    7.3.6 Cascading VG-LNA, RF Divider, Mixer, and Baseband Amplifier....... 133
7.4 Conclusion 134
Acknowledgment 134
References... 134

7.1 INTRODUCTION

In this chapter, we present the design of a fully integrated radio frequency (RF) front end [1] for a low-cost, low direct current (DC) power consumption, high-sensitivity millimeter-wave receiver operating in the V-band around 60 GHz. The technology used in this design was IHP's silicon germanium (SiGe) heterojunction bipolar transistor (HBT) 0.13 μm BiCMOS process. The proposed RF front end is well suited for battery-operated biomedical applications. The designed receiver will talk to the implanted body sensors and collect physiological data for health care monitoring. To realize the receiver, HBTs are preferred over metal oxide semiconductor field-effect transistors (MOSFETs) because of their high gain, high transitional frequencies (f_T), f_{max}, low noise figure, and low power consumption. The designed RF front end consumes around 100 mW of DC power and has a wide dynamic range to handle weak signals in the presence of high

109

interferers. The receiver has a good 1-dB compression making it linear in the frequency range of 57–64 GHz and has a voltage gain control to ensure that the analog-to-digital converter (ADC) does not saturate. This receiver operates at 60 GHz to exploit the high security that this license-free band offers and its performance in this band makes it an apt choice for transmitting sensitive biological data.

The rest of the chapter is organized as follows. The chapter begins with a brief introduction. Section 7.2 gives a background of millimeter wave wireless systems and discusses the possible future applications in the millimeter wave bands. Section 7.3 discuss the RF front-end receiver architecture suitable for health care applications and subsequently explains the low-noise amplifier (LNA) fundamentals such as gain, stability circle, noise figure, and gain trade-off, etc. This section also elaborates the design steps for the LNA and details the approach used for the designs. In addition to that, Section 7.3 explains the basics of single-balanced active mixers, active RF dividers, baseband amplifiers, and the complete integration to realize an RF front-end receiver covering a 57–64 GHz frequency band. Finally, the chapter is concluded by a summary of accomplishments.

7.2   MILLIMETER WAVE WIRELESS SYSTEMS

The millimeter wave regime ranges from the frequencies of 30 GHz to 300 GHz and has been researched extensively because of the following advantages:

1. Increased data rates
2. Higher bandwidth
3. Higher security
4. Reduced physical cables required for transmission
5. Smaller antenna size

Higher data rates, bandwidth, and especially higher security [2–3] make transmission and reception at millimeter waves an attractive option. Since signals at 60 GHz undergo a lot of attenuation (as high as 16 dB/km at sea level) due to atmospheric oxygen absorbing the electromagnetic radiation, the signal cannot travel very far from the point of transmission. Also, the line of sight (LOS) propagation loss at 60 GHz is very high and is given by the formula [4]

$$Path \ loss \ (PL) = 92.5(dB) + 20\log_{10} f (GHz) + 20\log_{10} d (km) + \alpha \ (dB) \qquad (7.1)$$

where, α is the constant that represents the attenuation due to atmospheric oxygen. Eavesdropping is avoided because Earth's atmosphere creates an "absorption shield" that prevents tapping from an Earth-based station. This also holds true for point-to-point communication because the stray radiation is absorbed by the oxygen in the surroundings allowing for secure communication [5]. Figure 7.1 shows the attenuation due to atmospheric oxygen for a range of millimeter-wave frequencies.

The two main bands available for research in the millimeter-wave frequencies are (1) centered about 60 GHz and (2) centered about 180 GHz.

The 60 GHz band is used for research into communication application despite suffering from attenuation due to oxygen. This band is not standardized and hence

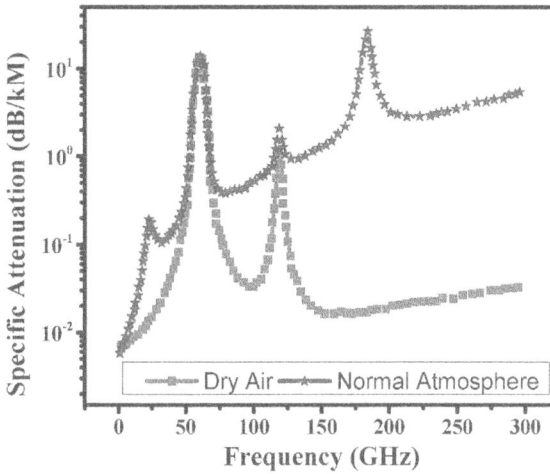

FIGURE 7.1   Atmospheric attenuation of electromagnetic waves due to dry air and normal atmospheric air.

the transmission ranges from 57 GHz to 66 GHz in Europe (9 GHz range) while it ranges from 57 GHz to 64 GHz in the United States (7 GHz range). The channels in the 60 GHz band are 2 GHz wide and have guard bands of 240 MHz and 120 MHz, respectively [6]. This is the band that will be in focus in this chapter and the proposed RF front end will be designed to operate in the frequency range from 57 GHz to 64 GHz.

The other bands in the millimeter-wave and their applications include [6]:

1. 76–77 GHz band: used for safety applications and automotive radar
2. 77–81 GHz band: ultra-wide band (UWB) short range radar
3. 77 GHz and 94 GHz: millimeter wave imaging
4. 40 GHz band: high-speed microwave data links
5. 71–76 GHz, 81–86 GHz, and 92–95 GHz band: point-to-point communication links

The frequency bands in the millimeter wave from 30 GHz to 300 GHz have been defined by the International Telecommunications Union (ITU) and have been summarized aptly in [6].

7.3   SIGE HBT RECEIVER FRONT-END ARCHITECTURE

7.3.1   RECEIVER ARCHITECTURE

One of the widely accepted receiver architectures is the super heterodyne or dual conversion-based receiver architecture [7] because it offers good sensitivity and selectivity. However, such receiver architecture suffers from high power consumption because it involves two separate synthesizers to generate the local oscillator (LO) signals. Therefore, to improve the power penalty, direct conversion receiver

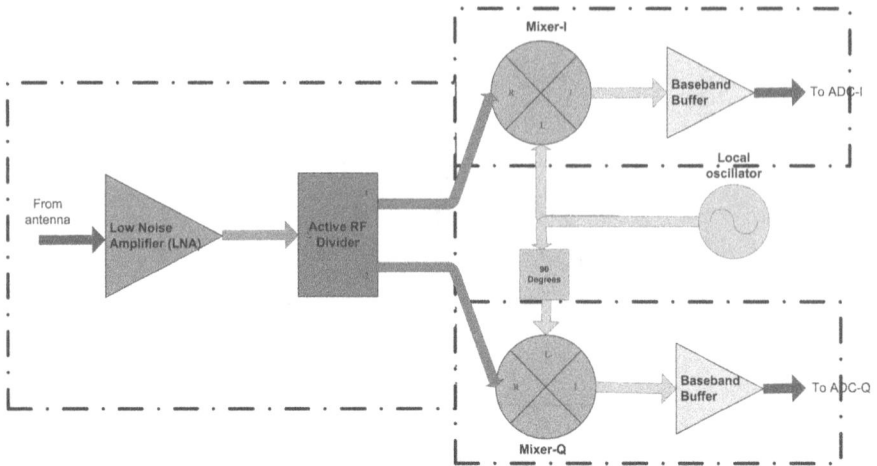

FIGURE 7.2   Direct conversion receiver architecture.

architectures were preferred as they require single-stage frequency conversion and hence, one frequency synthesizer for the same. Figure 7.2 shows the direct-conversion receiver architecture [8]. The first stage is an LNA whose main function is to receive a signal from antenna and amplify with minimal noise and distortion. Following the LNA is a quadrature mixer that demodulates the incoming RF signal to baseband in phase (I) and quadrature phase (Q) signals with the help of an LO tuned to the same frequency as the RF signal. The baseband signal after getting amplified in the baseband amplifier goes to the ADC and finally to the baseband signal processor for further processing.

The drawback associated with this kind of architecture is getting acceptable phase noise performance in the quadrature oscillator at the millimeter wave. The design of a receiver is heavily dependent on the design of an LNA, with a high gain, high linearity, and a low noise figure of the entire receiver architecture. Moreover, the attention has been made on the design of the down converter including the high gain LNA chain and the mixer operating at millimeter wave frequencies.

7.3.2   Low Noise Amplifier

The LNA constitutes the first component in an RF receiver, after the antenna. Hence, it contributes significantly to the overall performance of the RF transceiver. The main function of the LNA is to amplify the signal of interest while contributing minimal noise. In most cases, the received signal would be buried in noise and therefore needs to be amplified with an addition of minimum noise before it can be further processed. Generally, an LNA comprises of both active and passive devices. Active devices are incorporated to achieve the desired gain while passive devices are incorporated for better matching to ensure maximum power transfer from the antenna to the RF front-end receiver [9]. At 60 GHz, HBTs are preferred over MOSFETs, because MOSFETs offer lower f_T compared to HBTs. Whereas passive devices have traditionally been realized using lumped components. However, inductors can also be realized using transmission lines at millimeter-wave, but lumped components are preferred because they can provide the adequate quality factor. Care must be taken to design these passive components at frequencies below their resonant frequency.

7.3.2.1 Design Considerations for an LNA

While designing an LNA, the following are the set of parameters that an LNA should ultimately achieve:

1. Minimum noise figure [10]
2. Obtaining gain beyond the minimum required value
3. Minimum power consumption
4. Reducing the input and output reflection coefficient below a certain value (< −10 dB) over the frequency range of interest
5. Maximum linearity (i.e., high IIP3)

Traditionally, the LNA design at millimeter wave frequencies involved biasing the transistor to obtain minimum noise figure. This was achieved by matching the real part of the optimum noise impedance with the real part of the input impedance of the transistor. However, this would lead to a compromise in the gain and input reflection coefficient to the amplifier.

This design procedure changes with the emergence of monolithic microwave integrated circuits (MMICs). Geometry and bias current of the transistor became design variables and the passive networks despite their huge size and losses (metal and substrate), in which degraded performance determined the cost and size of the LNA.

Here, we will describe LNA design in two major steps: active matching and passive component matching

7.3.2.1.1 Active Matching

In this phase, we are trying to equate the optimum noise impedance with the source impedance. Since the noise figure of an amplifier is given by the expression [9]

$$F = F_{MIN} + \frac{R_n}{G_S} |Y_S - Y_{sopt}|^2 \tag{7.2}$$

it is apparent that to reduce the noise figure of the LNA, the signal source admittance Y_S must be equal to the optimum source admittance Y_{sopt}. Thus, to minimize the noise figure of the LNA, the following can be done: (1) the minimum noise figure must be as low as possible and (2) the optimal source admittance should be equal to or close to the signal source admittance.

F_{MIN} can be reduced by proper sizing and biasing of the transistor. In HBTs, the current density is directly proportional to the frequency and is configuration dependent (the cascode stage has more current density than single HBT). On plotting the NF_{MIN} versus frequency, the optimum current density point (J_{OPT}) can be obtained and this is the current at which the LNA should be biased.

The next step is to size the transistor. This is done so that the optimal noise impedance can be made equal to the signal source impedance. The noise parameters of the HBT can be expressed by the formulas [9]

$$R_n = \frac{R_{HBT}}{Nl_E} \tag{7.3}$$

$$G_u = G_{HBT}\omega^2 Nl_E \tag{7.4}$$

$$G_{cor} = G_{C,HBT}\omega Nl_E \tag{7.5}$$

$$B_{cor} = B_{HBT}\omega Nl_E \tag{7.6}$$

where, N = number of emitter stripes; l_E = emitter length; and R_{HBT}, $G_{u,HBT}$, $G_{C,HBT}$, and B_{HBT} are technology constants.

These noise parameters can also be obtained from the simulation of the process design kit (PDK) components.

The noise impedance can also be expressed as a function of the transistor size, emitter length, number of fingers, and number of transistors through the formula [9]

$$Z_{sopt}(HBT) \approx \frac{f_{Teff}}{f * Nl_E * g_{meff}}\left[\sqrt{\frac{g_m}{2}(r_E + R_b)} + j\right] = Z_0 + jX_{sopt} \tag{7.7}$$

$$Nl_E = \frac{f_{Teff}}{f * Z_0 * g_{meff}}\left[\sqrt{\frac{g_m}{2}(r_E + R_b)}\right] \tag{7.8}$$

Since the minimum noise figure increases with increase in emitter width in HBTs, devices with minimum emitter width are used in the design of an LNA amplifier.

7.3.2.1.2 Passive Component Matching

Traditional impedance matching is done by using passive elements. Resistors are often avoided as they contribute to the noise figure. Capacitor and inductors are purely reactive components and hence do not contribute to the noise figure. Transformers and transmission lines can also be used for matching.

The imaginary part of the optimum noise figure is tuned out using capacitors and inductors or other reactive elements leading to only the minimum noise figure. A negative reactive feedback is also used to match the input impedance without degrading the noise figure.

7.3.2.2 Design Methodology for Cascode LNA

1. The V_{CE} of the transistor is chosen with an aim to maximize linearity.
2. The transistor is biased at the optimum current density point (J_{OPT}).
3. Set the emitter length and number of fingers for the transistor (HBT) that result in the best noise figure.

4. The size of the transistor should be such that the $r_E(Z_{OPT}) = Z_0$. The parasitics that accompany the transistors should be identified and made a part of the matching network or tuned out using appropriate pure reactive components (capacitor/inductor).
5. A degenerate inductance L_E is added which helps in setting the real part of the input impedance equal to Z_0. The value of L_E can be determined by using the formula

$$L_E = \frac{Z_0 - R_b - r_E}{2\pi f_T} \qquad (7.9)$$

6. An inductor is added to the base L_B to tune out the imaginary part of the input impedance and make Z_{opt} equal to Z_{in}.
7. The output of the LNA is designed such that maximum power transfer takes the place of the range of frequency for which the LNA is designed.
8. Add the bias circuitry without affecting noise figure.

7.3.2.3 LNA Simulation Results and Observations

7.3.2.3.1 DC I-V characteristics and DC biasing

The first step in the design of an LNA is to perform DC biasing. The DC I-V characteristics of the SiGe NPN transistor are found using the circuit and simulation software. The voltage source to be used for this design is 1.3 V and the base current is approximately 14 µA. From Figure 7.3, it is apparent that in the base current of 14 µA and a V_{CE} of 1.3 V, the collector current is 8 mA and the device power consumption will be approximately 10 mW.

Therefore, the DC biasing of the cascode structure is made in such a way that the common emitter amplifier of the structure has a base current of approximately 14 µA. The common base amplifier is biased in a way that the HBT is in soft saturation. The values of the voltages and currents can be observed from Figure 7.4.

7.3.2.3.2 f_T Measurement

Since the f_T of the cascode structure is unknown, it can be found out using the circuit simulation software by plotting the f_T versus the collector current as shown in Figure 7.5. An emitter degenerate inductor is added to the cascode stage whose value is approximately 40 pH to model the parasitics that may be encountered when designing the layout of the LNA. The emitter inductor also acts as a series-series feedback and helps improve stability. On observing Figure 7.5, it is found that at a collector current of 8 mA, the f_T is approximately 490 GHz. Therefore, the DC biasing of the cascode structure performs well at our frequency of interest (i.e., 60 GHz).

7.3.2.3.3 Input and Output Impedance Matching

After the f_T has been measured, a series capacitor and resistor of suitable values are connected in a shunt feedback to stabilize the network. A setup similar to the one used to measure the f_T of the network is used to plot the stability circles and measure

BJT DC Collector Current vs. Collector-Emitter Voltage
Use with BJT_curve_tracer Schematic Template

m1
VCE=1.300
IC.i=0.008
IBB=1.400E-5

Values at bias point indicated by marker m1.
Move marker to update.

VCE	Device Power Consumption, Watts
1.300	0.010

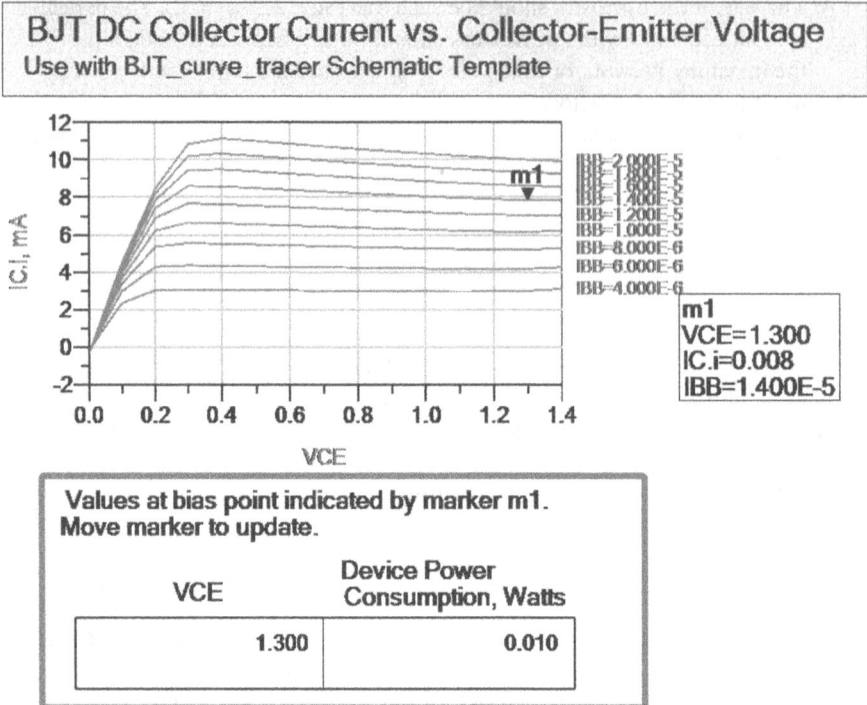

FIGURE 7.3   DC I-V characteristics of the NPN HBT.

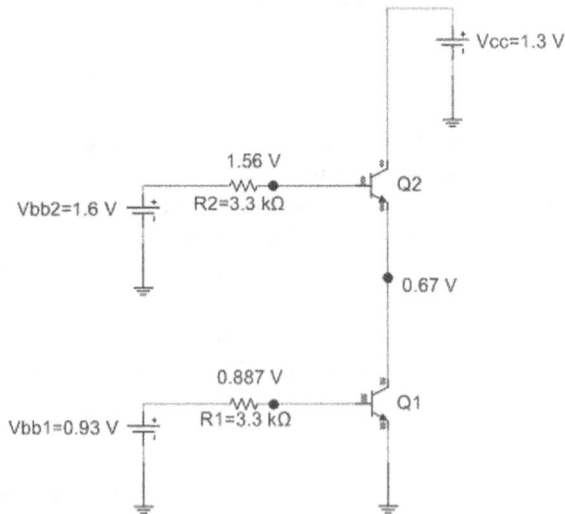

FIGURE 7.4   DC biasing of the cascode structure.

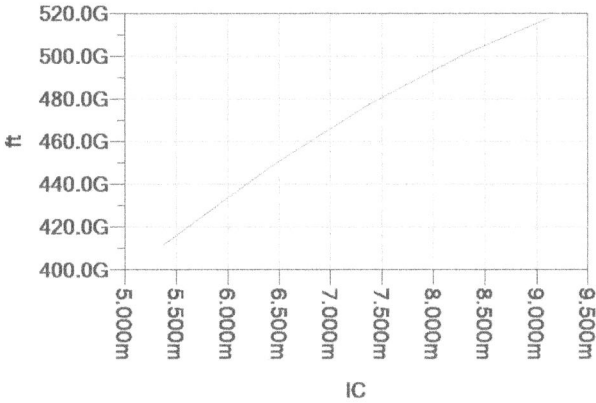

FIGURE 7.5 f_T versus collector current (I_C).

the k (Rollett factor) value. Figure 7.6 shows the point on the Smith chart that is chosen for impedance matching at the input so that the network meets the required specifications such as a gain of at least 10 dB and a noise figure of less than 6 dB.

The matching is performed on the Smith chart, as shown in Figure 7.7, and an inductor/capacitor of appropriate value is added in a series/shunt to the input and output of the network. Matching results only in the real part of the impedance remaining and has a value close to 50 Ω while the imaginary part of the impedance is cancelled out.

After carrying out the simulation on the Smith chart, an inductor of 61 pH was added in series to the input side of the LNA to achieve matching.

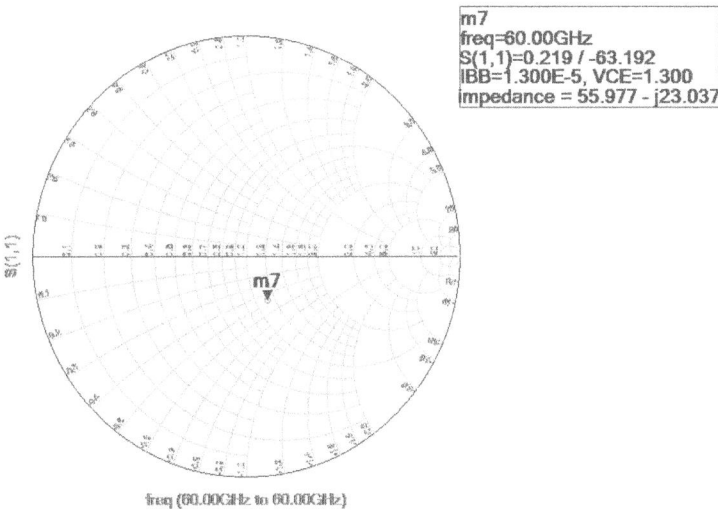

m7
freq=60.00GHz
S(1,1)=0.219 / -63.192
IBB=1.300E-5, VCE=1.300
impedance = 55.977 - j23.037

freq (60.00GHz to 60.00GHz)

FIGURE 7.6   Smith chart indicating input impedance matching point.

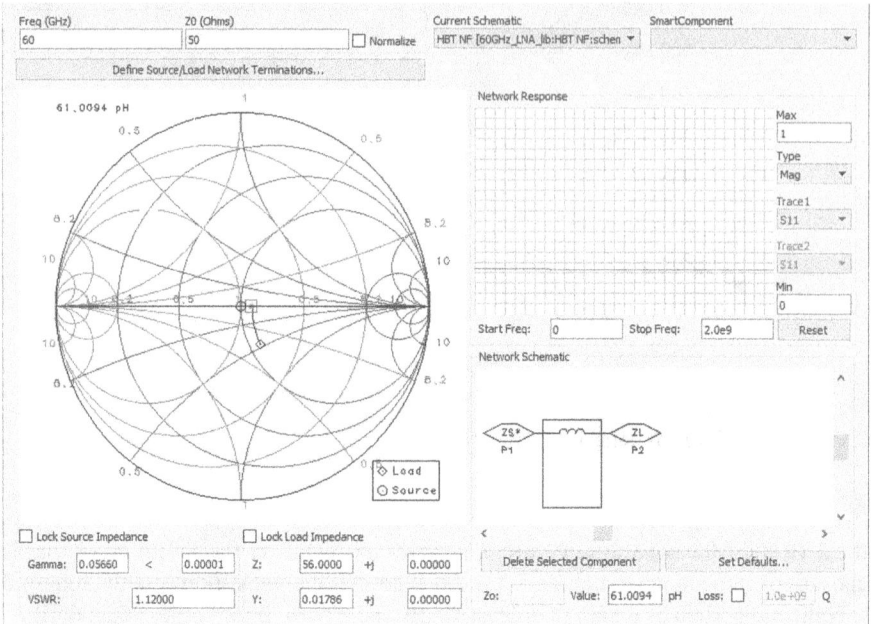

FIGURE 7.7 Matching input impedance using a Smith chart.

7.3.2.3.4 Single-Stage LNA Circuit

Once the values for the series inductor at the input and series capacitor at the output of the circuit had been determined, the entire LNA circuit was designed as shown in Figure 7.8. The inductors are replaced by transmission lines and inductors designed by the Muehlhaus RFIC Inductor Toolkit. The inductor on the output of the network with a value of 180 pH was replaced by a spiral inductor designed with the help of the Inductor Toolkit, whereas the remaining inductors were replaced by transmission lines. The inductor at the output of the LNA (collector terminal of the common gate HBT) was split into two inductors so as to improve the gain while giving up bandwidth in the process. The inductor at the emitter of the common emitter HBT was used to model the parasitics that are encountered while doing the layout of the LNA.

The values of all these passive components were slightly changed through manual tuning to improve the performance of the LNA by observing the S-parameters plot, shown in Figure 7.9. The plot for the noise figure exhibited by the LNA is shown in Figure 7.10.

From the S-parameter plot, it can be concluded that the single stage LNA achieves a gain of approximately 11 dB throughout the band and its reflection coefficients lie well below 10 dB in our frequency of interest (57–64 GHz).

The noise figure plot indicates that the LNA achieves a noise figure of approximately 2.6 dB at 60 GHz with a fluctuation of ±0.7 dB. Thus, the LNA does not add much noise to the received signal. The S-parameters and noise figure values are summarized in Table 7.1.

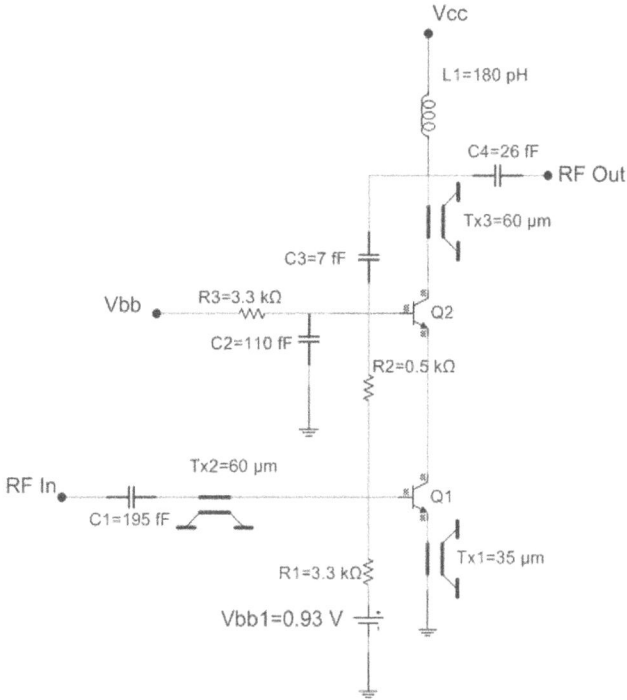

FIGURE 7.8   Single-stage cascode LNA with design parameters.

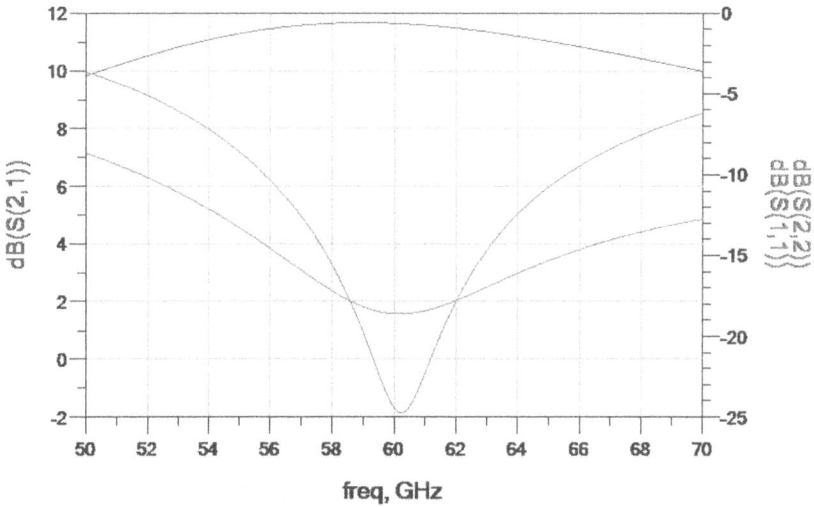

FIGURE 7.9   S parameters of the single-stage LNA.

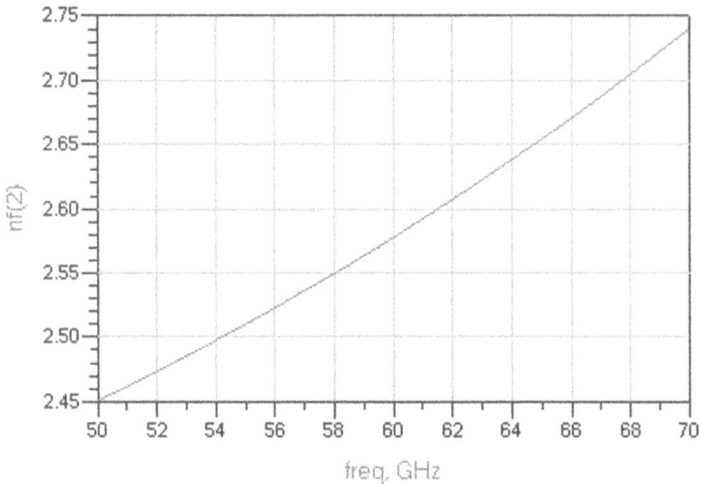

FIGURE 7.10   Plot of noise figure versus frequency for single-stage LNA.

TABLE 7.1

Simulated Parameters of Single-Stage LNA

Frequency Range	Gain (RF to Baseband)	Input Reflection Coefficient (S_{11})	Output Reflection Coefficient (S_{22})	Noise Figure
50–70 GHz	11.646 dB	−18.609 dB	−24.459 dB	2.578

7.3.2.3.5   Two-Stage LNA Circuit

The single-stage LNA designed earlier was then cascaded by merely duplicating itself and adding it to the output of the first stage LNA. Two inductors in the first stage LNA were realized using spiral inductors designed and optimized using the Inductor Toolkit and the remaining inductors were realized using transmission lines whose values were adjusted to ensure they were realizable on layout. The values of the various inductors and capacitors, especially at the input, output, and the inter stage of the two-stage LNA were manually tuned so that the two-stage LNA was stable and achieved the required specifications including a 20-dB gain. Figure 7.11 shows the cascaded two-stage LNA.

The S-parameters and noise figure of the two-stage LNA can be observed in Figures 7.12 and 7.13.

From Figure 7.12, it can be inferred that the two-stage LNA achieves a much higher gain than that achieved using a single stage. The reflection coefficients are similar to that exhibited by the single-stage LNA with both input and output reflection coefficient well below 10 dB in the frequency band of 57–64 GHz.

From Figure 7.13, it can be inferred that the noise figure slightly increases when compared to the noise figure encountered in the single-stage LNA. This is because the first stage contributes the maximum noise and subsequent stages have little

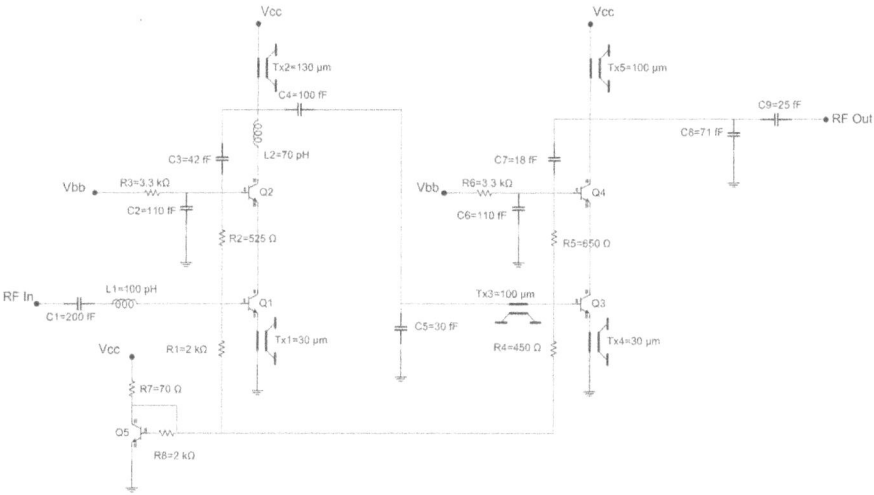

FIGURE 7.11   Two-stage cascode LNA with design parameters.

impact on the noise figure if the gain of the first stage is considerable. This is the
very same principle of Friis law. The S-parameters and noise figure values are summarized in Table 7.2.

7.3.2.3.6   Two-Stage VG-LNA

The two-stage cascode LNA was slightly modified by adding a voltage control
to each stage of the LNA to be able to vary the gain of the LNA [11]. The structure of the VG-LNA looks similar to that of a differential amplifier as shown in
Figure 7.14.

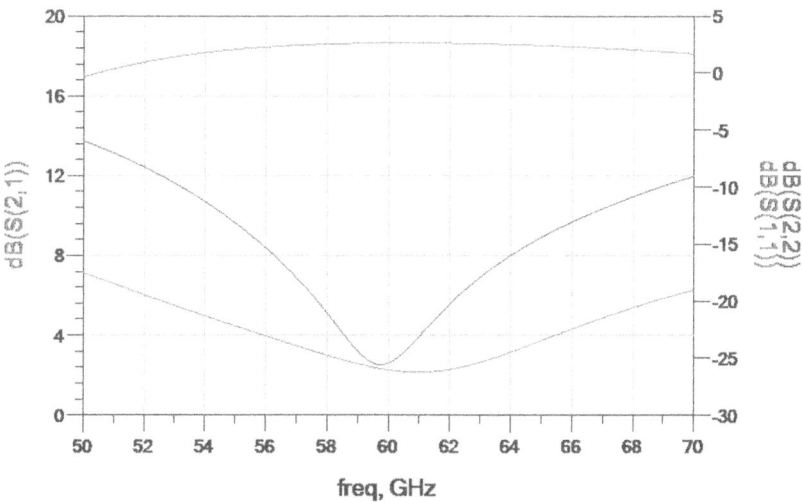

FIGURE 7.12   S parameters of the two-stage cascode LNA.

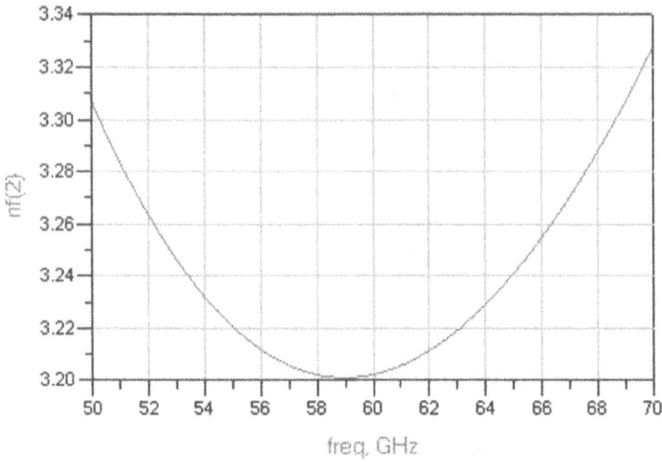

FIGURE 7.13   Plot of noise figure versus frequency for two-stage LNA.

TABLE 7.2

Simulated Parameters of Two-Stage LNA

Frequency Range	Gain (RF to Baseband)	Input Reflection Coefficient (S_{11})	Output Reflection Coefficient (S_{22})	Noise Figure
50–70 GHz	18.713 dB	−27.52 dB	−23.849 dB	3.275

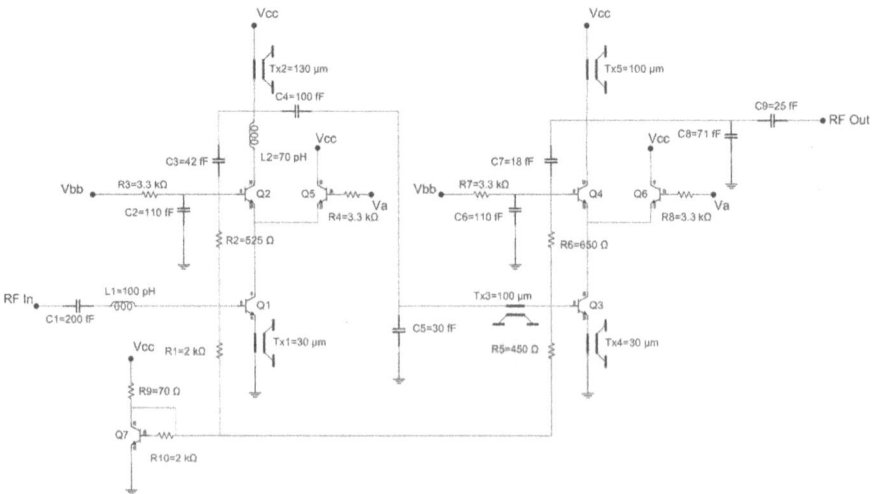

FIGURE 7.14   Two-stage VG-LNA cascode structure.

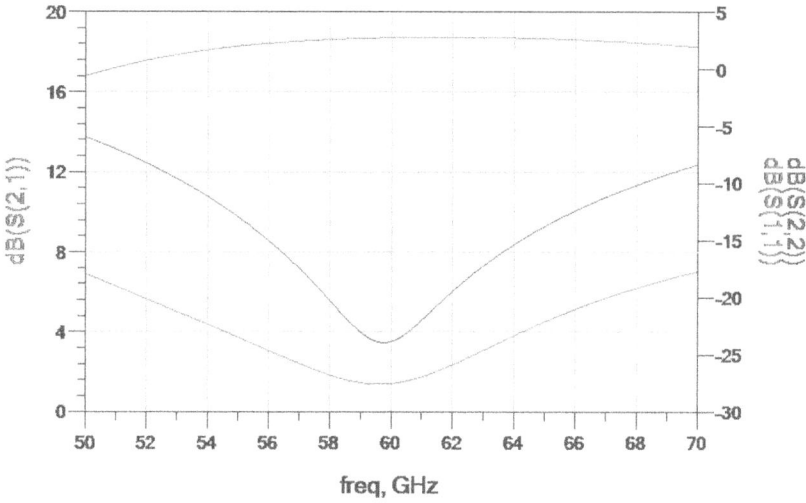

FIGURE 7.15   S parameters of the two-stage VG-LNA.

The S-parameters, variation in gain by varying gain control voltage, P1 dB, and noise figure of the two-stage LNA can be observed in Figures 7.15–7.18.

The S parameter plot of the two-stage VG-LNA is similar to that of the two-stage LNA. The only change is that the gain can be controlled by varying the gain control voltage with the maximum gain being exhibited at 1.4 V and the minimum gain being exhibited at 1.75 V. The addition of the voltage gain control helps avoid saturation of the ADC.

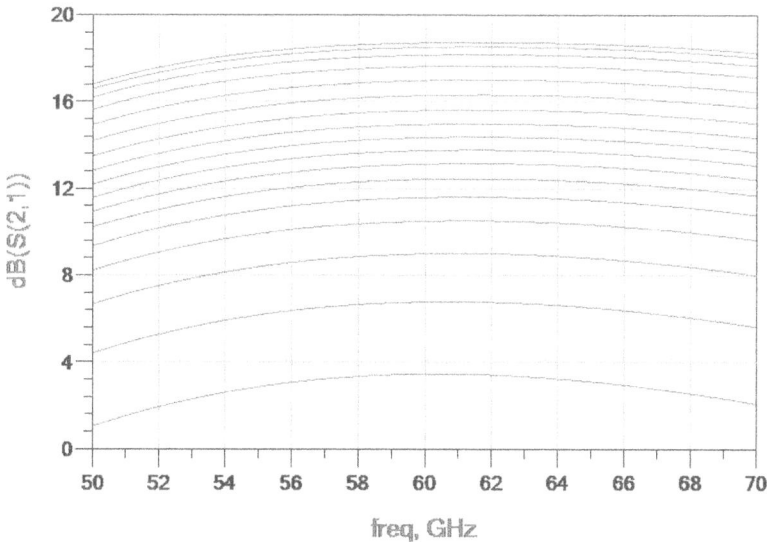

FIGURE 7.16   Variation of gain by varying the gain control voltage from 1.4 V to 1.75 V.

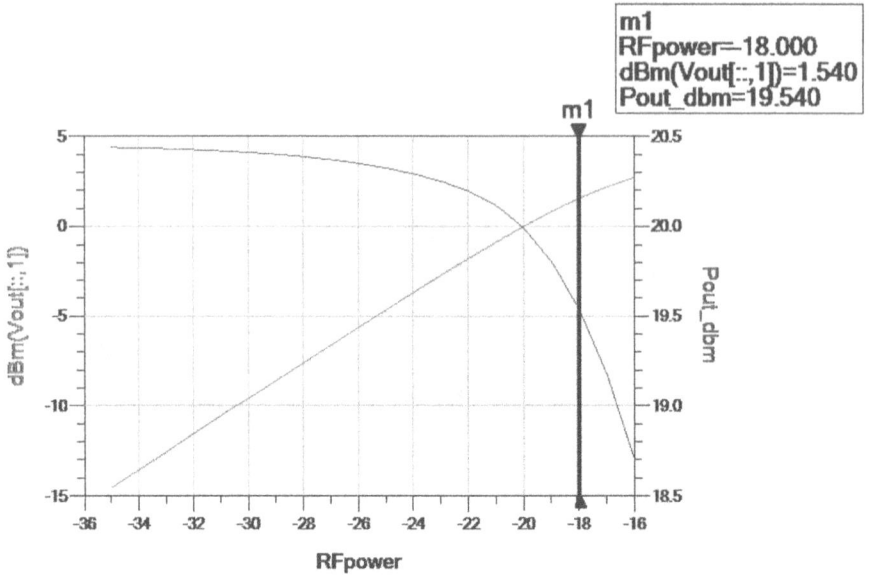

FIGURE 7.17   Plot of output power versus RF power to measure the 1 dB compression point.

The 1-dB compression point for the two-stage VG-LNA is found to be 1.5 dB, which indicates that the LNA is linear in our frequency range of interest.

The noise figure of the two-stage VG-LNA has only slightly increased when compared to the two-stage LNA and this may be attributed to the losses due to addition of two HBTs (Q5 and Q6) and their resistance used to provide gain control voltage. The S-parameters, noise figure, gain due to varying AGC voltage and 1-dB compression point values are summarized in Table 7.3.

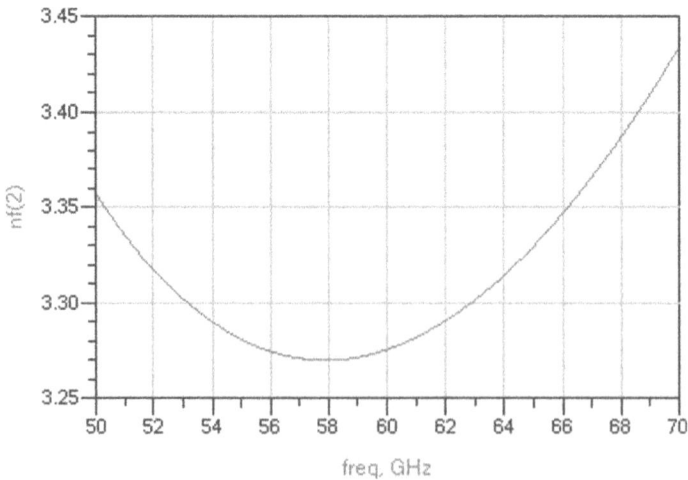

FIGURE 7.18   Plot of noise figure versus frequency for two-stage VG-LNA.

TABLE 7.3

Simulated Parameters of Two-Stage VG-LNA

Frequency Range	Gain (RF to Baseband)	Input Reflection Coefficient (S_{11})	Output Reflection Coefficient (S_{22})	Rollett Factor	Noise Figure	Gain due to V_{AGC} (1.4–1.75 V)	1 dB Compression Point
50–70 GHz	18.713 dB	−27.52 dB	−23.849 dB	7.463	3.275	18.713– 3.442 dB	1.54 dBm

7.3.3 RF DIVIDER

An RF divider was designed and its circuit is shown in Figure 7.19. The RF divider is cascaded after the VG-LNA and just before the I and Q phase mixer to divide the power equally between both the mixers. The inductor and capacitor were used for matching at the input and output port. The RF divider has a differential configuration. Similar to the LNA, the HBTs have been biased using a resistor and a DC source (1.6 V). The inductor at the input and the emitter of the HBT at the input are realized using transmission lines and whereas, the inductors at the output of the RF divider are realized using inductors designed using the Inductor Simulation Toolkit. The inductance at the emitter end of the HBT at the input port is used to model the parasitics. The current mirror designed is similar to the one used in the VG LNA.

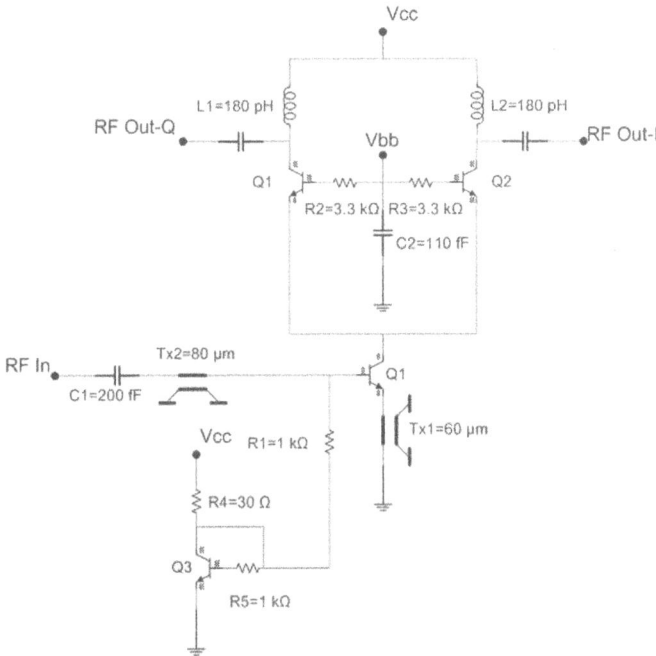

FIGURE 7.19   Schematic of the RF divider with design parameters.

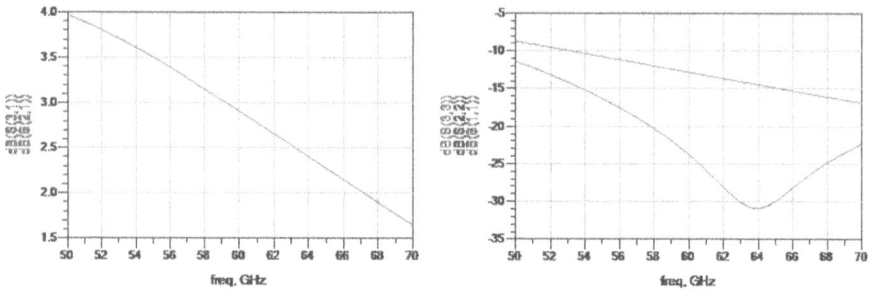

FIGURE 7.20   S parameters of the RF divider.

Figures 7.20 and 7.21 show the S-parameters and noise figure of the simulated RF divider. The following observations can be made from the S parameters plot of the divider.

The S-parameter plot indicates that while we designed the RF divider with the purpose of just dividing power equally among both the I and Q phase mixer, the RF divider also provides a gain of around 3 dB. However, this gain may drop when the RF divider is realized on a layout.

While the noise figure of the RF divider is higher than that exhibited by the two-stage VG-LNA, this will have little impact on the overall noise figure of the receiver. This is because the gain obtained from the two stage VG-LNA is high and will result in the noise figure due to the RF divider being neglected (refer to the Friis formula). The S-parameters and noise figure values are summarized in Table 7.4.

7.3.4   MIXER

A mixer is a three-port device that performs the function of frequency translation (multiplication of two signals) in the communication systems, both at the transmitter

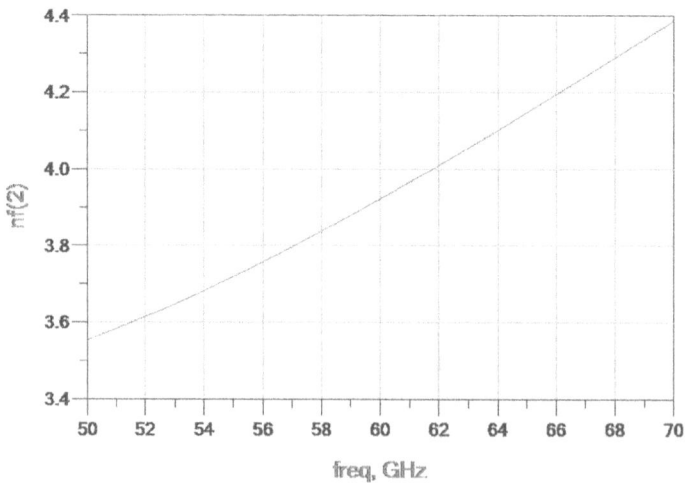

FIGURE 7.21   Plot of noise figure versus frequency for the RF divider.

TABLE 7.4

Simulated Parameters of RF Divider

Frequency Range	Gain (RF to Baseband)	Input Reflection Coefficient (S_{11})	Output Reflection Coefficient (S_{22} and S_{33}) (from Each Port)	Noise Figure
50–70 GHz	2.908 dB	−12.836 dB	−23.799 dB (at 60 GHz)	3.922

and receiver. In the transmitter section, it is used as an up converter, which translates the low frequency data signal to a high frequency (carrier frequency), which is then transmitted through the antenna. In the receiver section, it is used as a downconverter, which separates the data signal from the carrier and performing frequency translation so that the data signal is shifted to a lower frequency. Since the focus of this report is on the receiver section, the mixer designed would be used for down conversion.

The input to a downconverter is an RF frequency, whereas the output obtained from the mixer is a baseband signal. Eq. (7.10) shows how the mixer multiplies two signals of arbitrary frequencies [7, 9].

$$A_1 cos(\omega_1 t) * A_2 cos(\omega_2 t) = \frac{A_1 A_1}{2} \left(cos(\omega_1 + \omega_2)t + cos(\omega_1 - \omega_2)t \right) \quad (7.10)$$

In the case of down conversion, the sum term is discarded through the use of a filter and the difference term is obtained. In the receiver end, one sideband around the LO signal contains the desired RF signal. The other sideband is termed as the image frequency. The image frequency, if not removed, is found to be difficult to differentiate from the RF signal once down conversion takes place. Hence, the image frequency is an undesirable signal that is removed using a band stop filter before down conversion takes place.

The mixer can be designed using a number of devices and exploiting their nonlinear characteristics such as diodes, HBT and BJT, and MOSFET or HEMT.

However, as in the case of the LNA, we will realize the mixer using HBTs.

7.3.4.1 Classification of a Mixer

7.3.4.1.1 Passive and Active Mixer

The mixer can be classified as [6]:

1. Passive mixer: These are mixers that involve the use of FETs or diodes being used as resistors. Passive mixers do not consume power.
2. Active mixer: These are mixers realized using devices such as transistors (MOSFET and HBT). Active mixers consume power.

7.3.4.1.2 Single-Balanced and Double-Balanced Mixers

The mixer can be realized using two different configurations:

1. Single-balanced mixer: This kind of configuration is so called because of the differential LO waveforms. The output of such a mixer is in a differential

configuration and a single-ended RF signal is the input. This is the configuration that is used to realize the I and Q phase mixers in this project.

2. Double-balanced mixer: This kind of configuration is realized using two single-balanced mixers. It also operates using a balanced/differential LO waveform such as the single-balanced mixer, but unlike it, the RF input is also applied in differential configuration. While a single-ended RF input can also be applied by grounding one of the terminals, this may lead to higher input referred noise.

7.3.4.2   Single-Balanced Gilbert Cell Mixer Topology

In the Gilbert cell topology, the AC differential current can be found by referring to Figure 7.22 and through the following derivation:

The current through the bipolar transistors Q1 and Q2 is

$$i_{C1} = I_S \left\{ \exp\left(\frac{V_{LO}}{2V_T}\right) - 1 \right\} \text{ and } i_{C2} = I_S \left\{ \exp\left(\frac{V_{LO}}{2V_T}\right) - 1 \right\} \tag{7.11}$$

From the above equations, we get

$$\frac{i_{C1} - i_{C2}}{i_{C2}} = \left\{ \exp\left(\frac{V_{LO}}{V_T}\right) - 1 \right\} \tag{7.12}$$

$$\frac{i_{C1} + i_{C2}}{i_{C2}} = \left\{ \exp\left(\frac{V_{LO}}{V_T}\right) + 1 \right\} \tag{7.13}$$

Also, from the diagram, we have

$$I_{TAIL} = i_{C1} + i_{C2} \tag{7.14}$$

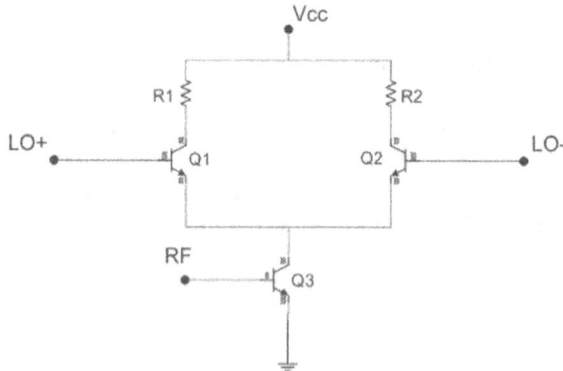

FIGURE 7.22   Single-balanced Gilbert cell mixer.

Substituting Eq. (7.14) in (7.13), and dividing Eqs. (7.12) and (7.11), we get

$$i_{C1} - i_{C2} = I_{TAIL} \tanh\left(\frac{V_{LO}}{2V_T}\right) \tag{7.15}$$

The expansion of tanh(x) is

$$\tanh(x) = x - \frac{x^3}{3} + \frac{2x^5}{15} - \frac{17x^7}{315} \tag{7.16}$$

Substituting the expansion of *tanh(x)* in Eq. (7.15), we get

$$i_{C1} - i_{C2} = I_{TAIL} \left\{ \frac{V_{LO}}{2V_T} - \frac{V_{LO}^3}{24V_T^3} + other \ higher \ order \ terms \right\} \tag{7.17}$$

Assuming the higher order terms are filtered, we are left with

$$i_{C1} - i_{C2} = I_{TAIL} \left\{ \frac{V_{LO}}{2V_T} \right\} \tag{7.18}$$

Also, the expression for I_{TAIL} is

$$I_{TAIL} = I_0 + g_m v_{RF} \tag{7.19}$$

where I_0 is the DC current that divides equally between both transistors Q1 and Q2. g_m is the trans conductance of transistor Q3 and $g_m = I_{CQ}/V_T$. Substituting the above expression obtained for I_{TAIL} in Eq. (7.18), we get

$$i_{C1} - i_{C2} = (I_0 + g_m v_{RF}) \left\{ \frac{V_{LO}}{2V_T} \right\} \tag{7.20}$$

Since the first term is lost due to filtering, therefore

$$i_{C1} - i_{C2} = \Delta i_C = \frac{g_m v_{RF} V_{LO}}{2V_T} \tag{7.21}$$

Referring to the single-balanced Gilbert cell topology and its derivations shown above, we design the I-phase mixer and an identical Q-phase mixer with the inputs to this mixer varying by an angle or phase of 90 degrees.

7.3.4.3 Single-Balanced Mixer

The mixer circuit was designed as shown in Figure 7.23. Just as in the case of the LNA, the values of the various inductors and capacitors were slightly adjusted through manual tuning to improve the performance of the mixer. Two capacitors

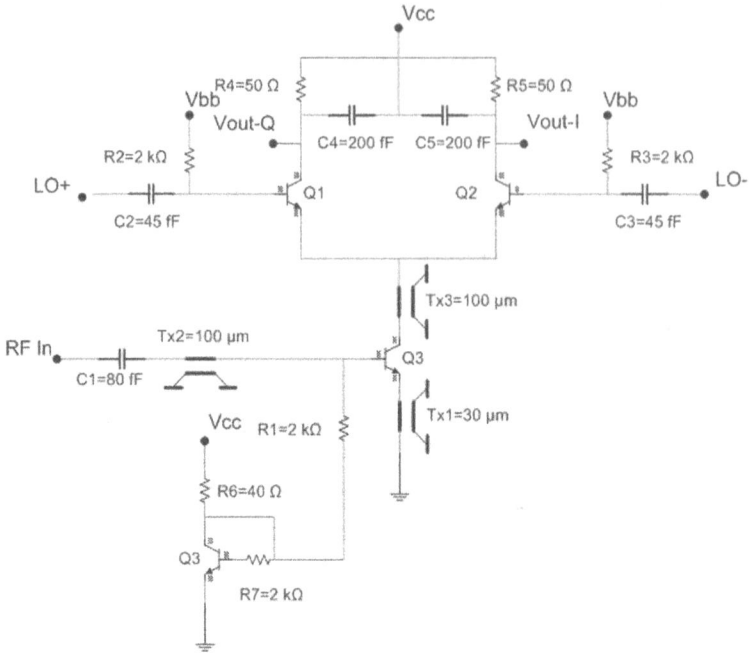

FIGURE 7.23   Schematic of the single-balanced mixer with design parameters.

have been placed in the output to form an RC low pass filter that would filter out the LO and RF frequencies leaving only the IF frequencies to pass through it. The inductor at the emitter of the common emitter HBT was used to model the parasitics. The current mirror and biasing used in the case of LNA has also been replicated and used in the mixer circuit to feed current to the mixer circuit and bias the HBTs, respectively.

The mixer was simulated and its output power (in dBm) and convergence gain (in dBm) plots can be observed in Figures 7.24 and 7.25. The following observations can be made from the plots of the mixer.

With an RF input signal having a power of −50 dBm, this signal on passing through the mixer would be amplified and the output signal would have a power level of −46.7 dBm. Therefore, the gain provided by the mixer is 3.3 dB. The mixer, like the RF divider, does not need to amplify the signal because the majority of the gain in the receiver is provided by the LNA and baseband amplifier. Table 7.5 indicates the output power and convergence gain obtained from the simulated single balanced mixer.

TABLE 7.5

Simulated Parameters of Single-Balanced Mixer

Frequency Range	Output Power (in dBm)	Convergence Gain (at 0 dBm LO Power)
50–70 GHz	−46.7 dBm	3.293 dBm

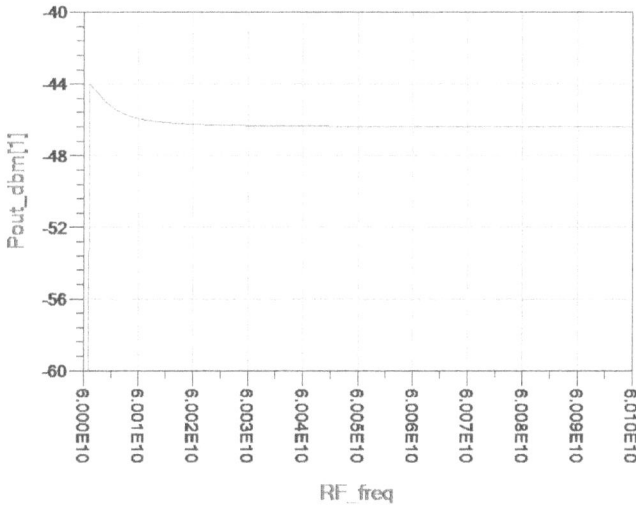

FIGURE 7.24   Output power (in dBm) from the single balanced mixer.

7.3.5   BASEBAND AMPLIFIER

The baseband amplifier was designed to improve or enhance the gain of the RF front end and its design is similar to the RF divider (i.e., it is also a differential amplifier with an emitter degenerate resistor). A 1.6 V DC source has been used as the power source and the capacitors have been placed parallel to the output resistors to form a low pass RC filter. Figure 7.26 shows the design of the baseband amplifier.

The baseband amplifier was simulated and the output power (in dBm) plot can be observed in Figure 7.27. The following observations can be made from the plot of the baseband amplifier.

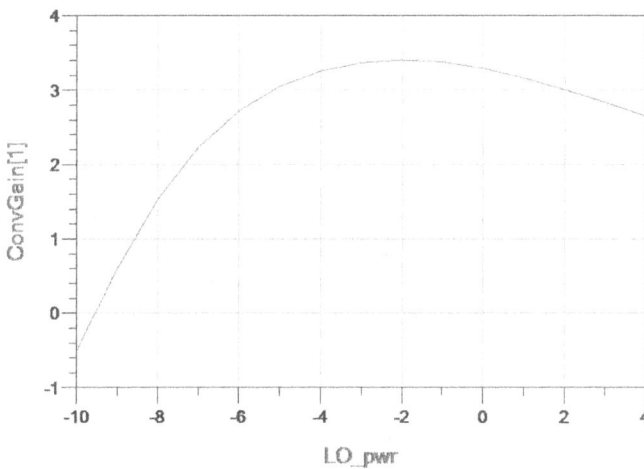

FIGURE 7.25   Output power (in dBm) versus LO power for the single-balanced mixer.

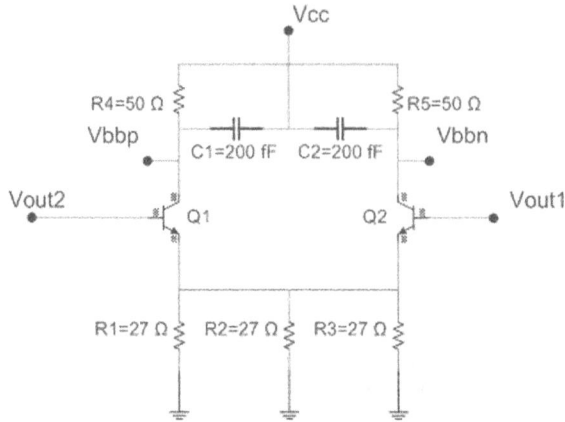

FIGURE 7.26   Schematic of the baseband amplifier with design parameters.

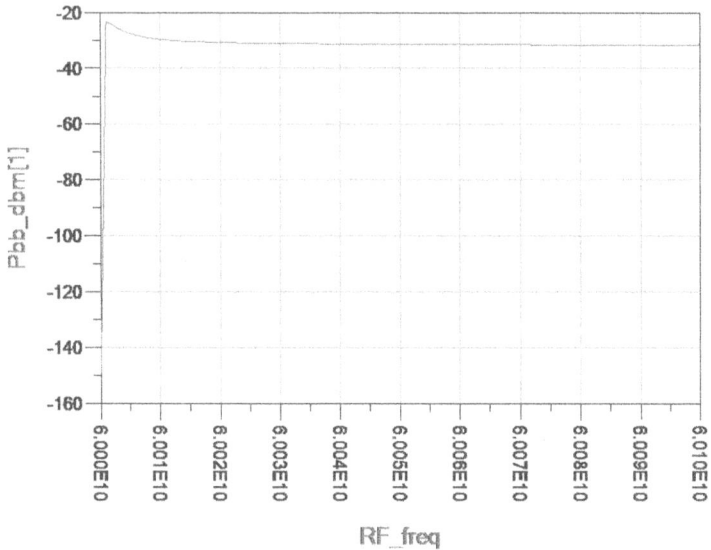

FIGURE 7.27   Plot of output power versus RF frequency for the baseband amplifier.

The baseband amplifier has provided an additional gain of 15 dB to the receiver, which would allow the receiver to detect received signals that have been attenuated and amplify them so that they can be sent to the ADC for further processing. Table 7.6 indicates the gain of the simulated baseband amplifier.

TABLE 7.6

Simulated Parameters of Baseband Amplifier

Frequency Range	Gain (RF to Baseband)
50–70 GHz	15.2 dB

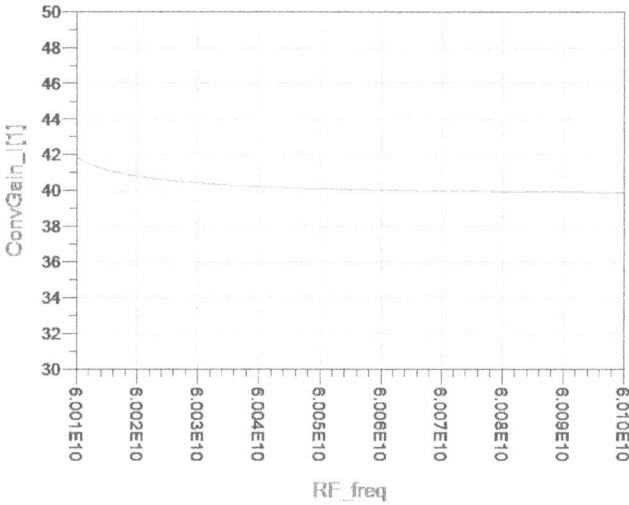

FIGURE 7.28   Output power (in dBm) from the RF front end and baseband amplifier.

7.3.6   CASCADING VG-LNA, RF DIVIDER, MIXER, AND BASEBAND AMPLIFIER

The entire RF front end, which comprises of the VG-LNA, RF divider, and mixer (I and Q phase), is cascaded with a baseband amplifier (I and Q phase) to realize an additional gain. The output voltage from the mixer stage is input to the baseband amplifier and controls the current flowing in the differential circuit. Figure 7.2 shows the block diagram for the same.

Figures 7.28 and 7.29 show the output power (in dBm) and convergence gain (in dBm) of the entire RF front end. The following observations can be made from the plots.

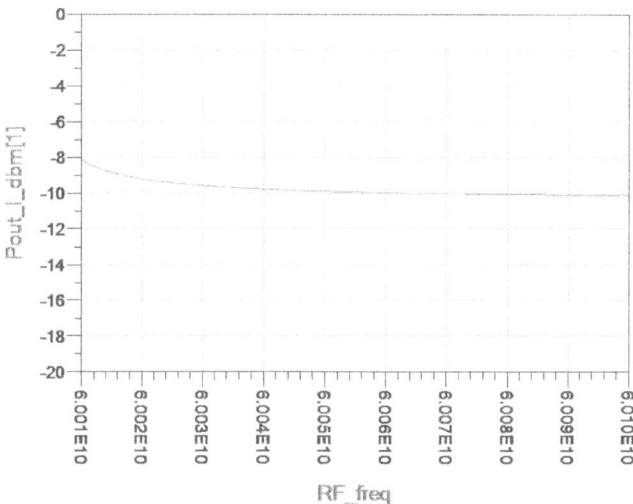

FIGURE 7.29   Convergence gain (in dBm) from the RF front end and baseband amplifier.

TABLE 7.7

Simulated Parameters of RF Front End

Frequency Range	Output Power (in dBm)	Convergence Gain
50–70 GHz	−10 dBm	40 dBm

Figures 7.28 and 7.29 clearly indicate that if an RF input is of -50 dBm, then the signal received at the output would have a power of −10 dBm. Thus, the entire RF front end has a gain of 40 dB, which is sufficient enough to amplify signals received that have been attenuated significantly. Table 7.7 indicates the output power and convergence gain of the entire RF front end.

7.4 CONCLUSION

In this chapter, the RF front end comprising of a two-stage VG-LNA, active RF divider, mixer (I and Q phase) and baseband amplifier (I and Q phase) was designed using an advanced design system and the IHP PDK SG13G2. First, the individual modules were designed using ideal circuit components and manual tuning was carried out to optimize the performance and meet the desired specifications. Further, the same design was realized using PDK components. The RF front end achieved has a noise figure of approximately 4.8 dB with a gain of 40 dB. The IIP_3 of the entire system is approximately 12 dBm and its power dissipation is 95 mW. Therefore, in this chapter, a mm-wave RF front end for a short-range communication (as in the case of biomedical sensors) capable of operating at 60 GHz has been designed and simulated.

ACKNOWLEDGMENT

This work was carried out by using IHP standard cell libraries and design kits provided by Europractice IC service.

REFERENCES

1. Iniewski, Krzysztof, ed. 2017. *Wireless technologies: circuits, systems, and devices.* Boca Raton, FL: CRC Press.
2. Chai, Yuan, Lianming Li, and Tiejun Cui. 2012. "Design of a 60 GHz LNA with 20 dB gain and 12 GHz BW in 65 nm LP CMOS," *in 2012 IEEE MTT-S International Microwave Workshop Series on Millimeter Wave Wireless Technology and Applications,* 1–4. IEEE.
3. Heydari, Babak, Mounir Bohsali, Ehsan Adabi, and Ali M. Niknejad. 2007. "Millimeter-wave devices and circuit blocks up to 104 GHz in 90 nm CMOS." *IEEE Journal of Solid-State Circuits* 42 (12): 2893–903.
4. Everything RF. "Free space path loss calculator." https://www.everythingrf.com/rf-calculators/free-space-path-loss-calculator.
5. Advantages of 60 GHz Unlicensed Wireless Communications - 60GHz Wireless Networks. 60ghz-wireless.com.

6. Božanić, Mladen, and Saurabh Sinha. 2018. *Millimeter-wave low noise amplifiers*. Springer.

7. Razavi, Behzad. 2012. *RF microelectronics*, 2. New York: Prentice Hall.

8. Razavi, Behzad. 2005. "A 60-GHz CMOS receiver front-end." *IEEE Journal of Solid-State Circuits* 41 (1): 17–22.

9. Voinigescu, Sorin. 2013. *High-frequency integrated circuits*. Cambridge, UK: Cambridge University Press.

10. Lai, Ivan Chee-Hong, and Minoru Fujishima. 2008. *Design and Modeling of Millimeter-wave CMOS Circuits for Wireless Transceivers: Era of Sub-100nm Technology*. Springer Science & Business Media.

11. Siao, Di-Sheng, Jui-Chih Kao, and Huei Wang. 2014. "A 60 GHz low phase variation variable gain amplifier in 65 nm CMOS." *IEEE Microwave and Wireless Components Letters* 24 (7): 457–59.

8 Gate-Overlap Tunnel Field-Effect Transistors (GOTFETs) for Ultra-Low-Voltage and Ultra-Low-Power VLSI Applications

Sanjay Vidhyadharan and Surya Shankar Dan

CONTENTS

8.1 Introduction 137
8.2 Current Conduction Mechanism in TFETs 138
8.3 Gate-Overlap Tunnel Field Effect Transistor (GOTFET) 140
    8.3.1 GOTFET Structure 140
    8.3.2 Fabrication Process for the GOTFETs 141
    8.3.3 GOTFET Device Characterization 143
    8.3.4 GOTFET Operation at High Frequencies 144
    8.3.5 Suppression of Ambipolar Currents in GOTFETs 145
    8.3.6 Operating Principle of the GOTFET 146
    8.3.7 Modeling Approach for the Proposed GOTFETs 147
    8.3.8 Comparison of the GOTFET Characteristics with the Other
          State-of-the-Art TFETs 149
8.4 Performance Benchmarking of GOTFETs 150
    8.4.1 Comparison of Power and Propagation Delays CGOT versus
          CMOS Inverter, NAND, NOR, and XOR Gates 150
    8.4.2 Comparison of Jitter in a CGOT and CMOS Inverter and Chain
          of Inverters 155
8.5 Summary 160
References 160

8.1 INTRODUCTION

The advancement in complementary metal oxide semiconductor (CMOS) technology during the last few decades has enabled scaling down of the metal oxide semiconductor field-effect transistor (MOSFET) feature size to below the 100 nm range. While the scaling of CMOS technology has helped in significantly increasing the packing density within the chip, it has also given rise to undesirable constraints on low threshold voltage V_{TH} and low power supply V_{DD}. The power supply voltage across the devices must be reduced proportionately with the reduction in feature size to maintain the electric fields inside the device within junction breakdown limits. The reduction in V_{DD} also necessitates a reduction of threshold voltage to keep the on-state currents I_{ON} at the desired levels. The device I_{ON} directly influences the speed of operation of the device. The threshold voltage V_{TH} is usually lowered through appropriate selection of the gate material and through dopants ion implantation in the channel. Lower V_{TH} results in high off currents I_{OFF}, thereby increasing the static power consumption in the CMOS circuits. With a large number of these nm-scale MOSFETs packed densely on a single chip, the static power consumption has increased significantly (Helms et al. 2004; Roy et al. 2003). CMOS technology also has a severe limitation of significant parasitic capacitance and poor I_{ON}:C_{GG} ratio, which adversely affects the circuit performance at high frequencies (Caka et al. 2007; Wei et al. 2011; Lin et al. 2009). Hence various beyond-CMOS technology devices are slowly finding widespread applications in the very large-scale integration (VLSI) domain, tunnel field effect transistors (TFETs) and carbon nanotube field-effect transistors (CNFETs) being among the forerunners.

TFETs have an inverse subthreshold slope (SS) of 24–30 mV/decade at room temperature, which is significantly steeper than the conventional MOSFETs, which have a theoretical minimum SS of 60 mV/decade at room temperature. Additionally, TFETs have two to three orders of magnitude lower I_{OFF} than the corresponding MOSFETs. The steep SS and low I_{OFF} of the TFETs make them attractive beyond-CMOS candidates for low-voltage and low-power VLSI applications. TFETs also have an added advantage of having the fabrication process similar to that of CMOS, which enables easy integration of TFETs with the latest CMOS technology.

8.2   CURRENT CONDUCTION MECHANISM IN TFETS

The current conduction in MOSFETs occurs due to the drift of majority carriers (electrons for nMOSFETs and holes for pMOSFETs) from the source to the drain region, under the influence of drain to source voltage V_{DS}. An applied gate bias voltage V_{GS} greater than the threshold voltage V_{TH} inverts the surface into a channel conducive for carrier conduction. TFETs work on a different mechanism. Instead of channel inversion, in TFETs, a very high electric field across the source-channel junction due to the applied gate bias voltage V_{GS}, results in band-to-band (BtB) tunneling from the valence band of the source to the conduction band of the channel as shown in the Figure 8.1.

Models for BtB tunneling current for direct bandgap materials have been reported in the literature (Kane 1961; Saurabh and Kumar 2017; Sze and Ng 2006). The quantum mechanical BtB generation rate is given by

$$G_{btb} = AF_y^\gamma exp\left(-\frac{B}{F_y}\right) \tag{8.1}$$

where, $\gamma = 2$ and 2.5 for direct and indirect phonon-assisted tunneling process, respectively. For direct BtB tunneling in direct bandgap materials,

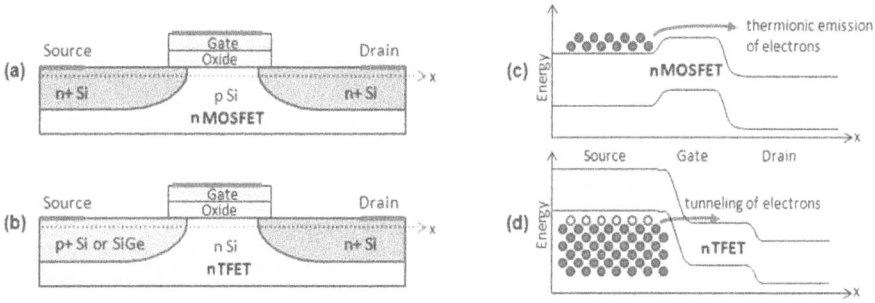

FIGURE 8.1 Schematic and current conduction mechanism: (a), (c) conventional nMOS-FET; (b), (d) the conventional nTFET, respectively.

material-dependent factors A and B have been modeled as Eq. (8.2) and (8.3), which are similar to the relations reported in the seminal work on the fundamental theory of BtB (Kane 1961; Kane 1960)

$$A = \left(\frac{g\pi}{9}\right)\left(\frac{q}{h}\right)^{\gamma}\sqrt{\frac{m_{red}}{E_g}} \tag{8.2}$$

$$B = \frac{\pi^2\sqrt{m_{red}}\,E_g^{3/2}}{qh} \tag{8.3}$$

where g denotes the degeneracy factor of the material used, h the Plank's constant, q is the elementary charge, E_g the bandgap. For phonon-assisted BtB tunneling in indirect bandgap materials, material-dependent factors A and B are given by (Kane 1961; Kane 1960)

$$A = \left(\frac{gm_{avg}^{3/2}}{\sqrt{2}}\right)\left(\frac{1+2N_{op}}{\rho\varepsilon_{op}}\right)D_{op}^2\left(\frac{q}{h}\right)^{\gamma}\left(\frac{m_{red}}{2E_g}\right)^{7/4} \tag{8.4}$$

$$B = \frac{4\pi\sqrt{m_{red}}\,(2E_g)^{3/2}}{3qh} \tag{8.5}$$

The reduced mass $m_{red} = (m_c^{-1} + m_v^{-1})^{-1}$ and the average mass $m_{avg} = (m_c + m_v)/2$, where

$$\frac{1}{m_c} = \frac{1}{2m_{red}} + \frac{1}{m_0} \text{ and } \frac{1}{m_v} = \frac{1}{2m_{red}} - \frac{1}{m_0} \tag{8.6}$$

with m_0 being the rest mass of electrons. Line tunneling occurs due to the electric field $\hat{y}F_y(x,y)$ (in the vertical $\hat{y}$ direction), and the total number of free carriers N_S generated within the source due to BtB is obtained from

$$N_S = 2W\iint_{x\ y} G_{btb}\,dy\,dx \tag{8.7}$$

where W denotes the device width, and a factor of 2 arises because this is a double-gated device. It is evident from Eq. (8.2) that a high electric-field at the tunneling junction results in a large BtB tunneling providing high I_{ON}. In TFETs, the source is heavily doped and the channel doping is maintained at near-intrinsic levels to ensure a large built-in electric-field across the source-channel junction.

8.3   GATE-OVERLAP TUNNEL FIELD EFFECT TRANSISTOR (GOTFET)

Recent studies (Ilatikhameneh et al. 2015; Schulte-Braucks et al. 2017; Gupta et al. 2015; Saurabh and Kumar 2017; Amir et al. 2016; Chander et al. 2015; Kao et al. 2012b; Liu et al. 2015; Strangio et al. 2018; Shrivastava 2017; Settino et al. 2017) have proposed TFET structures with high-κ dielectric materials as the gate insulator, gate overlapping the source, double-gated structure, and silicon germanium (SiGe) substrate, which yield significantly higher I_{ON} than conventional TFET structures. Unlike conventional TFET designs (Dan et al. 2012), these devices can out-perform CMOS devices at the same technology node. The following subsections present the device structure, characteristics, and circuit performance of a novel 45-nm channel length ultra-low-power GOTFET device. The I_{ON} of the GOTFET is twice as that of the MOSFET, while the I_{OFF} is one-two orders of magnitude lower than the 45 nm MOSFET, channel widths of both devices being the same (Vidhyadharan et al. 2019a, 2019b, 2020c, 2019d, 2020a, 2020b; Yadav et al. 2019a, 2019b, 2020a, 2020b).

8.3.1   GOTFET STRUCTURE

Figure 8.2a shows the Gate-Overlap TFET structure and Figure 8.2c and d show the 2-D BtB generation contours for nGOTFET and pGOTFET, respectively. Table 8.1 lists the doping concentrations of the materials used in both nGOTFET and pGOT-FET devices. The unique double-gates overlapping the entire source structure of the

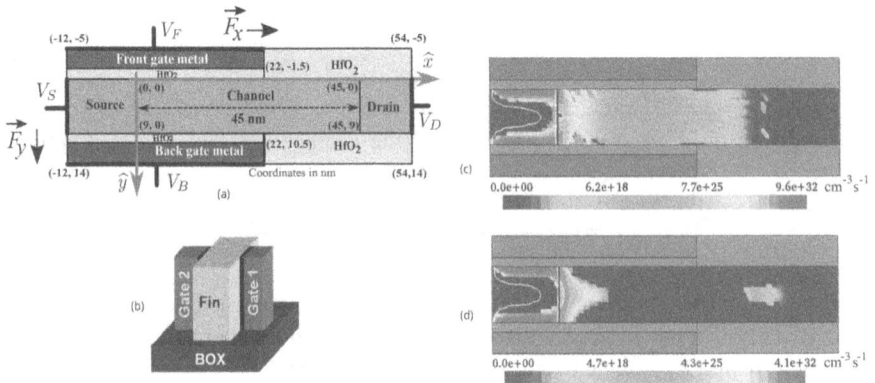

FIGURE 8.2   **(a)** GOTFET structure; **(b)** 3D structure of the proposed GOTFET; **(c)** electron; **(d)** hole BtB generation in nGOTFET and pGOTFET devices, respectively.

TABLE 8.1

Materials, Doping, and Layer Details of the nGOTFET and pGOTFET Devices

	nGOTFET			pGOTFET			Layer details* (nm)	
Region	Material	Doping (/cm³)	Type	Material	Doping (/cm³)	Type	Length	Thickness
Source	$Si_{0.15}Ge_{0.85}$	10^{20}	p^+	$Si_{0.2}Ge_{0.8}$	10^{20}	n^+	12	9
Channel	Undoped Si	—	—	$Si_{0.2}Ge_{0.8}$	5×10^{17}	p	45	9
Drain	Si	10^{20}	n^+	$Si_{0.2}Ge_{0.8}$	5×10^{17}	p	9	9
Oxide	HfO_2	—	—	HfO_2	—	—	34	1.5
Gate	Al	—	—	Mo	—	—	34	3.5

*Width $W = 1$ μm is kept as a design variable for circuit implementation.

GOTFET results in an extremely high BtB tunneling within the source region from valence band of the inner core of the source region to the conduction band of the source surface underneath the gate stack, on application of a positive V_{GS}, resulting in a significantly higher I_{ON}. The reverse-biased p-i-n regions of the TFET restricts the I_{OFF} to extremely low levels, when $V_{GS} = 0$.

Gate metal with the appropriate work function helps in obtaining the desired V_{TH}, which results in higher I_{ON}. Aluminum (Al), with a work function of $\Phi_m = 4.1$ eV, yields the best device characteristics for nGOTFET, while molybdenum (Mo), with $\Phi_m = 4.53$ eV, is ideal for the pGOTFET device. The GOTFETs have a 45 nm long and 9 nm thick channel. The gate oxide thickness t_{ox} is 1.5 nm and the gate metal is 3.5 nm thick. The I_{ON} of GOTFET can be further increased by reducing the t_{ox}; however, t_{ox} less than 1.2 nm may lead to dielectric breakdown as has been reported in Intel's process (Chau et al. 2003). Physical 1.2 nm HfO_2 is already in use in the 90 nm CMOS technology, and 0.8 nm physical HfO_2 has also been produced (Chau et al. 2003; Harame 2004). The fabricated pMOSFET and nMOSFET with 1.2 nm HfO_2 gate oxide showed very high drive performance in terms of I_{Dsat} with the right V_{TH}, exhibiting extremely low gate leakage and hence have been recommended for all future high-performance CMOS devices (Chau et al. 2003).

8.3.2 FABRICATION PROCESS FOR THE GOTFETs

There are several reports on TFET fabrication (Ramaswamy and Kumar 2017; Chang et al. 2013; Dewey et al. 2011; Ashita and Rafat, 2018; Kao et al. 2012a, 2012b) which are tabulated in Table 8.2 for a comparison with the proposed GOTFET. Kao et al. (2012a, 2012b) reported characterized data on fabricated devices, which are quite similar to the GOTFET, under the name "Gate over Source only TFET" (GoSoTFET). The GoSoTFET, as the name indicates, had the gate-stack only over the source region. In a GOTFET, the gate overlap on the source region extends to nearly half of the channel region to facilitate surface conduction as found in standard MOSFETs to enhance I_{ON} at higher biases. The GOTFETs behave exactly as the

TABLE 8.2

Comparison of the Proposed GOTFET with the State-of-the-Art in Simulated and Fabricated TFETs.

Parameter	Unit	Simulated TFETs			Fabricated TFETs					GOTFET (this work)
		(R1)	(R2)	(R3)	(R4)	(R5)	(R6)	(R7)	(R8)	
Material	–	SiGe	Si	Si	Si	Si	Si	InGaAs	InGaAs	SiGe
L_{ch}	nm	50	50	150	1000	1000	100	100	150	45
Max V_{DS}	V	1	1	1	1.2	0.5	1.1	0.3	0.5	1
Max V_{GS}	V	1	1	2	1.5	0.5	2	1.2	1.5	1
I_{ON}	µA/µm	6.2	22	141	0.068	0.45	1.4	6.4	135	**720**
I_{OFF}	pA/µm	0.00015	0.41	110	0.019	1.7	0.1	62	50	**0.52**
$I_{ON}:I_{OFF}$	µA/µA	10^{20}	10^8	10^6	10^6	10^5	10^7	10^5	10^4	**10^9**
SS	mV/dec	unreported	62	unreported	85.4	34	46	77	169	**30**

Note: Best figures are highlighted in bold.

Sources: (R1): Kumar et al. (2017), (R2): Horst et al. (2019), (R3): Safa et al. (2017), (R4): Huang et al. (2018), (R5): Ramaswamy and Kumar (2017), (R6): Chang et al. (2013), (R7): Dewey et al. (2011), and (R8): Ashita et al. (2018).

GoSoTFETs under lower V_{GS} biases, as indicated in Figure 8.2. The GOTFET can be fabricated following the procedure summarized below, as reported previously in Schmidt et al. (2014), Aspar et al. (1997), Bruel et al. (1995), and Bruel et al. (1997).

1. Preparation of the SOI substrate.
2. Active area of p+ source region with boron (B) doped silicon germanium ($Si_{0.15}Ge_{0.85}$) with 85% Ge mole fraction.
3. Atomic layer deposition (ALD) of 1.5 nm high-κ hafnium oxide (HfO_2) as the gate dielectric.
4. 30 nm HfO_2 is deposited as a hard mask for gate sidewalls or spacers for source and drain sides.
5. Gate deposition using atomic vapor deposition (AVD).
6. n+ Si drain region formed by arsenic (As) ion implantation.

The GOTFET structure is fundamentally a double-gate FDSOI structure, and there are several fabrication techniques to develop such structures. One of the most popular and reliable modern techniques is the proprietary "Smart-Cut" technique patented by SOITEC (Schmidt et al. 2014; Aspar et al. 1997; Bruel et al. 1995; Bruel et al. 1997). Using this standard SOI technique, one can fabricate structures such as GOTFET with nano-scale precision, with front and back-gate terminals. Hence, in real applications, with the Smart-Cut technique, we discard the bottom of the wafer with the flip-and-bond process. This process will result in a symmetrical double-

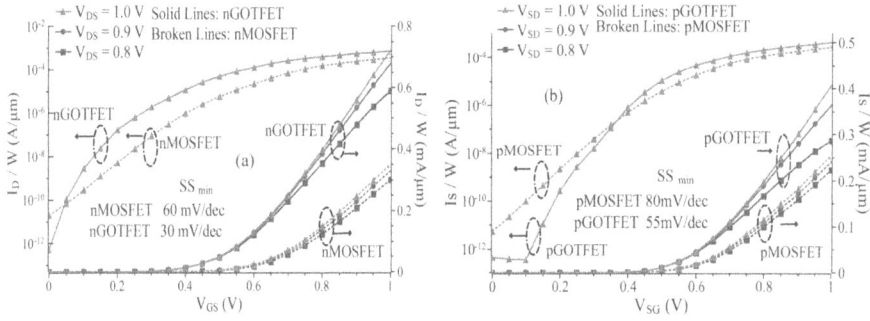

FIGURE 8.3 Comparison of device characteristics of GOTFETs versus MOSFETs: (a) drain current I_D variation with gate voltage of the n-channel devices for increasing V_{DS} biases; (b) source current I_S variation with gate voltage of the p-channel devices for increasing V_{SD} biases.

gated structure shown in Figure 8.2. However, a cheaper alternative is the 3D structure shown in Figure 8.2(b) that can be considered as a top view rather than a vertical cross section. The front and back gates will turn out to be the left (Gate 2) and right (Gate 1) sides of a vertical SiGe channel (annotated as the "Fin") whose height will be the effective width of the device as shown in Figure 2b. As one can easily identify, this structure is the basis of the typical FinFET structure, the current device technology in the industry driving all the state-of-the-art ICs.

8.3.3 GOTFET DEVICE CHARACTERIZATION

$I_D - V_{DS}$ and $I_D - V_{GS}$ characteristics of the nGOTFET and $I_S - V_{SD}$ and $I_S - V_{SG}$ characteristics of the pGOTFET are shown in Figures 8.3 and 8.4, respectively. The MOSFET characteristics have also been shown in Figures 8.3 and 8.4 on the same axes along with the GOTFET characteristics, to illustrate the superior performance

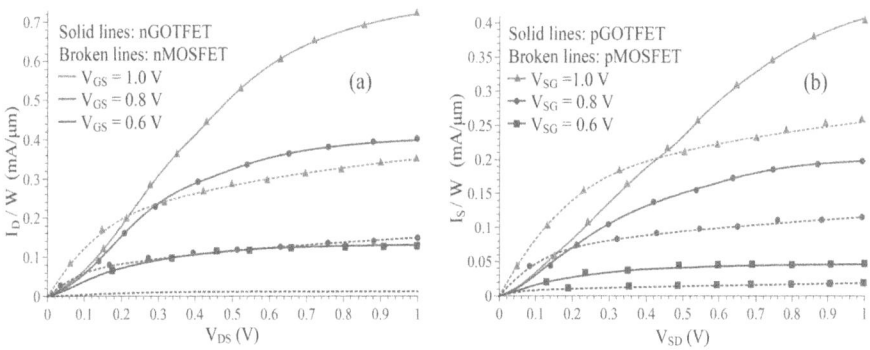

FIGURE 8.4 Comparison of device characteristic GOTFET versus MOSFET: (a) $I_D - V_{DS}$ characteristics of the n-channel devices for increasing V_{GS} biases; (b) $I_S - V_{SD}$ characteristics of the p-channel devices for increasing V_{SG} biases.

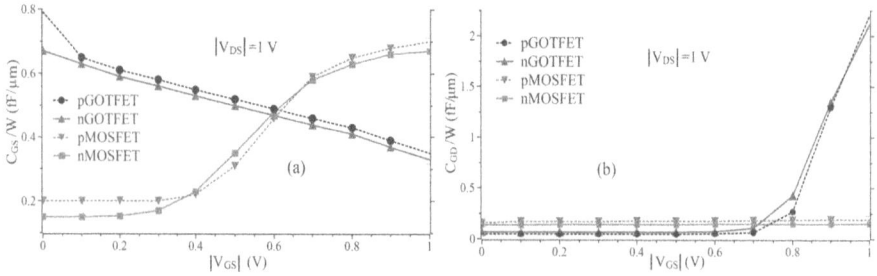

FIGURE 8.5 (a) $C_{GS} - V_{GS}$; (b) $C_{GD} - V_{GS}$ variation with gate bias of the n- and p-channel devices.

of 45 nm GOTFETs with respect to the 45 nm industry-standard CMOS devices (Cadence 2008). The variation of C_{GS}, C_{GD} and C_{GG} ($C_{GS} + C_{GD}$) capacitances for $V_{DS} = 1$ V and V_{GS} sweep from 0 to 1 V are shown in Figures 8.5 and 8.6. The C_{GG} of the GOTFETs and MOSFETs are almost the same in the range $V_{GS} = 0 - 0.8$ V; however the C_{GG} of the GOTFETs increases steeply beyond $V_{GS} = 0.8$ V, due to better electrostatic coupling between drain and channel as a result of reduced channel to drain potential. Figure 8.7 compares the transconductance g_m of nGOTFET and pGOTFET devices with the corresponding MOSFETs under deep saturation ($V_{DS} \geq V_{GS}$) bias conditions. The g_m of the GOTFETs is higher than the analogous MOSFETs as seen in Figure 8.7, for the entire range of V_{GS} range of $0 - 1$ V, which enables faster operation of the GOTFETs in circuits.

8.3.4 GOTFET OPERATION AT HIGH FREQUENCIES

The unity-gain bandwidth or transition frequency f_T is calculated using Eq. (8.8) from the device characteristics extracted from synopsys TCAD, shown in Figure 8.8.

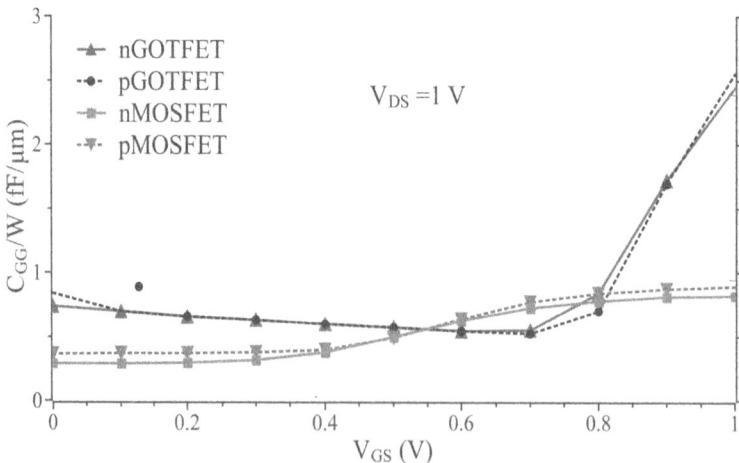

FIGURE 8.6 Total gate capacitance C_{GG} variation of the n- and p-channel GOTFET and MOSFET devices for V_{GS} change from 0 to 1 V.

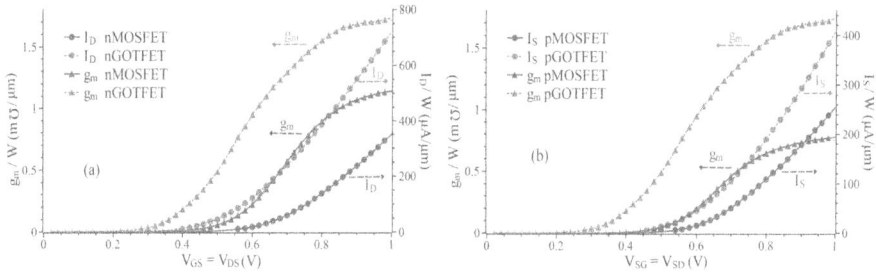

FIGURE 8.7 Transconductance g_m for: **(a)** n-; **(b)** p-channel GOTFET versus MOSFET, respectively at $V_{GS} = V_{DS}$ (under deep saturation bias condition).

The decrease in f_T of the GOTFETs for V_{GS} greater than 0.8 V is due to the increase in the C_{GD} beyond $V_{GS} = 0.8V$.

$$f_T = \frac{g_m}{2\pi(C_{GS} + C_{GD})} \tag{8.8}$$

Figure 8.8b shows the f_T characteristics observed while carrying out circuit simulation of the GOTFET in cadence EDA tool. The proposed nGOTFETs and pGOTFETs have almost one order of magnitude higher f_T (263 GHz and 244 GHz, respectively) than the analogous 45 nm nMOSFETs and pMOSFETs (28.80 GHz and 24.40 GHz, respectively) as shown in Figure 8.8b. GOTFETs have over 19–20 times higher $I_{ON} : I_{OFF}$ ratios than the analogous MOSFETs, while the gate capacitance of the GOTFETs are of the same order of magnitude (fF/μm) as that of MOSFETs. In addition, GOTFETs have significantly higher transconductance g_m than MOSFETs under low bias conditions experienced during switching.

8.3.5 SUPPRESSION OF AMBIPOLAR CURRENTS IN GOTFETS

Conventional TFETs have relatively high undesirable drain currents for negative gate biases. This phenomenon in TFETs is commonly known as "ambipolar behavior." To

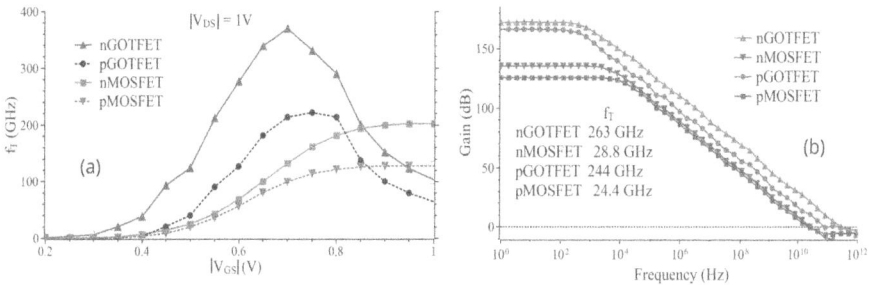

FIGURE 8.8 **(a)** Unity-gain BW f_T characteristics of the nGOTFET and pGOTFET versus nMOSFET and pMOSFET derived from the device characteristics extracted from TCAD; **(b)** Unity-gain BW f_T characteristics of the nGOTFET and pGOTFET versus nMOSFET and pMOSFET obtained through circuit simulation.

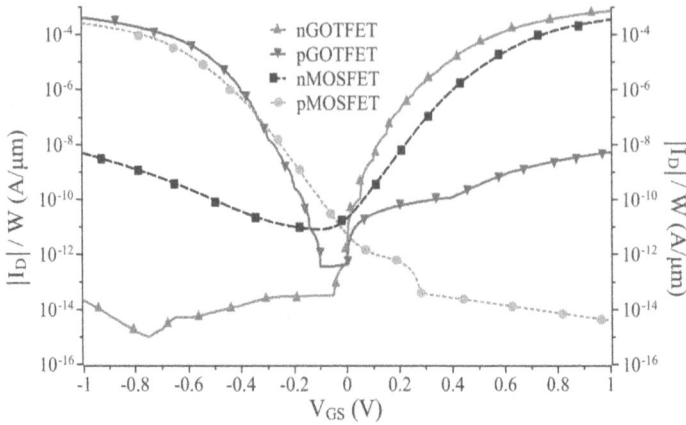

FIGURE 8.9 I_D vs. V_{GS} plot of GOTFETs under $V_{DS} = 1$ V, indicating negligible ambipolar currents comparable to MOSFETs.

limit ambipolar currents, the GOTFETs have a gate stack, which does not extend until the vicinity of the drain region. The gate overlaps the source completely, while it is terminated 22 nm short of the channel-drain junction. This enables high BtB tunneling from source to the channel regions, on the application of $V_{GS} \geq V_{TH}$ voltage, while it prevents tunneling from the drain region to the channel region on the application of negative gate-source bias. Thus the GOTFETs have significantly reduced ambipolar currents as compared to the conventional tunnel FETs. The $I_D - V_{GS}$ plot for V_{GS} range of −1 to +1 V at $V_{DS} = 1$ V is shown in Figure 8.9.

8.3.6 OPERATING PRINCIPLE OF THE GOTFET

The Gate-Overlap Tunnel FET is an advanced TFET that functions on the fundamental physical principle of BtB generation, but in a different and more effective manner than the conventional TFETs. The GOTFET structure facilitates vertical BtB tunneling from the *central core* region of the p+ doped source to the *surface* region of the source, under the influence of V_{GS}, since the gate stack overlaps the source region. In conventional TFETs, the BtB tunneling takes place from the source region to the channel region, due to the lateral electric field $\hat{x}F_x(x, y)$ across source-channel junction on application of forward gate bias. In contrast, BtB in a GOTFET occurs within the source region due to the vertical field $\hat{y}F_y(x, y)$ resulting from the gate stack overlapping the source. When a +ve voltage is applied at the gate terminals of GOTFET, the *electrons* from the valence band of the inner core region of the p+ source tunnel directly into the conduction band of the surface region of the p+ source as shown in Figure 8.10.

Figure 8.10 shows the energy-band diagram along the vertical $\hat{y}$ direction. When a $+V_{GS}$ is applied, the electrons from the valence band of the bulk/central core region of the source will tunnel into the conduction band of the surface region of the source, near the gate oxide. This effectively inverts the surface of the source region into an

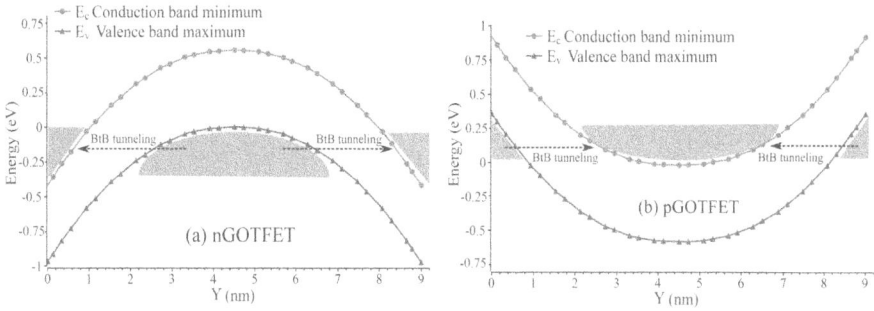

FIGURE 8.10 Energy band diagram and BtB tunneling in the source region of: **(a)** n-; **(b)** p-channel GOTFET extracted from synopsys TCAD.

effective n+ region. The GOTFET in this condition exhibits characteristics similar to that of a double-gate (DG) MOSFET with an n+ source, a p-channel, and an n+ drain. The I_{ON} current levels of GOTFETS are of the same order as that of the DG MOSFETs, unlike the low I_{ON} of conventional TFETs. The GOTFET combines the advantages of the TFETs and MOSFETs into a single effective device.

The higher I_{ON} can also be explained by an in-depth analysis of the maximum number of carriers available for conduction.The atomic density of silicon is in 10^{22}/cm³. Hence, the source region of a silicon MOSFET can have an maximum doping concentration of 10^{22} dopant atoms /cm³ since higher doping concentration 10^{20}/cm³ can disrupting the semiconductor properties of Si. This puts a limit on the maximum free holes available for conduction to 10^{22}/cm³, even at highly elevated temperatures. In GOTFETs, the free electrons for conduction are provided by the valence band of source inner core. Since the density of electrons in the valence band in the p+ source will be practically infinite, the on-currents of GOTFETs are significantly higher than MOSFETs. The source acts as a *super-source* providing 10^{32}/cm³ free electrons from the valence band for conduction through vertical BtB tunneling, unlike the MOSFETs which can have a maximum of only 10^{20}/cm³ free electrons.

8.3.7 Modeling Approach for the Proposed GOTFETs

The BtB generation rate G_{btb} in TFET is computed by the synopsys TCAD utilizing the compact model equations given by Kumar et al. (2017), Safa et al. (2017), Horst et al. (2019), and Kao et al. (2012a)

$$G_{BtB} = AF_y^{\gamma} exp\left(-B / F_y\right) \tag{8.9}$$

where, $\gamma = 2$ and 2.5 for direct and indirect phonon-assisted tunneling process, respectively. For direct BtB tunneling in direct bandgap materials, material-dependent factors *AA* and *BB* are given by

$$A = \left(\frac{g\pi}{9}\right)\left(\frac{q}{h}\right)^{\gamma}\sqrt{\frac{m_{red}}{E_G}} \tag{8.10}$$

$$B = \frac{\pi^2 \sqrt{m_{red}} \, E_G^{3/2}}{qh} \tag{8.11}$$

For phonon-assisted BtB tunneling in indirect bandgap materials, material-dependent factors A and BAB are given by

$$A = \left(\frac{gm_{avg}^{3/2}}{\sqrt{2}}\right)\left(\frac{1+2N_{op}}{\rho\varepsilon_{op}}\right)D_{op}^2\left(\frac{q}{h}\right)^\gamma\left(\frac{m_{red}}{2E_G}\right)^{7/4} \tag{8.12}$$

$$B = \frac{4\pi\sqrt{m_{red}}\,(2E_G)^{3/2}}{3qh} \tag{8.13}$$

The reduced mass $m_{red} = \left(m_c^{-1} + m_v^{-1}\right)^{-1}$ and average mass $m_{avg} = (m_c + m_v)/2$ where

$$\frac{1}{m_c} = \frac{1}{2m_{red}} + \frac{1}{m_0} \quad \& \quad \frac{1}{m_v} = \frac{1}{2m_{red}} - \frac{1}{m_0} \tag{8.14}$$

m_0 being the rest mass of electrons. The line tunneling occurs due to the vertical field $\hat{y}F_y$ (in the vertical $\hat{y}$ direction), so the total number of free carriers N_S generated within the source due to BtB is obtained from

$$N_S = 2W \iint_{x\ y} G_{BtB}dydx \tag{8.15}$$

where, WW denotes the width of the device and a factor of 2 is included because this is a double gate device. Current I_D is modeled using the standard diffusion transport along the channel (in the lateral x direction) using

$$I_D = (\text{cross-sectional area}) \cdot qN_S v_{dr} \tag{8.16}$$

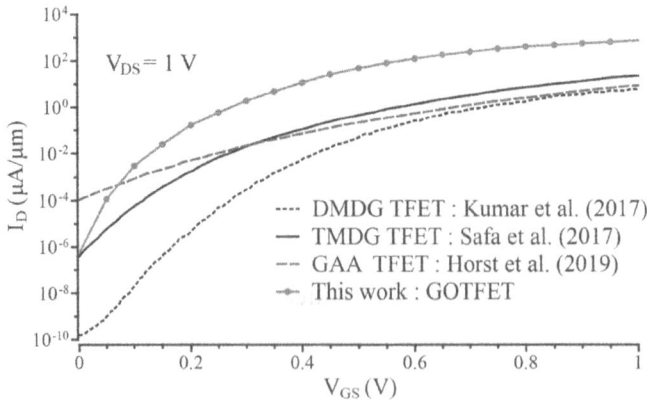

FIGURE 8.11 Benchmarking the performance of GOTFET with other state-of-the-art TFETs. (Reported in the literature: Kumar et al. 2017; Safa et al. 2017; Horst et al. 2019.)

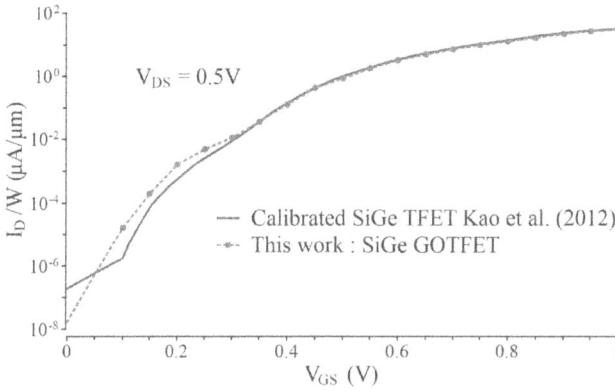

FIGURE 8.12 Benchmarking the GOTFET performance with fabricated SiGe TFET. (From Kao et al. 2012a.)

8.3.8 COMPARISON OF THE GOTFET CHARACTERISTICS WITH THE OTHER STATE-OF-THE-ART TFETs

The performance of GOTFET is compared with some other state-of-the-art tunnel FETs reported in the literature [Gate All Around (GAA) TFET (Horst et al. 2019), Dual Metal Double Gate (DMDG) TFET (Kumar et al. 2017), and Triple Metal Double Gate (TMDG) TFET (Safa et al. 2017)] and fabricated SIiGE TFET (Kao et al. 2012b) in Figures 8.11 and 8.12. The structure of these state-of-the-art TFETs is shown in Figure 8.13.

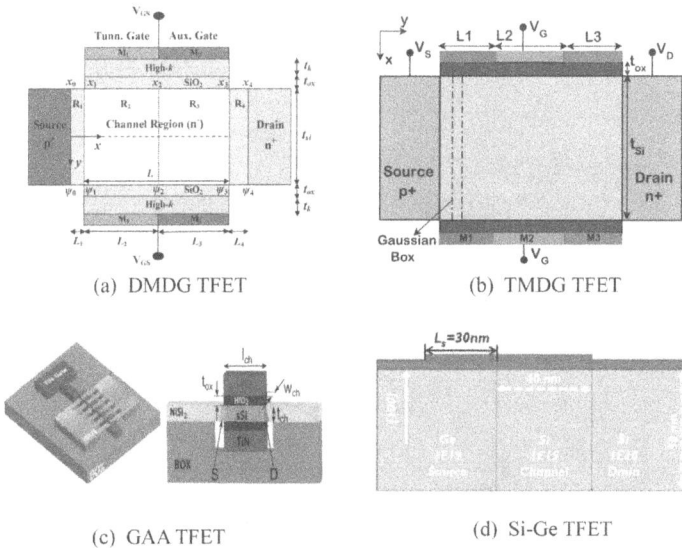

(a) DMDG TFET

(b) TMDG TFET

(c) GAA TFET

(d) Si-Ge TFET

FIGURE 8.13 Structure of: **(a)** DMDG TFET (Kumar et al. 2017); **(b)** TMDG TFET (Safa et al. 2017); **(c)** GAA TFET (Horst et al. 2019); **(d)** fabricated SiGe TFET (Kao et al. 2012a).

8.4   PERFORMANCE BENCHMARKING OF GOTFETS

8.4.1   COMPARISON OF POWER AND PROPAGATION DELAYS CGOT VERSUS CMOS INVERTER, NAND, NOR, AND XOR GATES

Due to steep subthreshold slope and low leakage currents, the GOTFET is a promising alternative for the MOSFET, especially when it comes to low power designs. Higher I_{ON} and steeper sub-threshold slope enables the CGOT based digital circuits to operate faster. At the same time, the lower I_{OFF} ensures lower static power consumption as compared to the same circuit implemented with CMOS devices. Figure 8.14 shows the schematic of the CGOT-based inverter circuit and its delay characteristics.

The performance of the CGOT inverter has been benchmarked with an identical CMOS-based inverter with the same aspect ratio. The circuit has been simulated at 1 GHz clock frequency with a load capacitance of 10 fF, as evident from Figure 8.14. The CGOT inverter operates 1.43 times faster than the corresponding CMOS inverter. Figure 8.15 shows the comparison of static power consumption of CGOT with a CMOS inverter.

The CGOT inverter consumes merely 0.009 times the power consumed by a corresponding CMOS inverter. Overall, a decrease of 99.45% in power delay product (PDP) can be achieved by replacing the CMOS devices with the proposed CGOT devices. Figure 8.16 shows the voltage transfer characteristics (VTC) curves of the CGOT inverter for supply voltage V_{DD} range 0.25–1 V. The CGOT inverter provides full rail-to-rail output even for the subthreshold operation region (until 0.25 V), thereby indicating the suitability of GOTFET devices for low-voltage applications. Figure 8.17 shows the variation in propagation delay of CGOT inverter as a function of the pGOTFET/nGOTFET transistor ratio β. Symmetrical delay characteristics are achieved for $\beta \approx 2$.

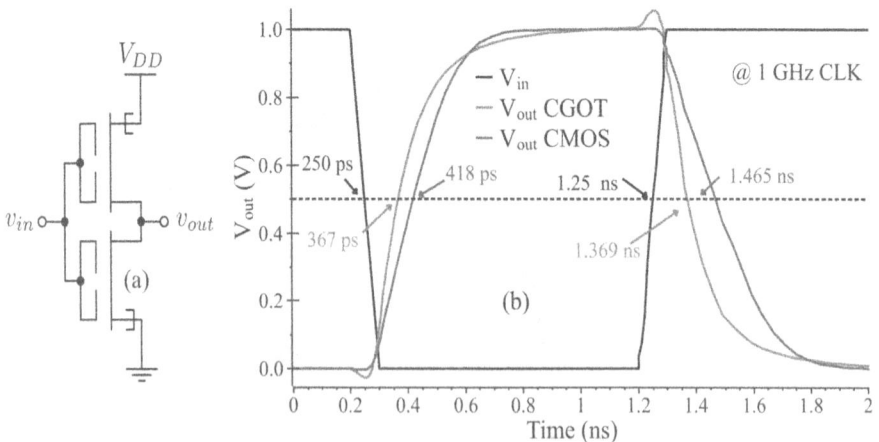

FIGURE 8.14   **(a)** Schematic of CGOT inverter; **(b)** delay comparison CGOT versus CMOS inverter.

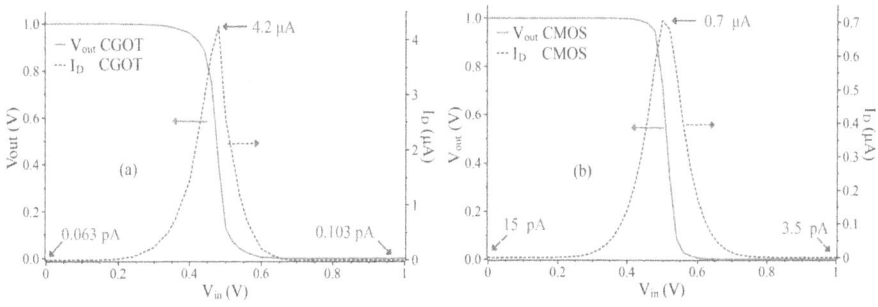

FIGURE 8.15 Comparison of static currents: **(a)** CGOT versus **(b)** CMOS inverter.

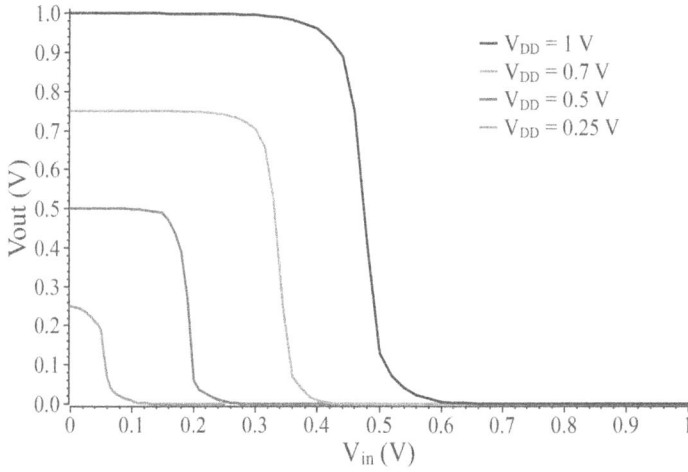

FIGURE 8.16 VTC curves of a CGOT inverter for supply voltage V_{DD} range $0.25-1$ V.

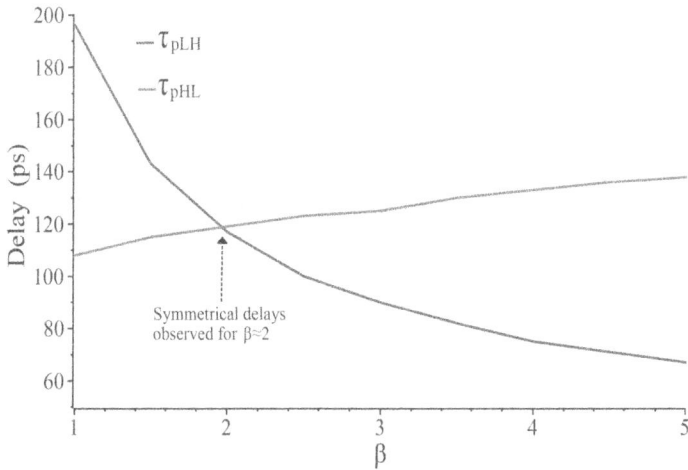

FIGURE 8.17 Variation in propagation delay of a CGOT inverter as a function of the pGOTFET/nGOTFET transistor ratio β.

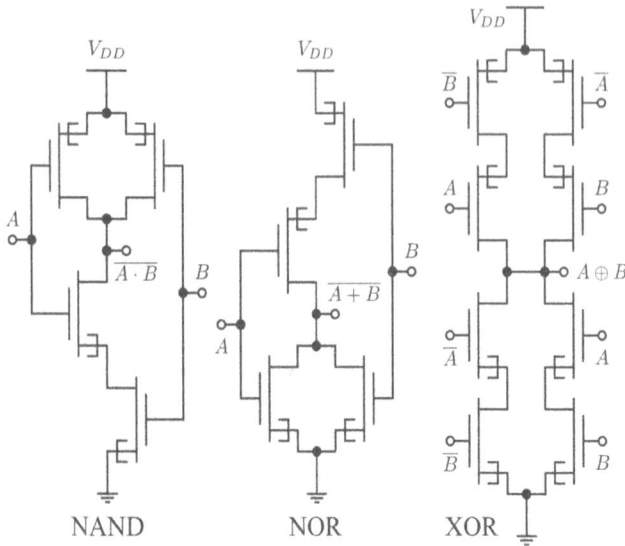

FIGURE 8.18   Schematic of CGOT digital circuits: 2-input: (a) NAND; (b) NOR; (c) XNOR gates.

Figure 8.18 shows the schematic of the CGOT 2-input NAND, NOR, and XNOR gates. Figures 8.19–8.21 show the comparison of delay and static power characteristics for NAND, NOR, and XNOR gates, respectively. The improvement in the performance parameters is summarized in Table 8.3.

Figures 8.22–8.25 show the total power consumption plot (including dynamic power) and transient response waveform of the inverter, NAND, NOR, and XNOR gates, respectively. The total power of the CGOT gates is 18–54% lower than

TABLE 8.3

Comparison the Performance Parameters of Inverter NAND, NOR, and XOR Gates Implemented with CGOT and CMOS Technologies at 10 fF Load Capacitance

Circuit parameter	Units	Inverter		2-i/p NAND		2-i/p NOR		2-i/p XNOR	
		CMOS	CGOT	CMOS	CGOT	CMOS	CGOT	CMOS	CGOT
Bias V_{DD}	V	1	1	1	1	1	1	1	1
Delay $\tau_{pd,LH}$	ps	168	117	184	106	201	146	331	223
Delay $\tau_{pd,HL}$	ps	215	119	254	148	337	157	290	182
Average delay	ps	191.5	118	219	127	269	151.5	310.5	202.5
Static I_{high}	pA	15	0.063	19	0.125	51	0.125	55.6	0.251
Static I_{low}	pA	3.5	0.103	9.4	0.206	5.4	0.165	48.8	0.413
Average P_{static}	pW	9.25	0.083	14.2	0.166	28.2	0.145	52.2	0.332
PDP ($\times 10^{-23}$)	J	177.1	0.98	310.98	2.1	758.6	2.2	1620.81	6.72
Decrease in PDP	%	99.45 %		99.32%		99.71%		99.58%	

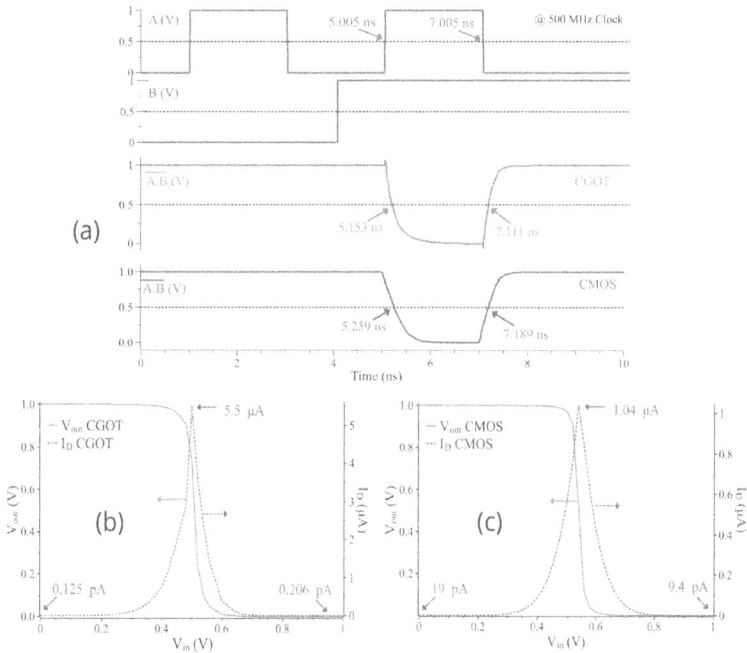

FIGURE 8.19 (a) Delay comparison CGOT versus CMOS NAND gates. Comparison of static currents (b) CGOT versus (c) CMOS NAND gates.

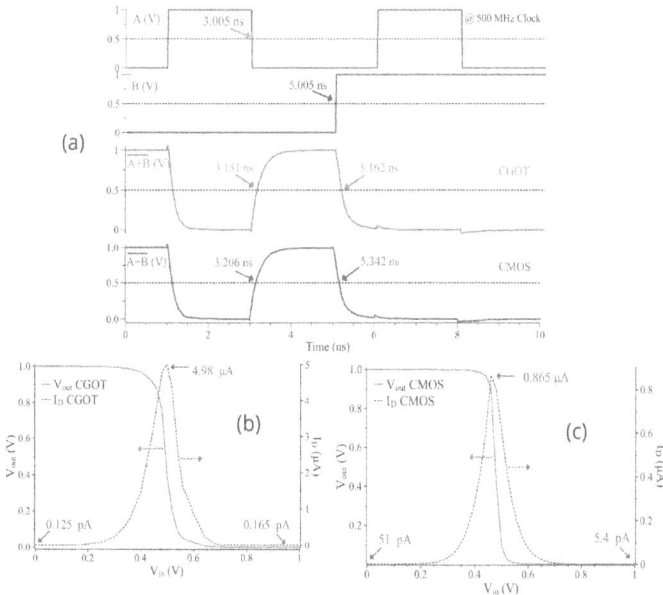

FIGURE 8.20 (a) Delay comparison CGOT versus CMOS NOR gates. Comparison of static currents (b) CGOT versus (c) CMOS NOR gates.

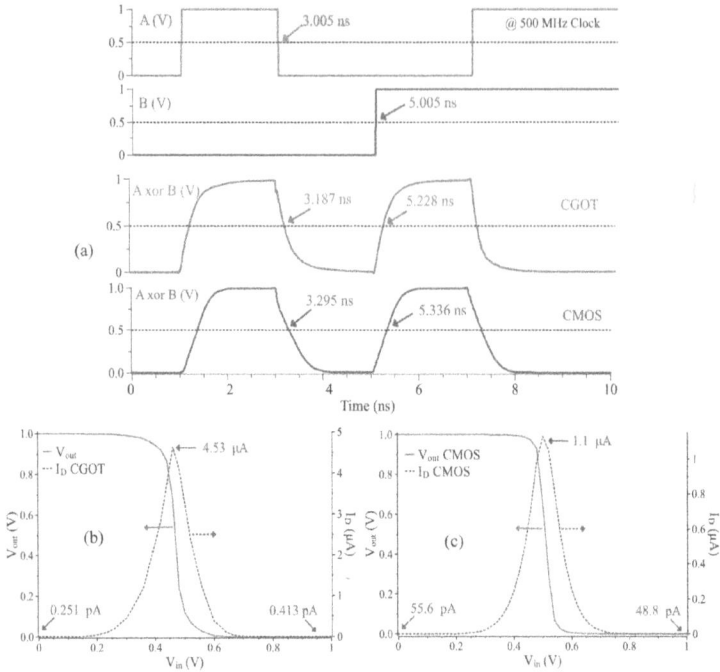

FIGURE 8.21   (a) Delay comparison CGOT versus CMOS XOR gates. Comparison of static currents (b) CGOT versus (c) CMOS XOR gates.

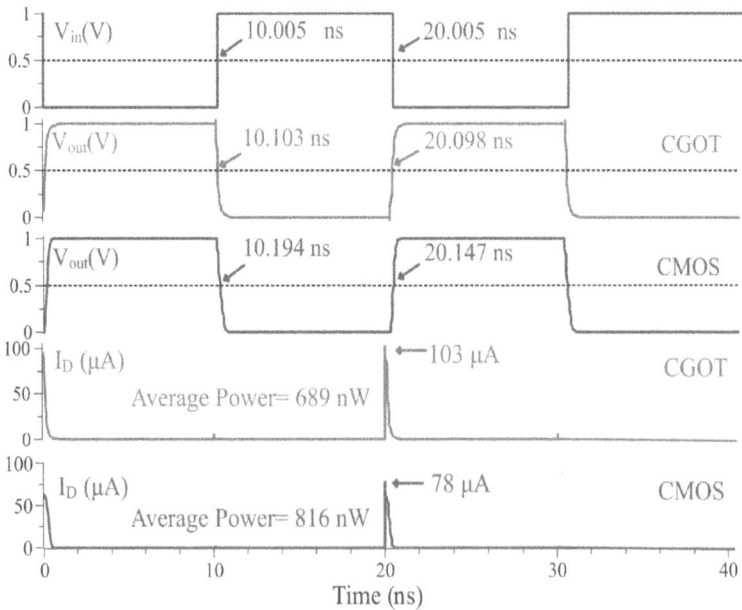

FIGURE 8.22   Total power consumption plot (including dynamic power) and transient response waveform CGOT versus CMOS inverter.

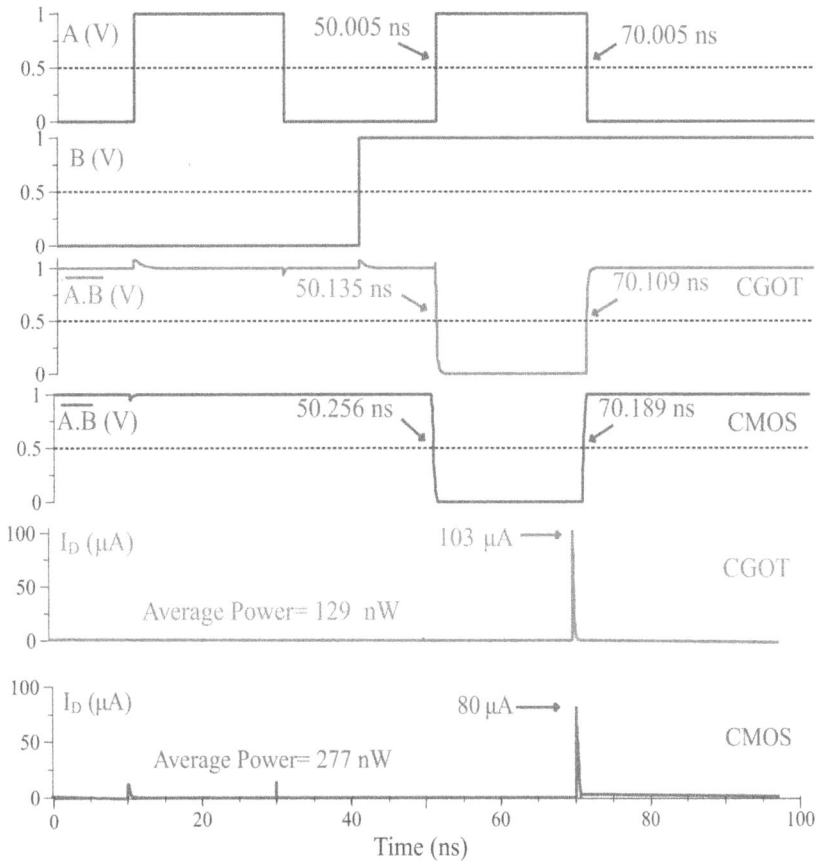

FIGURE 8.23 Total power consumption plot (including dynamic power) and transient response waveform CGOT versus CMOS NAND.

corresponding CMOS gates on account of lower I_{OFF} of GOTFETs. However, the peak short circuit current is higher in CGOT circuits due to the higher I_{ON} of GOTFETs. Innovative circuit modifications have been proposed in GOTFET-based VLSI circuits in the subsequent chapters to keep the total power of CGOT circuits significantly lower than corresponding conventional CMOS circuits.

The simulation plot showing VTC dependence on input pattern (Rabaey et al. 2016) for the two-input NAND and NOR gates are shown in Figure 8.26. The simulation plot showing delay dependence on input pattern (Rabaey et al. 2016) for the two-input NAND gates are shown in Figures 8.27 and 8.28 and that of NOR gates are shown in Figures 8.29 and 8.30.

8.4.2 COMPARISON OF JITTER IN A CGOT AND CMOS INVERTER AND CHAIN OF INVERTERS

Jitter is the variation of the clock edge from its ideal instance. Clock jitter is usually caused by the clock generating circuit, noise, power supply fluctuations,

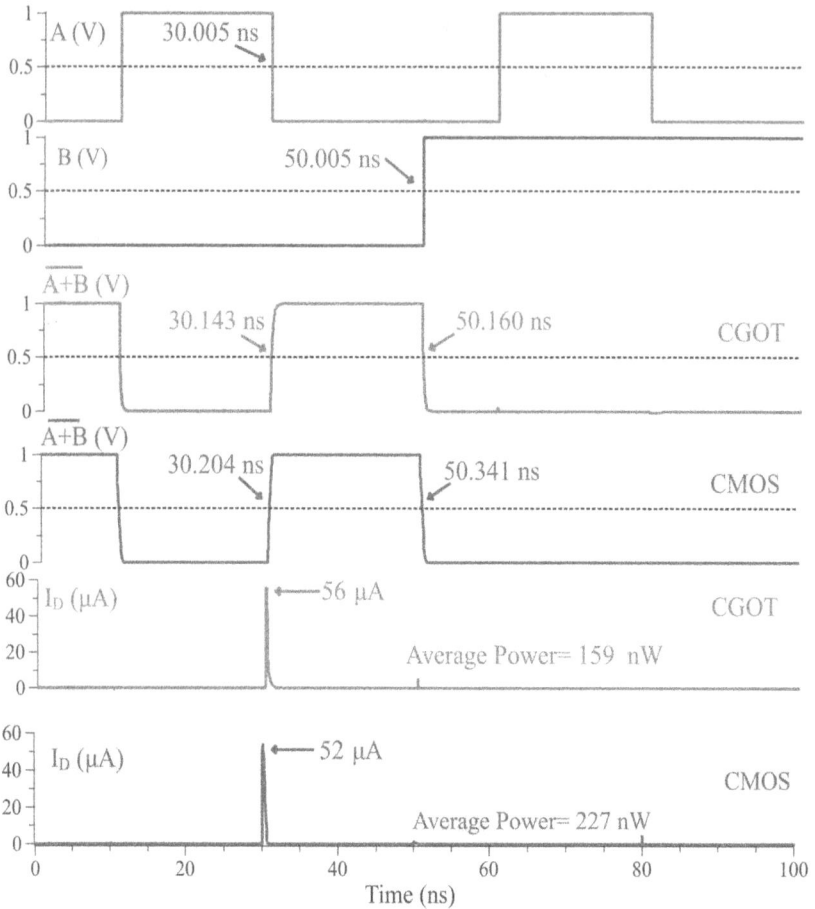

FIGURE 8.24 Total power consumption plot (including dynamic power) and transient response waveform CGOT versus CMOS NOR.

and interference from adjacent components. The eye diagram of the inverter (with a 1fF load capacitance) and a three-stage chain of inverters (with load capacitance 100 fF), implemented with CGOT and CMOS devices, are shown in Figures 8.31 and 8.32, respectively. It is evident from the plots that the jitters in the analogous CMOS and CGOT circuits are almost identical, indicating that the CGOT circuits are as immune to jitter as its counterpart CMOS circuit. However, the exact jitter estimation is impossible without physical designs, which are currently not possible with TFETs due to the unavailability of a standard TFET technology library. Nevertheless, the jitter issues can be taken care of using rigorous static timing analysis (STA) of the data-path and clock-path propagation delays.

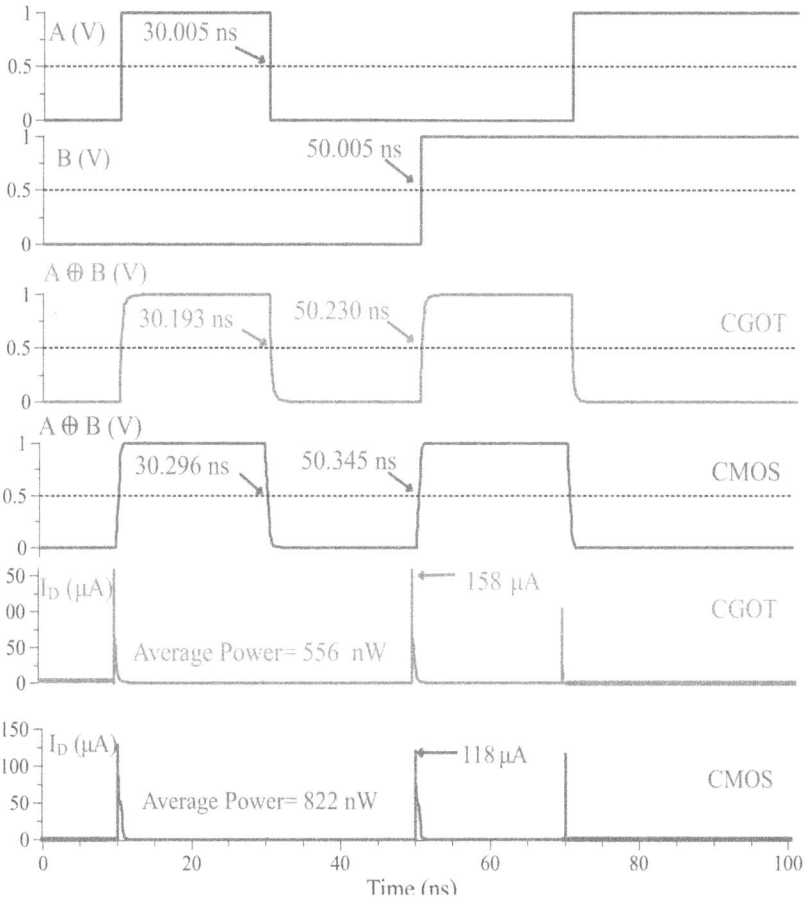

FIGURE 8.25   Total power consumption plot (including dynamic power) and transient response waveform CGOT versus CMOS XOR.

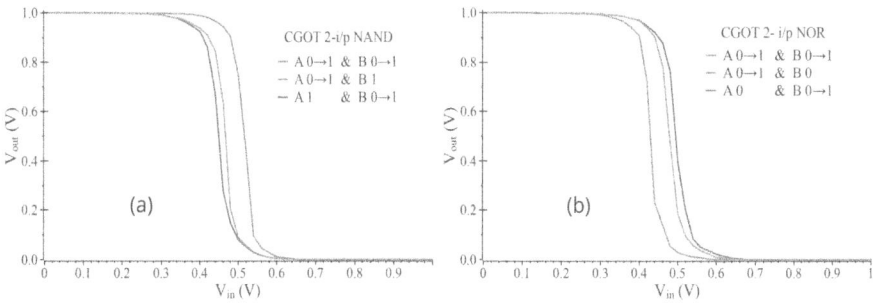

FIGURE 8.26   VTC dependence on input pattern for the CGOT: (a) 2-input NAND; (b) 2-input NOR.

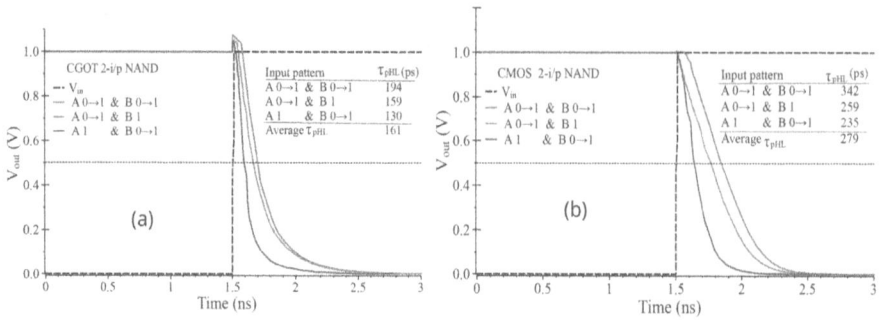

FIGURE 8.27 $\tau_{pd,HL}$ dependence on input pattern for the 2-input NAND at 10 fF load capacitance: **(a)** CGOT; **(b)** CMOS.

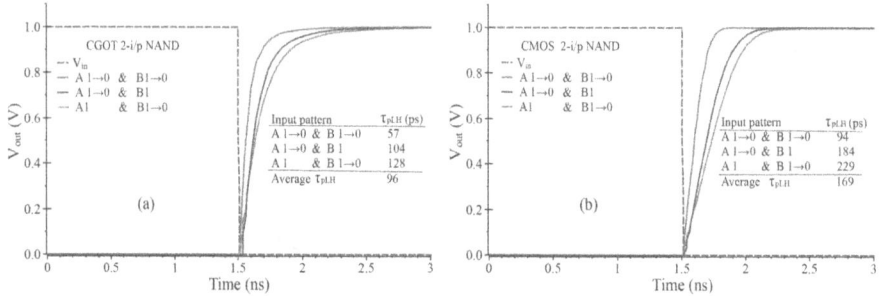

FIGURE 8.28 $\tau_{pd,LH}$ dependence on input pattern for the 2-input NAND at 10 fF load capacitance: **(a)** CGOT; **(b)** CMOS.

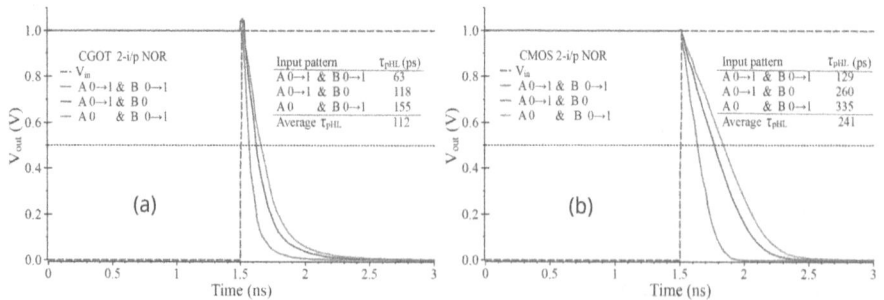

FIGURE 8.29 $\tau_{pd,HL}$ dependence on input pattern for the 2-input NOR at 10 fF load capacitance: **(a)** CGOT; **(b)** CMOS.

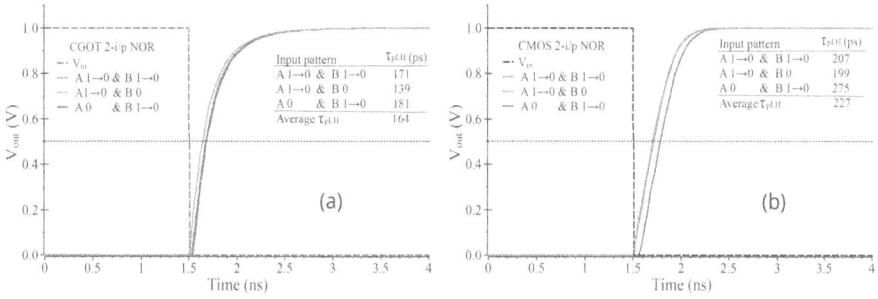

FIGURE 8.30 $\tau_{pd,LH}$ dependence on input pattern for the 2-input NOR at 10 fF load capacitance: **(a)** CGOT; **(b)** CMOS.

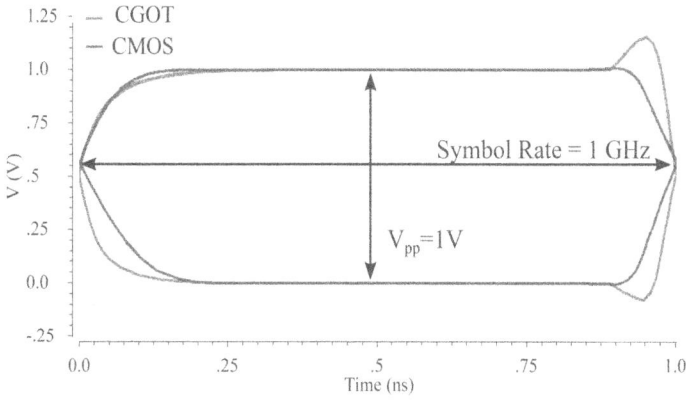

FIGURE 8.31 Eye diagram for the CGOT versus CMOS inverter with a load capacitance of 1 fF.

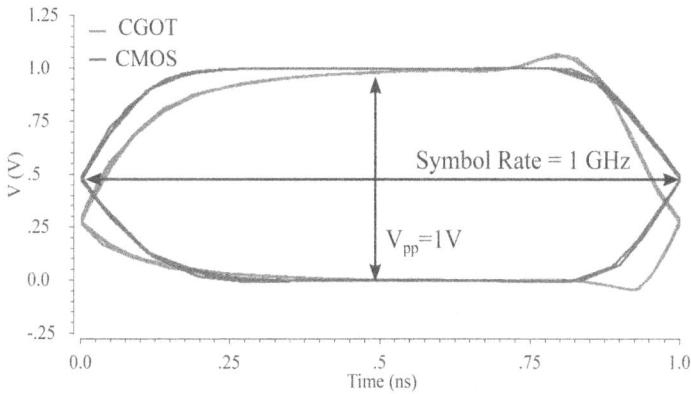

FIGURE 8.32 Eye diagram for the CGOT versus CMOS three-stage chain of inverters with a load capacitance of 100 fF.

8.5 SUMMARY

The performance benchmarking of the GOTFETs in standard digital building blocks such as inverter, NAND, NOR, and XNOR gates with the same circuits implemented with 45 nm MOSFET devices indicates that the two to three orders lower I_{OFF} of the GOTFET devices as compared to the CMOS devices results in a significant reduction in static and overall power consumption (92–94% reduction in total power) of the digital gates. The GOTFETs have almost twice the I_{ON} of analogous CMOS devices, and this helps in reducing the propagation delays in these basic digital circuits by 45–50%. Overall, a reduction of 98–99% in PDP has been observed in CGOT digital circuits as compared to the corresponding CMOS-based digital circuits.

REFERENCES

Amir, M. F., Trivedi, A. R., and Mukhopadhyay, S. 2016. Exploration of Si/Ge tunnel FET bit cells for ultra-low power embedded memory. *IEEE Journal on Emerging and Selected Topics in Circuits and Systems* 6 (2): 185–97.

Ashita, Loan S. A., and Rafat, M. 2018. A high-performance inverted-C tunnel junction FET with source-channel overlap pockets. *IEEE Transactions on Electron Devices* 65 (2): 763–68.

Aspar, B., Bruel, M., Moriceau, H., Maleville, C., Poumeyrol, T., Papon, A. M., Claverie, A., Benassayag, G., Auberton-Herve, A. J., and Barge, T. 1997. Basic mechanisms involved in the Smart-CutÂ® process. *Microelectronic Engineering* 36 (1): 233–40.

Bruel, M., Aspar, B., and Auberton-Herve, A.-J. 1997. Smart-cut: A new silicon on insulator material technology based on hydrogen implantation and wafer bonding*1. *Japanese Journal of Applied Physics* 36 (3S): 1636.

Bruel, M., Aspar, B., Charlet, B., Maleville, C., Poumeyrol, T., Soubie, A., Auberton-Herve, A., Lamure, J., Barge, T., Metral, F., and Trucchi, S. 1995. "Smart cut: a promising new SOI material technology," *in 1995 IEEE International SOI Conference Proceedings*, 178–79.

Cadence. 2008. "Cadence collaborates with common platform and ARM to deliver 45-nm RTL-to-GDSII reference flow." https://www.semiconductoronline.com/doc/cadence-collaborates-with-common-platform-and-0001.

Caka, N., Zabeli, M., Limani, M., and Kabashi, Q. 2007. Impact of MOSFET parameters on its parasitic capacitances. Proceedings of the 6th WSEAS International Conference on Electronics, Hardware, Wireless and Optical Communications, Corfu Island, Greece, February 16–19, 2007. https://www.researchgate.net/publication/235955632_Impact_of_MOSFET_parameters_on_its_parasitic_capacitances.

Chander, S., Bhowmick, B., and Baishya, S. 2015. Heterojunction fully depleted SOI-TFET with oxide/source overlap. *Superlattices and Microstructures* 86: 43–50.

Chang, H., Adams, B., Chien, P., Li, J., and Woo, J. C. S. 2013. Improved subthreshold and output characteristics of source-pocket Si tunnel FET by the application of laser annealing. *IEEE Transactions on Electron Devices* 60 (1): 92–96.

Chau, R., Datta, S., Doczy, M., Kavalieros, J., and Metz, M. 2003. Gate dielectric scaling for high-performance CMOS: From SiO2 to high-K, in *Extended Abstracts of International Workshop on Gate Insulator (IEEE Cat. No.03EX765)*, 124–26.

Dan, S., Biswas, A., Le Royer, C., Grabinski, W., and Ionescu, A. 2012. A novel extraction method and compact model for the steepness estimation of FDSOI TFET lateral junction. *IEEE Electron Device Letters* 33 (2): 140–42.

Dewey, G., Chu-Kung, B., Boardman, J., Fastenau, J. M., Kavalieros, J., Kotlyar, R., Liu, W. K., Lubyshev, D., Metz, M., Mukherjee, N., Oakey, P., Pillarisetty, R.,

Radosavljevic, M., Then, H. W., and Chau, R. 2011. Fabrication, characterization, and physics of III–V heterojunction tunneling field effect transistors (H-TFET) for steep sub-threshold swing, in *2011 International Electron Devices Meeting*, 33.6.1–33.6.4.

Gupta, Ajay, M., Narang, R., and Saxena, M. 2015. Analysis of cylindrical gate junctionless tunnel field effect transistor (CG-JL-TFET), in *2015 Annual IEEE India Conference (INDICON)*, 155.

Harame, D. L. 2004. *SiGe–materials, Processing, and Devices: Proceedings of the First International Symposium*. The Electrochemical Society.

Helms, D., Schmidt, E., and Nebel, W. 2004. Leakage in CMOS circuits. An introduction. In Macii, E., Paliouras, V., and Koufopavlou, O., eds., *Integrated Circuit and System Design. Power and Timing Modeling, Optimization and Simulation,* Lecture Notes in Computer Science, 17–35, Berlin, Heidelberg. Springer.

Horst, F., Farokhnejad, A., Zhao, Q., Iñíguez, B., and Kloes, A. 2019. 2-D physics-based compact DC modeling of double-gate tunnel-FETs. *IEEE Transactions on Electron Devices* 66 (1): 132–138.

Huang, P., Tanamoto, T., Goto, M., Takenaka, M., and Takagi, S. 2018. Investigation of electrical characteristics of vertical junction Si n-type tunnel FET. *IEEE Transactions on Electron Devices* 65 (12): 5511–17.

Ilatikhameneh, H., Tan, Y., Novakovic, B., Klimeck, G., Rahman, R., and Appenzeller, J. 2015. Tunnel field-effect transistors in 2-D transition metal dichalcogenide materials. *IEEE Journal on Exploratory Solid-State Computational Devices and Circuits*, 1: 12–18.

Kane, E. 1960. Zener tunneling in semiconductors. *Journal of Physics and Chemistry of Solids* 12 (2): 181–88.

Kane, E. O. 1961. Theory of tunneling. *Journal of Applied Physics* 32 (1): 83–91.

Kao, K. H., Verhulst, A. S., Vandenberghe, W. G., Soree, B., Groeseneken, G., and Meyer, K. D. 2012a. Direct and indirect band-to-band tunneling in germanium-based TFETs. *IEEE Transactions on Electron Devices* 59 (2): 292–301.

Kao, K., Verhulst, A. S., Vandenberghe, W. G., Soree, B., Magnus, W., Leonelli, D., Groeseneken, G., and Meyer, K. D. 2012b. Optimization of gate-on-source-only tunnel FETs with counter-doped pockets. *IEEE Transactions on Electron Devices* 59 (8): 2070–77.

Kumar, S., Goel, E., Singh, K., Singh, B., Singh, P. K., Baral, K., and Jit, S. 2017. 2-D analytical modeling of the electrical characteristics of dual-material double-gate TFETs with a SiO2/HfO2 stacked gate-oxide structure. *IEEE Transactions on Electron Devices* 64 (3): 960–68.

Lin, Jyi-Tsong, Chih-Hao Kuo, Tai-Yi Lee, Yi-Chuen Eng, Tzu-Feng Chang, Po-Hsieh Lin, Hsuan-Hsu Chen, Chih-Hung Sun, and Hsien-Nan Chiu 2009. Improving reliability and diminishing parasitic capacitance effects in a vertical transistor with embedded gate, in *2009 16th IEEE International Symposium on the Physical and Failure Analysis of Integrated Circuits*, 75–78.

Liu, H., Saripalli, V., Narayanan, V., and Datta, S. 2015. *III-V Tunnel FET Model*. https://nanohub.org/resources/21015/download/PennState_IIIV_TFET_VerilogAModel_1.0.0_Manual.pdf.

Rabaey, J. M., Chandrakasan, A., and Nikolic, B. 2016. *Digital Integrated Circuits: A Design Perspective*, 2nd ed. Pearson Education India.

Ramaswamy, S. and Kumar, M. J. 2017. Double gate symmetric tunnel FET: Investigation and analysis. *IET Circuits, Devices Systems* 11 (4): 365–70.

Roy, K., Mukhopadhyay, S., and Mahmoodi-Meimand, H. 2003. Leakage current mechanisms and leakage reduction techniques in deep-submicrometer CMOS circuits. *Proceedings of the IEEE* 91 (2): 305–27.

Safa, S., Noor, S. L., and Khan, Z. R. 2017. Physics-based generalized threshold voltage model of multiple material gate tunneling FET structure. *IEEE Transactions on Electron Devices* 64 (4): 1449–54.

Saurabh, S. and Kumar, M. J. 2017. *Fundamentals of Tunnel Field-Effect Transistors*. Boca Raton, FL: CRC Press, Taylor & Francis Group.

Schmidt, M., Schäfer, A., Minamisawa, R. A., Buca, D., Trellenkamp, S., Hartmann, J.-M., Zhao, Q.-T., and Mantl, S. 2014. Line and point tunneling in scaled Si/SiGe heterostructure TFETs. *IEEE Electron Device Letters* 35 (7): 699–701.

Schulte-Braucks, C., Pandey, R., Sajjad, R. N., Barth, M., Ghosh, R. K., Grisafe, B., Sharma, P., von den Driesch, N., Vohra, A., Rayner, G. B., Loo, R., Mantl, S., Buca, D., Yeh, C., Wu, C., Tsai, W., Antoniadis, D. A., and Datta, S. 2017. Fabrication, characterization, and analysis of Ge/GeSn heterojunction p-type tunnel transistors. *IEEE Transactions on Electron Devices* 64 (10): 4354–362.

Settino, F., Lanuzza, M., Strangio, S., Crupi, F., Palestri, P., Esseni, D., and Selmi, L. 2017. Understanding the potential and limitations of tunnel FETs for low-voltage analog/mixed-signal circuits. *IEEE Transactions on Electron Devices* 64 (6): 2736–43.

Shrivastava, M. 2017. Drain extended tunnel FET-A novel power transistor for RF and switching applications. *IEEE Transactions on Electron Devices* 64 (2): 481–87.

Strangio, S., Settino, F., Palestri, P., Lanuzza, M., Crupi, F., Esseni, D., and Selmi, L. 2018. Digital and analog TFET circuits: Design and benchmark. *Solid-State Electronics* 146: 50–65.

Sze, Simon M. and Kwok K. Ng. 2006. *Physics of Semiconductor Devices*. Boca Raton, FL: CRC Press, Taylor & Francis Group.

Vidhyadharan, S., Ramakant, R., Vidhyadharan, A. S., Shyam, A. K., Hirpara, M. P., and Dan, S. S. 2019a. An efficient design approach for implementation of 2 bit ternary flash ADC using optimized complementary TFET devices, in *2019 32nd International Conference on VLSI Design and 2019 18th International Conference on Embedded Systems (VLSID)*, 401–406.

Vidhyadharan, S., Yadav, R., Akhilesh, G., Gupta, V., Ravi, A., and Dan, S. S. 2019b. Part II: Benchmarking the performance of optimized TFET-based circuits with the standard 45 nm CMOS technology using device & circuit co-simulation methodology, in *The Physics of Semiconductor Devices*, 619–628, Springer International Publishing.

Vidhyadharan, S., Yadav, R., Hariprasad, S., and Dan, S. S. 2019c. A nanoscale gate-overlap tunnel FET (GOTFET) based improved double tail dynamic comparator for ultra-low-power VLSI applications. *Springer Analog Integrated Circuits & Signal Processing* 101: 109–117. https://doi.org/10.1007/ s10470-019-01487-x.

Vidhyadharan, S., Dan, S. S., Abhay, S., Yadav, R., and Hariprasad, S. 2020a. Novel gate-overlap tunnel FET based innovative ultra-low-power ternary flash ADC. *Integration* 73: 101–113.

Vidhyadharan, S., Dan, S. S., Yadav, R., and Hariprasad, S. 2020b. A novel ultra-low-power gate overlap tunnel FET (GOTFET) dynamic adder. *International Journal of Electronics*, 1–19. Taylor & Francis eprint: https://doi.org/10.1080/00207217.2020. 1740800.

Vidhyadharan, S., Yadav, R., Hariprasad, S., and Dan, S. S. 2020c. An advanced adiabatic logic using Gate Overlap Tunnel FET (GOTFET) devices for ultra-low power VLSI sensor applications. *Analog Integrated Circuits and Signal Processing* 102 (1): 111–123.

Wei, L., Boeuf, F., Skotnicki, T., and Wong, H. P. 2011. Parasitic capacitances: Analytical models and impact on circuit-level performance. *IEEE Transactions on Electron Devices* 58 (5): 1361–1370.

Yadav, R., Dan, S. S., Vidhyadharan, S., and Hariprasad, S. 2020a. Innovative multi-threshold gate-overlap tunnel FET (GOTFET) devices for superior ultra-low power digital, ternary and analog circuits at 45-nm technology node. *Journal of Computational Electronics* 19 (1): 291–303.

Yadav, R., Dan, S. S., Vidhyadharan, S., and Hariprasad, S. 2020b. Suppression of ambipolar behavior and simultaneous improvement in RF performance of gate-overlap tunnel field effect transistor (GOTFET) devices. *Silicon*. https://doi.org/10.1007/s12633-020-00506-1

Yadav, R., Vidhyadharan, S., Akhilesh, G., Gupta, V., Ravi, A., and Dan, S. S. 2019a. Part I: Optimization of the tunnel FET device structure for achieving circuit performance better than the current standard 45 nm CMOS technology, in *The Physics of Semiconductor Devices*, 611–18, Springer International Publishing.

Yadav, R., Vidhyadharan, S., Shyam, A. K., Hirpara, M. P., Chaudhary, T., and Dan, S. S. 2019b. Novel low and high threshold TFET based NTI and PTI cells benchmarked with standard 45 nm CMOS technology for ternary logic applications, in *2019 32nd International Conference on VLSI Design and 2019 18th International Conference on Embedded Systems (VLSID)*, 419–24.

9 The Role of PMU for Frequency Stability in Hybrid Power Systems

Renuka Loka, Alivelu M. Parimi,
and P. Shambhu Prasad

CONTENTS

9.1 Introduction 165
9.2 Comparison Between SCADA and PMU Measurements for Frequency
    Stability in Hybrid Power Systems 166
9.3 PMU Installations and Research across the Globe 168
9.4 General Applications of PMU 169
9.5 PMUs in Frequency Stability 170
    9.5.1 Frequency Stability Applications of PMU 171
    9.5.2 Controller Aspects of PMUs in Frequency Stability 172
9.6 Future Scope of PMU Applications for Frequency Stability of Hybrid
    Power Systems 173
9.7 Conclusion 173
References 173

9.1 INTRODUCTION

The frequency stability of conventional power systems, consisting of predominantly thermal power generation, can be obtained by using primary frequency control and load frequency control [1]. In conventional power systems, by maintaining real power balance, the frequency can be maintained near nominal values [2]. Recently, many microgrids and large renewable generations have gradually become a part of the power system for which the suitability of traditional controls employed in conventional systems has to be analyzed.

When conventional power systems operate along with microgrids and large renewable generations, the dynamic behavior of the systems becomes unpredictable. This type of hybrid power system (HPS) needs to assess the system frequency on a real-time basis. This allows the controllers to take appropriate and fast-acting controls such that frequency is quickly restored to nominal values without sustained frequency oscillations.

In microgrids, frequency regulation is one of the important criteria for their seamless deployment [3]. Due to intrinsic features of microgrids, such as low inertia, high R/X ratio, uncertainties in renewable energy source, and so on, frequency stability becomes more challenging both in the grid-connected and islanded mode of operation [4]. The frequency instability that causes low-frequency oscillating modes, if not damped, may collapse the system and lead to blackout [5]. Damping of frequency oscillations requires different levels of optimally designed robust controllers to ensure frequency stability.

The effects of poor damping on system stability in power systems, which may result from improper control actions, can be well interpreted from a few events that occurred in the past. The US and Canadian grid failure in 2003, due to undamped frequency oscillations and failure of real-time contingency analysis is one of the examples [6]. Other instances of power outage include the Mexico and US grid failure in 2011[7], Indian grid failure in 2012 [8], and the US grid failure in 2017 [9]. These power outages cause huge economic loss to the system [10] and take time to restore.

The mentioned power outages could have been avoided if the system data had been acquired at a faster rate. Appropriate fast-acting control actions are needed to minimize the effect of such failures, which in turn calls for real-time dynamic measurements. Measurements in power systems are acquired using two main methods: supervisory control and data acquisition (SCADA) systems and phasor measurement units (PMUs).

Ultimately, real-time data acquisition for fast-acting controls is an important requirement. Therefore, the following section discusses determining the apt method for acquiring real-time data.

The organization of this chapter is as follows. A comparison on SCADA and PMU is presented in section 9.2. General applications of PMU measurements in HPSs is discussed in section 9.3. A review on applications of PMUs for power system frequency stability is provided in section 9.4. A brief discussion on controller aspects are given in section 9.5, followed by future scope in section 9.6 and the conclusion in section 9.7.

9.2 COMPARISON BETWEEN SCADA AND PMU MEASUREMENTS FOR FREQUENCY STABILITY IN HYBRID POWER SYSTEMS

In conventional power systems, SCADA systems were popularly employed in enhancing system reliability and improving performance [11]. They collect data from the sensors that are further processed to perform power system monitoring. The SCADA systems were moderately efficient in improving frequency regulation. These systems were often used at the secondary level of the hierarchical control strategy of power systems [12].

However, SCADA systems suffer from drawbacks such as low sampling rate (2–4 samples/cycle) and lack of synchronization [13]. Low sampling rate leads to failure in identifying low-frequency oscillations and poor monitoring especially for dynamic systems such as power systems. Moreover, SCADA systems are more vulnerable to security threats and high communication delays, which can disrupt the operation of power system controls [14], which may jeopardize the dynamic stability of the system.

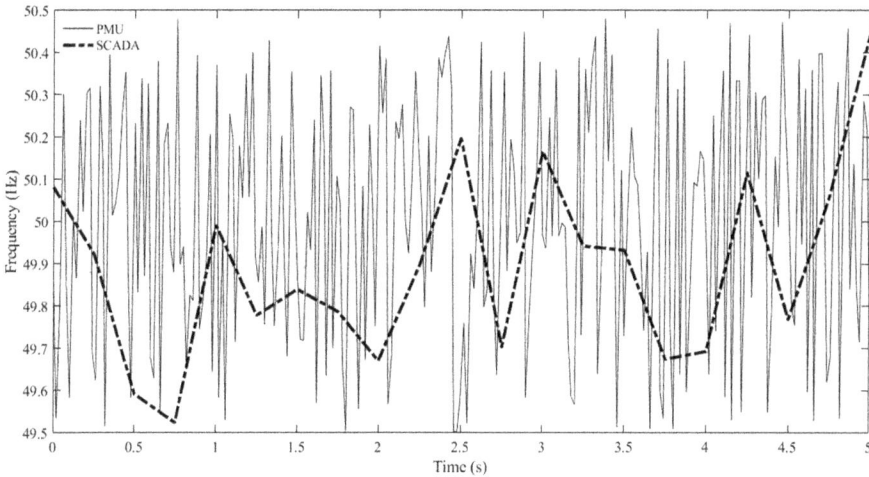

FIGURE 9.1. Dynamic frequency measurement data simulated for PMUs and SCADA systems.

Microgrids suffer from dynamic uncertainty in the generation and thus the controls need more precise measurements to achieve frequency stability. Real-time dynamic measurements with higher sampling rates are needed to provide solutions to dynamic stability problems in HPSs, for which PMUs are considered as the most efficient choice. The advancement of synchrophasor technology-based PMUs provided us with promising solutions to overcome the drawbacks of SCADA systems [15]. PMUs provide real-time dynamic measurements up to the level of 50–60 samples per second, which are found to help enhance power system dynamic stability [16].

An example of depicting frequency measurement data, which is obtained from SCADA systems and PMU measurements, has been plotted as shown in Figure 9.1. It can be observed that PMU measurements are more accurate and defined even for shorter time intervals, which are as less as 20 ms when compared with the measurements that are obtained from SCADA systems. Considering the PMU measurement data that can be seen from the graph, loss of important information that may be crucial for HPS controllers to maintain frequency stability is significantly less than that of SCADA.

On the other hand, SCADA measurements provide less accurate measurements at various points of time. It can be seen that the dynamics obtained from PMU measurements cannot be obtained from SCADA. The loss of system dynamics results in poor control actions, which may sometimes result in unanticipated dynamic stability problems of HPSs.

Another added advantage of a PMU is that it processes the signal and synchronizes the data with a Global Positioning System (GPS). Unlike SCADA systems, the synchronized time-stamping of data helps in identifying any frequency event occurring at a particular time frame, which is critical for frequency monitoring purposes. As per IEEE Standards, synchrophasor technology allows frequency as low as 0.005 Hz and the maximum rate of change of 0.4 Hz [17]. The data quality in

PMUs can be enhanced using various techniques and offer more secure gateways when compared to SCADA systems [18]. PMUs use cellular networks for faster communication, which can significantly reduce the communication delay problems [19], which makes PMUs superior to that of SCADA systems. Based on the advantages, PMU applications related to the conventional power system and microgrids have provided substantial information on the utilization of PMU data in HPSs [20]–[25].

It can be concluded that to perform robust frequency control of HPSs, dynamic measurements obtained from PMUs are superior to that of steady-state measurements obtained from SCADA systems. PMUs installed at various locations across the globe are already in operation and many future installations have also been planned, whose details are given in the following section.

9.3   PMU INSTALLATIONS AND RESEARCH ACROSS THE GLOBE

Given the importance of PMU measurements, power grids in different regions across the globe have PMUs installed and operating at various strategic locations of the grid [26]. As a part of the North American Synchrophasor Initiative (NASPI) [27], more than 1,700 PMUs have been installed. NASPI recognized the significance of PMUs and developed solutions to various applications of PMUs such as phase angle monitoring, in collaboration with different organizations including Power System Operations Corporation Ltd, India (POSOCO) [28].

In, India, around 62 PMUs have been installed in the Northern, Southern, and Western regions of the power system network [24], and a plan to install more PMUs all over the country for various applications such as wide area measurement systems (WAMS) has been initiated [29, 30]. Around 470 PMUs have been installed and many projects based on synchrophasor technology have been initiated in the Pennsylvania-New Jersey-Maryland Interconnection (PJM) [31].

Frequency quality was assessed using the PMU measurements in the Nordic grid [32] and around five PMUs were installed with a proposal to install more units [26]. China has installed around 3,000 PMUs, while Russia has installed 45 PMUs and Europe has installed around 30 PMUs for real-time data monitoring [26].

The number of installations across the globe is depicted in Figure 9.2 as per the current known status and a steady increase is expected in the number of installations over the years owing to the trend toward smart grids in various parts of the world.

Many countries have initiated research projects in different areas of synchrophasor technology and its applications. To name a few, the frequency monitoring network FNET/GridEye in North America [33], NASPI, synchrophasor as a part of energy innovations in PJM [34], Unified Real-Time Dynamic State Measurement (URTDSM) as a part of National Smart Grid Mission (NSGM), India [35], Smart Transmission Grids Vision for Europe who has noted the importance of PMU measurements in dynamic state estimation in its research interests [36], Statnett projects in wide-area monitoring protection and control such as synchrophasor-based automatic real-time control (SPARC), and the Nordic Early Warning Early Prevention System (NEWEPS) in Norway [37] are some of the active research projects in various parts.

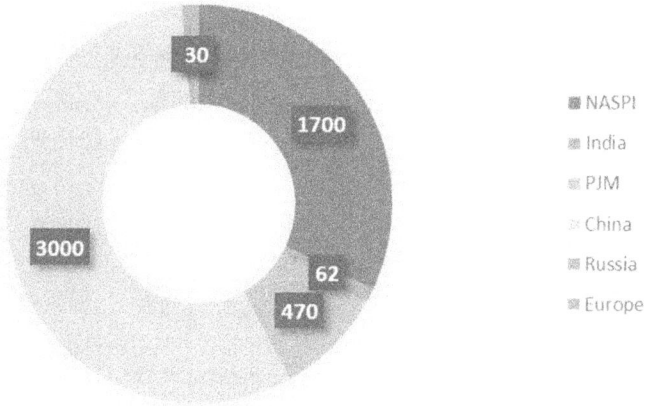

FIGURE 9.2 Number of PMUs installed across various parts of the world.

To successfully operate HPSs, PMUs can be installed at optimal locations to get the required measurement data. For this purpose, future PMU installations and application research are given great importance.

Although PMUs have been widely installed and used across the globe, more research needs to be done for effectively utilizing the PMU data in any specific HPS application. Previous studies related to conventional power systems as well as microgrids discussed in the following section provide a brief idea about the numerous applications of PMUs in the HPS.

9.4 GENERAL APPLICATIONS OF PMU

PMU measurements can be used for different power system applications at various stages of generation, transmission, and distribution. The broad applications of PMUs in modern HPSs including microgrids (MG) have been described in Table 9.1. based on the literature available. Major contributions of various works about the individual application have been listed out. A few works have provided solutions for both frequency stability and voltage stability problems. PMU data can also be used for other applications such as event detection and control, security of power systems, and so on. It is to be noted that applications other than frequency stability are provided for the general understanding of the reader, whose detailed discussion is beyond the scope of this chapter.

Frequency stability, among all other PMU applications, can be identified as the challenging problem in HPSs because additional control loops are involved. Moreover, because of RESs, system frequency changes more abruptly unlike conventional power systems, where frequency measurements from PMUs can be found useful. The following section provides the applications of PMUs in frequency stability

TABLE 9.1

Broad Applications of PMUs in Modern Power Systems

Sl No.	PMU Application	Major Contributions	References
1	Voltage Stability	Reporting rate effects on secondary control, voltage collapse proximity index calculation, coordination of resources	[38]–[40]
2	Monitoring and Protection	Software models for PMUs, automation of PMU data-driven microgrid monitoring and dynamics, standards implementation for synchronization	[41]–[43] [44]–[48]
3	Resilience	Distributed control	[49]
4	State Estimation	Online estimation techniques, real-time data processing	[50]–[52]
5	Power Quality	Harmonic calculation and tracking	[44]
6	Frequency Stability	Frequency monitoring, inter-area oscillations, rate of change of frequency, effect of communication latency aspects	[38], [40], [46], [53]–[62]

9.5 PMUS IN FREQUENCY STABILITY

Frequency stability in HPSs has both aspects of conventional power systems and microgrids. The system is prone to greater imbalances in active and reactive powers, which may result in frequency stability problems [1]. The penetration of RESs cannot produce power without having fluctuations throughout the day [1, 59] and these power deviations can greatly influence the frequency deviations in modern power systems. Such systems including microgrids suffer from the following challenges:

- Large power swings associated with the intermittent nature of RESs
- Large frequency deviations because of large disturbances in load or generation
- Need for real-time measurements to ensure proper frequency controls

It is the need of the hour to provide solutions to these challenges without causing any stability issues in the HPSs. Synchrophasor measurements can be used for obtaining real-time dynamic frequency measurements, which are essential for obtaining the system frequency response to maintain frequency stability. The controllers in HPSs are provided with different types of inputs depending on the controller design.

Regarding the successful operation of controllers, the data related to different input parameters are obtained from PMUs. This data is utilized for various PMU applications in obtaining the frequency stability of HPSs. The different types of PMUs and their applications to frequency stability are discussed in the following subsection.

9.5.1 FREQUENCY STABILITY APPLICATIONS OF PMU

Depending on the location and their application, PMUs can be classified into four different categories. The types of PMUs along with their respective roles in the frequency stability of the modern power system are depicted in Figure 9.3. Different major applications of these PMUs are listed about frequency stability.

PMUs are installed at generation or transmission substations to get frequency measurements to understand the wide-area dynamics [37] as well as inter-area oscillations. Distribution level phasor measurement units (DPMU) [63] and microphasor measurement units (μPMU) [64] are used at distribution levels for obtaining frequency stability. DPMU and μPMU can also be used for microgrid frequency stability applications. FNET/GridEye has gained a lot of attention for its ease of installation and usage, less cost, and better frequency monitoring with a large potential for research in various applications [65]. FNET can be used at either the transmission level or at the distribution level. In FNET, frequency disturbance recorders (FDR) are employed for monitoring purposes. FDRs are a special type of PMU that are more economical than the traditional PMU. Using FNET, event detection can be performed and the location of the event can also be obtained apart from continuous monitoring of frequency stability parameters such as area control error (ACE) and rate of change of frequency (RoCoF) [33].

In implementing these applications for frequency stability [15] of modern power systems, we require mainly communication infrastructure [19] and efficient data processing methods [66]. Communication latency was identified as one of the problems in efficiently utilizing PMU measurements [46] for distributed primary

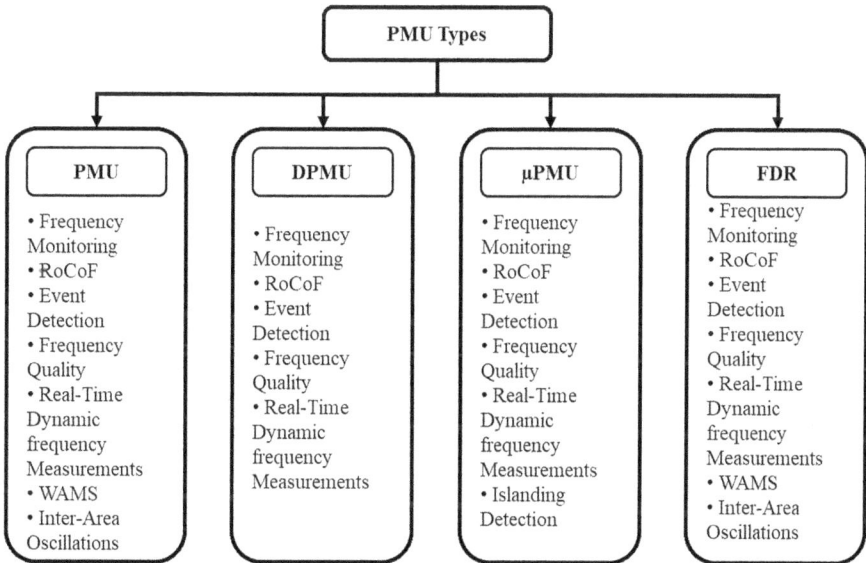

FIGURE 9.3   Role of different types of PMUs in frequency stability.

control where fast communication-based wide-area networks [67] can provide solutions.

Frequency monitoring and event detection are made more complex because of the strong coupling between active and reactive powers [68] in active distribution networks (ADN), which are growing components of modern power systems. Thus, frequency monitoring and RoCoF monitoring become a significant part of the dynamic stability of power systems.

Data obtained from wide-area measurements have to be carefully analyzed [69] for various reasons such as event detection, post-event scenario reconstruction, and to take prompt actions to avoid unnecessary tripping of relays, involuntary load shedding, or sometimes blackouts. The latest data processing-related areas such as big data [70] and intelligent tools such as deep learning [71] are enabling technologies for PMU frequency stability applications.

The measurement data obtained from PMUs on different frequency parameters are communicated with the controllers employed in HPSs. This data helps design optimal and robust frequency controllers suitable for HPSs. Subsequently, controller aspects are discussed in the next subsection.

9.5.2 CONTROLLER ASPECTS OF PMUs IN FREQUENCY STABILITY

Different types of controllers are employed for frequency stability in microgrids and conventional power systems. Droop control, distributed control, agent-based control, state feedback control, linear quadratic regulators, and hierarchical timescales-based control [72] are some examples. Each of these controls requires dynamic frequency measurements for their operation that can be obtained using different types of application-specific PMUs.

Local measurements are needed to establish a sparse communication-based distributed frequency control in microgrids [49] for which DPMUs or μPMUs can provide the required frequency and real and reactive power measurements.

Emergency frequency controls operate based on predefined frequency deviation thresholds or by considering RoCoF limits as per the frequency standards [60]. PMUs can provide the required data for the emergency controllers to operate effectively. As a part of emergency controls, generator tripping can be initiated depending on the over-frequency condition or because of the inter-area oscillations [1], whose tripping actions can be initiated manually by monitoring the frequency data from PMUs in WAMS applications or automatically by communicating the PMU measurement-based frequency parameters to a trip relay [2].

For wide-area control, different controllers such as local controllers, regional controllers, and central controllers are employed, which act through the aggregated measurements obtained from local PMUs and regional PMUs to achieve fast-acting controls for frequency stability of HPSs [57].

The PMU data can be processed, aggregated, or tuned depending on the requirement of the controller employed for achieving optimal frequency control. Based on the drawbacks of existing methods of PMU applications in obtaining frequency stability, the future scope is discussed in the following section.

9.6 FUTURE SCOPE OF PMU APPLICATIONS FOR FREQUENCY STABILITY OF HYBRID POWER SYSTEMS

Considering the role of PMUs in frequency stability and the controller aspects, the future scope of using PMUs in frequency stability applications is as follows:

1. FDRs have not been given sufficient attention for microgrid frequency applications. FDRs in microgrids can be a potential research area, especially for islanding detection and RoCoF monitoring applications.
2. Communication latency associated with PMUs is one of the problems in providing distributed primary control in microgrids or for distributed energy resources in modern power systems. This calls for the attention of the research community, either to develop communication latency resilient control architectures and/or deploy high-speed communication networks.
3. An increase in the number of microgrids and ADNs alters the conventional power system dynamics, which indicates a strong need for maintaining frequency stability through visualization of the grid dynamics using synchrophasor/FDR data.
4. PMU data can help obtain not only situational awareness but alert the operator for a possible frequency event, which is possible through advanced technologies such as big data and deep learning, which are newly emerging research areas in synchrophasor data applications.

The controllers designed for frequency stability need to overcome the drawbacks of certain PMU aspects such as high cost of monitoring, latency, and effective data handling, which sets the direction for PMU application-oriented future research works.

9.7 CONCLUSION

In this chapter, the importance of PMUs in the frequency stability of HPSs was discussed. A comparison of PMUs with SCADA systems on different parameters such as sampling rate, data quality, security, and latency has been done to indicate the importance of PMUs in designing frequency controllers for the HPSs. The current status of various PMU installations across the globe has also been provided. This chapter also discussed the generic applications of PMUs along with different types of PMUs and their application related to frequency stability. Various insights on different controller aspects for frequency stability through PMU applications were presented. An attempt has been made to bring out the future scope of PMU applications in HPSs concerning frequency stability.

REFERENCES

1. H. Bevrani, *Robust power system frequency control.* Cham: Springer International Publishing, 2014.
2. P. Kundur, *Power system stability and control* McGraw-Hill, Inc., New York, US, 1994.

3. "Grid-connected renewable energy sources," in *Microgrid dynamics and control*, John Wiley & Sons, Ltd, 2017, pp. 1–68.

4. "Microgrid technology and engineering application–1st edition," June 28, 2020. https://www.elsevier.com/books/microgrid-technology-and-engineering-application/li/978-0-12-803598-6.

5. H. Haes Alhelou, M. Hamedani-Golshan, T. Njenda, and P. Siano, "A survey on power system blackout and cascading events: Research motivations and challenges," *Energies*, vol. 12, no. 4, Art. no. 4, Feb. 2019, doi: 10.3390/en12040682.

6. P. Pourbeik, P. S. Kundur, and C. W. Taylor, "The anatomy of a power grid blackout— Root causes and dynamics of recent major blackouts," *IEEE Power Energy Magazine*, vol. 4, no. 5, Art. no. 5, Sep. 2006, doi: 10.1109/MPAE.2006.1687814.

7. E. C. Portante, S. F. Folga, J. A. Kavicky, and L. T. Malone, "Simulation of the September 8, 2011, San Diego blackout," in *Proceedings of the Winter Simulation Conference 2014*, Dec. 2014, pp. 1527–1538, doi: 10.1109/WSC.2014.7020005.

8. Y. Tang, G. Bu, and J. Yi, "Analysis and lessons of the blackout in Indian power grid on July 30 and 31, 2012," vol. 32, pp. 167–174, Sep. 2012.

9. I. Dobson and D. E. Newman, "Cascading blackout overall structure and some implications for sampling and mitigation," *International Journal of Electrical Power Energy Systems*, vol. 86, pp. 29–32, Mar. 2017, doi: 10.1016/j.ijepes.2016.09.006.

10. A. Rose, G. Oladosu, and S.-Y. Liao, "Business interruption impacts of a terrorist attack on the electric power system of Los Angeles: Customer resilience to a total blackout," *Risk Analysis*, vol. 27, no. 3, Art. no. 3, 2007, doi: 10.1111/j.1539-6924.2007.00912.x.

11. F. M. Enescu and N. Bizon, "SCADA applications for electric power system," in *Reactive Power Control in AC Power Systems: Fundamentals and Current Issues*, N. Mahdavi Tabatabaei, A. Jafari Aghbolaghi, N. Bizon, and F. Blaabjerg, Eds. Cham: Springer International Publishing, 2017, pp. 561–609.

12. H. Bevrani, "Frequency response characteristics and dynamic performance," in *Robust Power System Frequency Control*, H. Bevrani, Ed. Cham: Springer International Publishing, 2014, pp. 49–69.

13. P. Novák, R. Šindelář, and R. Mordinyi, "Integration framework for simulations and SCADA systems," *Simulation Modelling Practice and Theory*, vol. 47, pp. 121–140, Sep. 2014, doi: 10.1016/j.simpat.2014.05.010.

14. D. Pliatsios, P. Sarigiannidis, T. Lagkas, and A. G. Sarigiannidis, "A survey on SCADA systems: Secure protocols, incidents, threats and tactics," *IEEE Communications Survey Tutor*, pp. 1–1, 2020, doi: 10.1109/COMST.2020.2987688.

15. M. Hojabri, U. Dersch, A. Papaemmanouil, and P. Bosshart, "A comprehensive survey on phasor measurement unit applications in distribution systems," *Energies*, vol. 12, no. 23, Art. no. 23, Nov. 2019, doi: 10.3390/en12234552.

16. G. C. Patil and A. G. Thosar, "Application of synchrophasor measurements using PMU for modern power systems monitoring and control," in *2017 International Conference on Computation of Power, Energy Information and Commuincation (ICCPEIC)*, Melmaruvathur, Mar. 2017, pp. 754–760, doi: 10.1109/ICCPEIC.2017.8290464.

17. "IEEE standard for synchrophasor measurements for power systems–Amendment 1: Modification of selected performance requirements," *IEEE Std C37118la-2014 Amend. IEEE Std C37118l-2011*, pp. 1–25, Apr. 2014, doi: 10.1109/IEEESTD.2014.6804630.

18. A. Sundararajan, T. Khan, A. Moghadasi, and A. I. Sarwat, "A survey on synchrophasor data quality and cybersecurity challenges, and evaluation of their interdependencies," *Journal of Modern Power Systems and Clean Energy*, vol. 7, no. 3, Art. no. 3, May 2019, doi: 10.1007/s40565-018-0473-6.

19. B. Appasani and D. K. Mohanta, "A review on synchrophasor communication system: communication technologies, standards and applications," *Protection and*

Control of Modern Power Systems, vol. 3, no. 1, Art. no. 1, Dec. 2018, doi: 10.1186/s41601-018-0110-4.

20. N. Mekki, F. Derbel, L. Krichen, and F. Strakosch, "PMU deployment for state estimation in smart grids," in *2016 13th International Multi-Conference on Systems, Signals Devices (SSD)*, Mar. 2016, pp. 211–216, doi: 10.1109/SSD.2016.7473763.

21. T. J. Overbye and J. D. Weber, "The smart grid and PMUs: Operational challenges and opportunities," in *IEEE PES General Meeting*, Jul. 2010, pp. 1–5, doi: 10.1109/PES.2010.5589269.

22. M. Farsadi, H. Golahmadi, and H. Shojaei, "Phasor measurement unit (PMU) allocation in power system with different algorithms," in *2009 International Conference on Electrical and Electronics Engineering—ELECO 2009*, Nov. 2009, p. I-396–I–400, doi: 10.1109/ELECO.2009.5355226.

23. M. U. USMAN and M. O. FARUQUE, "Applications of synchrophasor technologies in power systems," *Journal of Modern Power Systems and Clean Energy*, vol. 7, no. 2, Art. no. 2, Mar. 2019, doi: 10.1007/s40565-018-0455-8.

24. H. Lee, Tushar B. Cui, A. Mallikeswaran, P. Banerjee, and A. K. Srivastava, "A review of synchrophasor applications in smart electric grid," WIREs Energy and Environment–Wiley Online Library, Jun. 29, 2020. https://onlinelibrary.wiley.com/doi/abs/10.1002/wene.223.

25. "Electric power system measurement—An overview," *ScienceDirect Topics*, Jun. 29, 2020. https://www.sciencedirect.com/topics/engineering/electric-power-system-measurement.

26. A. G. Phadke and T. Bi, "Phasor measurement units, WAMS, and their applications in protection and control of power systems," *Journal of Modern Power Systems and Clean Energy*, vol. 6, no. 4, pp. 619–629, Jul. 2018, doi: 10.1007/s40565-018-0423-3.

27. "Synchrophasor technology fact sheet," *NASPI*, Oct. 2014

28. "Using synchrophasor data for phase angle monitoring", *NASPI Control Room Solutions Task Team Paper*, May 2016.

29. "Unified real time dynamic state measurement", *A Report by Power Grid Corporation of India LTD*, Gurgaon, Feb 2012.

30. P. K. Agarwal, V. K. Agarwal, and H. Rathour, "Application of PMU-based information in the Indian power system," *International Journal of Emergency Electrical Power Systems*, vol. 14, no. 1, pp. 79–86, May 2013, doi: 10.1515/ijeeps-2013-0019.

31. "A resilient grid," *IEEE Power Energy Magazine*, vol. 18, no. 4, pp. 1–1, Jul. 2020, doi: 10.1109/MPE.2020.2967914.

32. Zhao Xu, J. Ostergaard, M. Togeby, and F. R. Isleifsson, "Evaluating frequency quality of Nordic system using PMU data," in *2008 IEEE Power and Energy Society General Meeting - Conversion and Delivery of Electrical Energy in the 21st Century*, Pittsburgh, PA, Jul. 2008, pp. 1–5, doi: 10.1109/PES.2008.4596468.

33. Department of Electrical Engineering and Computer Science, the University of Tennessee et al., "Recent developments of FNET/GridEye—A situational awareness tool for smart grid," *CSEE Journal of Power Energy Systems*, vol. 2, no. 3, pp. 19–27, Sep. 2016, doi: 10.17775/CSEEJPES.2016.00031.

34. "PJM learning center—synchrophasors." https://learn.pjm.com/energy-innovations/synchrophasors.aspx.

35. I. S. Jha, S. Sen, and R. Kumar, "Smart grid development in India—A case study," http://www.iitk.ac.in/npsc/Papers/NPSC2014/1569993451.pdf.

36. L. Vanfretti, D. Van Hertem, and J. O. Gjerde, "Smart transmission grids vision for Europe: Towards a realistic research agenda," in *Smart Grid Applications and Developments*, D. Mah, P. Hills, V. O. K. Li, and R. Balme, Eds. London: Springer London, 2014, pp. 185–220.

37. "Wide area monitoring protection and control," *Statnett.* https://www.stat-nett.no/en/about-statnett/research-and-development/our-prioritised-projects/wide-area-monitoring-protection-and-control.

38. E. De Din, G. Lipari, A. Angioni, F. Ponci, and A. Monti, "Effect of the reporting rate of synchrophasor measurements for distributed secondary control of AC microgrid," in *2017 IEEE International Workshop on Measurement and Networking (M N)*, Sep. 2017, pp. 1–6, doi: 10.1109/IWMN.2017.8078378.

39. R. R. Micky, R. Lakshmi, R. Sunitha, and S. Ashok, "Assessment of voltage stability in microgrid," in *2016 International Conference on Electrical, Electronics, and Optimization Techniques (ICEEOT)*, Chennai, India, Mar. 2016, pp. 1268–1273, doi: 10.1109/ICEEOT.2016.7754887.

40. M. A. Aftab, S. M. Suhail Hussain, V. Kumar, T. S. Ustun, and I. Ali, "IEC 61850 communication assisted synchronization strategy for microgrids," in *2018 IEEE 13th International Conference on Industrial and Information Systems (ICIIS)*, Rupnagar, India, Dec. 2018, pp. 401–406, doi: 10.1109/ICIINFS.2018.8721427.

41. Jae-Duck Lee, Seong-Joon Lee, Jeong-Hyo Bae, and Dae-Yun Kwon, "The PMU interface using IEC 61850," in *2013 International Conference on ICT Convergence (ICTC)*, JEJU ISLAND, Korea (South), Oct. 2013, pp. 1125–1128, doi: 10.1109/ICTC.2013.6675573.

42. M. Sanduleac, C. Chimirel, and M. Paun, "PMU Orchestrator as a solution for managing microgrid monitoring with 5G communication," in *2019 54th International Universities Power Engineering Conference (UPEC)*, Bucharest, Romania, Sep. 2019, pp. 1–3, doi: 10.1109/UPEC.2019.8893555.

43. K. V. S. Baba, S. R. Narasimhan, N. L. Jain, A. Singh, R. Shukla, and A. Gupta, "Synchrophasor based real time monitoring of grid events in Indian power system," in *2016 IEEE International Conference on Power System Technology (POWERCON)*, Wollongong, Australia, Sep. 2016, pp. 1–5, doi: 10.1109/POWERCON.2016.7753936.

44. M. Chakir, I. Kamwa, and H. Le Huy, "Extended C37.118.1 PMU algorithms for joint tracking of fundamental and harmonic phasors in stressed power systems and microgrids," *IEEE Transactions on Power Delivery*, vol. 29, no. 3, Art. no. 3, Jun. 2014, doi: 10.1109/TPWRD.2014.2318024.

45. S. Kumar, N. Das, and S. Islam, "Performance monitoring of a PMU in a microgrid environment based on IEC 61850-90-5," in *2016 Australasian Universities Power Engineering Conference (AUPEC)*, Brisbane, Australia, Sep. 2016, pp. 1–5, doi: 10.1109/AUPEC.2016.7749356.

46. S. Liu and X. Wang, "Analysis of frequency control in microgrids with multiple phasor measurement unit delays," in *2019 7th Workshop on Modeling and Simulation of Cyber-Physical Energy Systems (MSCPES)*, Apr. 2019, pp. 1–5, doi: 10.1109/MSCPES.2019.8738800.

47. N. K. Sharma and S. R. Samantaray, "Assessment of PMU-based wide-area angle criterion for fault detection in microgrid," *IET Generation Transmission Distribution*, vol. 13, no. 19, Art. no. 19, Oct. 2019, doi: 10.1049/iet-gtd.2019.0027.

48. Y. Bansal and R. Sodhi, "A half-cycle fast discrete orthonormal S transform based protection-class μPMU," *IEEE Transactions on Instrumentation and Measurement*, pp. 1–1, 2020, doi: 10.1109/TIM.2020.2980339.

49. J. Chen and Q. Zhu, "A game-theoretic framework for resilient and distributed generation control of renewable energies in microgrids," *IEEE Transactions on Smart Grid*, vol. 8, no. 1, pp. 285–295, Jan. 2017, doi: 10.1109/TSG.2016.2598771.

50. C. Lin, W. Wu, and Y. Guo, "Decentralized robust state estimation of active distribution grids incorporating microgrids based on PMU measurements," *IEEE Transactions on Smart Grid*, vol. 11, no. 1, Art. no. 1, Jan. 2020, doi: 10.1109/TSG.2019.2937162.

51. I. Ali, M. Huzaifa, O. Ullah, M. A. Aftab, and M. Z. Anis, "Real time microgrid state estimation using phasor measurement unit," in *2019 International Conference on Power Electronics, Control and Automation (ICPECA)*, Nov. 2019, pp. 1–6, doi: 10.1109/ICPECA47973.2019.8975460.
52. S. V. Hareesh and K. Shanti Swarup, "Dynamic state estimation of synchronous generator based distributed energy resource in autonomous microgrid," in *2019 8th International Conference on Power Systems (ICPS)*, Jaipur, India, Dec. 2019, pp. 1–6, doi: 10.1109/ICPS48983.2019.9067347.
53. O. Antoine, P. Janssen, Q. Jossen, and J.-C. Maun, "A laboratory microgrid for studying grid operations with PMUs," in *2013 IEEE Power & Energy Society General Meeting*, Vancouver, BC, 2013, pp. 1–5, doi: 10.1109/PESMG.2013.6672891.
54. C. Jamroen, N. Kesorn, A. Pichetjamroen, and S. Dechanupaprittha, "Impact of communication delays on PEVs charging power control for frequency stabilization in remote microgrid," in *2017 IEEE PES Asia-Pacific Power and Energy Engineering Conference (APPEEC)*, Bangalore, Nov. 2017, pp. 1–6, doi: 10.1109/APPEEC.2017.8308954.
55. Y. Yan, D. Shi, D. Bian, B. Huang, Z. Yi, and Z. Wang, "Small-signal stability analysis and performance evaluation of microgrids under distributed control," *IEEE Transactions on Smart Grid*, vol. 10, no. 5, Art. no. 5, Sep. 2019, doi: 10.1109/TSG.2018.2869566.
56. A. Srivastava and S. K. Parida, "Frequency and voltage data processing based feeder protection in medium voltage microgrid," in *2019 IEEE PES Innovative Smart Grid Technologies Europe (ISGT-Europe)*, Bucharest, Romania, Sep. 2019, pp. 1–5, doi: 10.1109/ISGTEurope.2019.8905705.
57. Q. Hong et al., "Design and validation of a wide area monitoring and control system for fast frequency response," *IEEE Transactions on Smart Grid*, vol. 11, no. 4, Art. no. 4, Jul. 2020, doi: 10.1109/TSG.2019.2963796.
58. C. Thilakarathne, L. Meegahapola, and N. Fernando, "Improved synchrophasor models for power system dynamic stability evaluation based on IEEE C37.118.1 reference architecture," *IEEE Transactions on Instruments and Measurement*, vol. 66, no. 11, Art. no. 11, Nov. 2017, doi: 10.1109/TIM.2017.2714558.
59. P. Mahish and A. K. Pradhan, "Synchrophasor data based distributed droop control in grid integrated wind farms to improve primary frequency response," in *2019 8th International Conference on Power Systems (ICPS)*, Jaipur, India, Dec. 2019, pp. 1–5, doi: 10.1109/ICPS48983.2019.9067554.
60. J. Tang, J. Liu, F. Ponci, and A. Monti, "Adaptive load shedding based on combined frequency and voltage stability assessment using synchrophasor measurements," *IEEE Transactions of Power Systems*, vol. 28, no. 2, Art. no. 2, May 2013, doi: 10.1109/TPWRS.2013.2241794.
61. P. Y. Kovalenko, M. D. Senyuk, V. I. Mukhin, and A. A. Korelina, "Synchronous frequency calculation based on synchrophasor measurements," in *2019 International Conference on Electrotechnical Complexes and Systems (ICOECS)*, Ufa, Russia, Oct. 2019, pp. 1–4, doi: 10.1109/ICOECS46375.2019.8949985.
62. L. Lugnani, D. Dotta, J. M. F. Ferreira, I. C. Decker, and J. H. Chow, "Frequency response estimation following large disturbances using synchrophasors," in *2018 IEEE Power Energy Society General Meeting (PESGM)*, Aug. 2018, pp. 1–5, doi: 10.1109/PESGM.2018.8586220.
63. Y. R. Rodrigues, M. Abdelaziz, and L. Wang, "D-PMU based secondary frequency control for islanded microgrids," *IEEE Transactions on Smart Grid*, vol. 11, no. 1, Art. no. 1, Jan. 2020, doi: 10.1109/TSG.2019.2919123.
64. L.-A. Lee and V. Centeno, "Comparison of μPMU and PMU," in *2018 Clemson University Power Systems Conference (PSC)*, Charleston, SC, USA, Sep. 2018, pp. 1–6, doi: 10.1109/PSC.2018.8664037.
65. "FNET Server Web Display." http://fnetpublic.utk.edu.

66. Z. Yang, H. Liu, T. Bi, Q. Yang, and A. Xue, "A PMU data recovering method based on preferred selection strategy," *Global Energy Interconnections*, vol. 1, no. 1, Art. no. 1, Jan. 2018, doi: 10.14171/j.2096-5117.gei.2018.01.008.
67. Q.-D. Ho and T. Le-Ngoc, "Smart grid communications networks: Wireless technologies, protocols, issues, and standards," in *Handbook of Green Information and Communication Systems*, Elsevier, 2013, pp. 115–146.
68. S. Fahad, A. Goudarzi, and J. Xiang, "Demand management of active distribution network using coordination of virtual synchronous generators," *IEEE Transactions on Sustainable Energy*, pp. 1–1, 2020, doi: 10.1109/TSTE.2020.2990917.
69. J. Chai et al., "Wide-area measurement data analytics using FNET/GridEye: A review," in *2016 Power Systems Computation Conference (PSCC)*, Jun. 2016, pp. 1–6, doi: 10.1109/PSCC.2016.7540946.
70. L. G. Meegahapola, S. Bu, D. P. Wadduwage, C. Y. Chung, and X. Yu, "Review on oscillatory stability in power grids with renewable energy sources: monitoring, analysis, and control using synchrophasor technology," *IEEE Transactions on Industrial Electronics*, pp. 1–1, 2020, doi: 10.1109/TIE.2020.2965455.
71. Y. Zhu, C. Liu, and K. Sun, "Image embedding of PMU data for deep learning towards transient disturbance classification," in *2018 IEEE International Conference on Energy Internet (ICEI)*, May 2018, pp. 169–174, doi: 10.1109/ICEI.2018.00038.
72. A. K. Sahoo, K. Mahmud, M. Crittenden, J. Ravishankar, S. Padmanaban, and F. Blaabjerg, "Communication-less primary and secondary control in inverter-interfaced AC microgrid: An overview," *IEEE Journal of Emerging and Selected Topics in Power Electronics*, pp. 1–1, 2020, doi: 10.1109/JESTPE.2020.2974046.

10 A Modular Zigbee-Based IoT Platform for Reliable Health Monitoring of Industrial Machines Using ReFSA

Amar Kumar Verma, Jaju Vedant Vinod, and Radhika Sudha

CONTENTS

10.1 Introduction 179
10.2 Methodology 180
10.3 Zigbee-Based Wireless Communication 181
    10.3.1 Zigbee Coordinator 181
    10.3.2 Zigbee Router 181
    10.3.3 End Device 181
10.4 Results and Discussion 181
    10.4.1 Confusion Matrix for Various Sets of Statistical Features 182
    10.4.2 Three-Statistical Features 182
    10.4.3 Six-Statistical Features 182
    10.4.4 Nine-Statistical Features 186
10.5 Conclusions 186
Acknowledgments 186
References 186

10.1 INTRODUCTION

Fault detection and diagnosis for reliable monitoring of industrial machinery's health condition is a growing technology in the world today. Integrating knowledge-based artificial intelligence (AI) techniques with current technologies will automate the entire process for detecting industrial faults (Verma et al. 2020; Radhika et al. 2010). Industrial machinery is an operating necessity for every manufacturing sector and is vulnerable to failure due to its continuous operation (Amar et al. 2020). The stator and rotor together introduce 45% of the total faults in an industrial machine (Ranjan et al. 2019). The EPRI and IEEE-IAS survey reveals that stator-related faults comprise 37% of overall failure, so current research focuses on stator-related faults that primarily identify stator-related inter-turn faults in their initial stage (Verma et al. 2019). Model, signature, and knowledge-based techniques are widely used in any industrial machine to identify failures (Vamsi et al. 2018). The signature-based approach uses various signals such as vibration, voltage, temperature, electromagnetic radiation, and current from industrial machines to identify the internal failures (Verma et al. 2018). Model-based methods are highly susceptible to variations in motor parameters. Although signature-based strategies can achieve high performance, the diagnosis of the short circuit between turns becomes challenging when dealing with developing faults. Hence, the diagnostic process based on the AI technique and the extraction of signatures will be reliable in the detection of stator faults and in the evaluation of severity (Angelo et al. 2009; Lie et al. 2018; Basu et al. 2010).

10.2 METHODOLOGY

Remote fault signature analyzer (ReFSA) hardware setup has been developed using AI computing technologies along with motor current signature analysis (MCSA). An ReFSA consists primarily of modular and virtual instruments such as a hall-effect transducer with its power circuitry, data acquisition module, and LabVIEW to remotely investigate and identify a faulty machine and predict its remaining useful life expectancy, thereby augmenting industry production efficiency. The stator current signature was obtained and analyzed from both healthy and faulty machines using computer-aided ReFSA to identify the machines' condition. A computer-aided data acquisition module uses networking features, high-level computer processing power that offers efficient and cost-effective measurement solutions, and is used to transfer data from sensors to the remote location wirelessly. The entire workflow of the ReFSA hardware setup is shown in Figure 10.1.

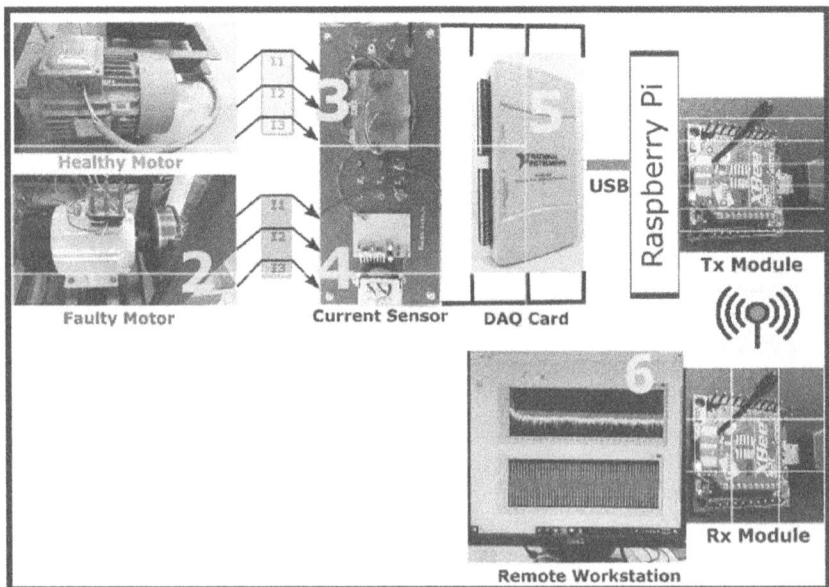

FIGURE 10.1　Block diagram for ReFSA workflow.

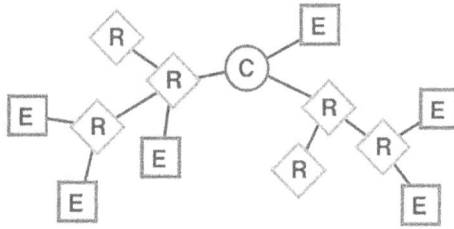

FIGURE 10.2   Mesh network topology of a Zigbee device.

10.3   ZIGBEE-BASED WIRELESS COMMUNICATION

A Zigbee-based communication system is a low-cost and low-power-consumption technology that can establish up to 1,500 meters of communication wirelessly at remote locations (Ali et al. 2019; Kuzminykh et al. 2017; Ramya et al. 2011; Ndih et al. 2016). The best data rate for periodic or intermediate two way is a 250-kbps data transfer between controllers and sensors. Bluetooth and WiFi-based technologies are the other wireless technologies that are most commonly available, capable of transferring data at a much faster rate but have a range of 100 meters and are also costly and consume more power (Sehawan et al. 2015). A Zigbee network structure mainly comprises three separate device types as shown in Figure 10.2.

10.3.1   ZIGBEE COORDINATOR

A Zigbee coordinator forms the network root. The mandatory node for all Zigbee networks has all the network information, including the keys, and acts as a trust center that plays a crucial role in security.

10.3.2   ZIGBEE ROUTER

This node can run an application function and also act as a relay station for other networked Zigbee devices.

10.3.3   END DEVICE

End devices have a minimal job of communicating with the parent nodes in such a way as to save battery power. The network is currently being implemented with only two nodes: one as the ground station coordinator and the other acting as an endpoint near the sensors. It uses the 802.15.4 protocol to achieve a higher point-to-point transmission speed, but this limits the network to be only a point-to-point, with no node mesh forming possibility.

10.4   RESULTS AND DISCUSSION

Few machine learning algorithms are used to identify industrial machines' internal failures, such as artificial neural networks (ANNs), support vector machines (SVMs), and random forest (RF) models. These classifier models have been trained on raw data using the experimental setup of both healthy and faulty motors. Statistical features have also been extracted from the raw experimental data and fed to the classifier algorithm for fault identification and fault severity evaluation. The selection of features is rendered using the F score of the analysis of variance test (ANOVA). It assigns a score to each feature, and this selects the best k features. This is computationally challenging due to the exponential increase in complexity, with an increase in feature number. Also, feed forward or backward feed algorithms are used, which are greedy approaches and do not guarantee the selection of the best subset. The feed forward algorithm starts with one feature, measures the performance based on some fixed performance parameters, and selects the one with the best performance and then adds one by one feature to that set.

10.4.1 CONFUSION MATRIX FOR VARIOUS SETS OF STATISTICAL FEATURES

We have trained three different models on the training data set and testing the performance of the model on a test data set. The range of data set features ranges from 3 features to 9 features out of a maximum of 15 features and their findings are thoroughly examined.

The confusion matrix with predicted label for different machine learning classifiers and average class-wise precision, recall, and bi-classifier F-measure are shown in Figures 10.3a–c and tabulated, respectively, in Tables 10.1–10.3. The pair plots for different sets of statistical features are correspondingly plotted in Figures 10.4–10.6.

10.4.2 THREE-STATISTICAL FEATURES

The best three-features under the proposed model are standard deviation, skewness, and root mean square (RMS) value, as per the ANOVA test.

10.4.3 SIX-STATISTICAL FEATURES

The best six-features under the proposed model are standard deviation, skewness, RMS value, max, peak-peak, and margin factor, as per the ANOVA test.

TABLE 10.1

Average Class-Wise Precision, Recall, and F-Measure of Bi-Classifier ANN

	Precision	Recall	F1-Score	Support
Healthy	0.97	0.93	0.95	1,202
Faulty	0.88	0.94	0.91	598
Avg/total		**0.9356**		**1,800**

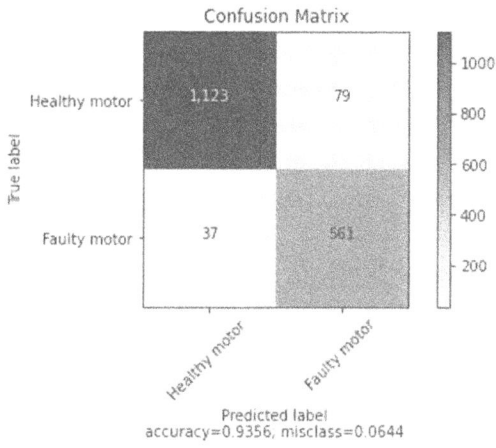

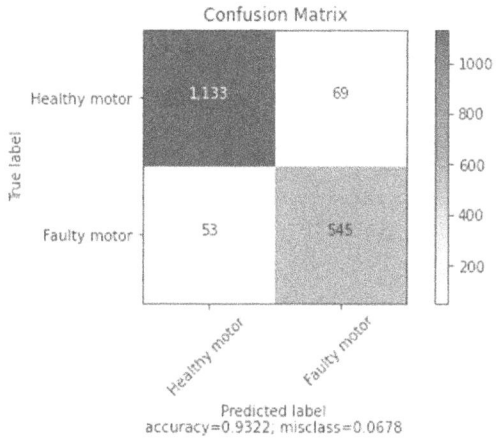

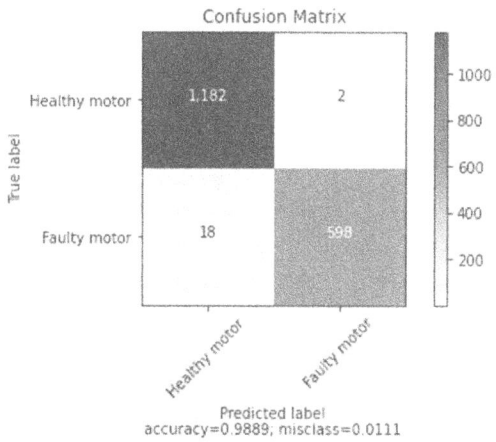

FIGURE 10.3 Confusion matrix with predicted label for various classifiers: **(a)** ANN; **(b)** SVM; **(c)** RF.

TABLE 10.2

Average Class-Wise Precision, Recall, and F-Measure of Bi-Classifier SVM

	Precision	Recall	F1-Score	Support
Healthy	0.96	0.94	0.95	1,202
Faulty	0.89	0.91	0.90	598
Avg/total		**0.9322**		**1,800**

TABLE 10.3

Average Class-Wise Precision, Recall, and F-Measure of Bi-Classifier RF

	Precision	Recall	F1-Score	Support
Healthy	0.98	0.93	0.96	1,202
Faulty	0.96	0.98	0.94	598
Avg/total		**0.9889**		**1,800**

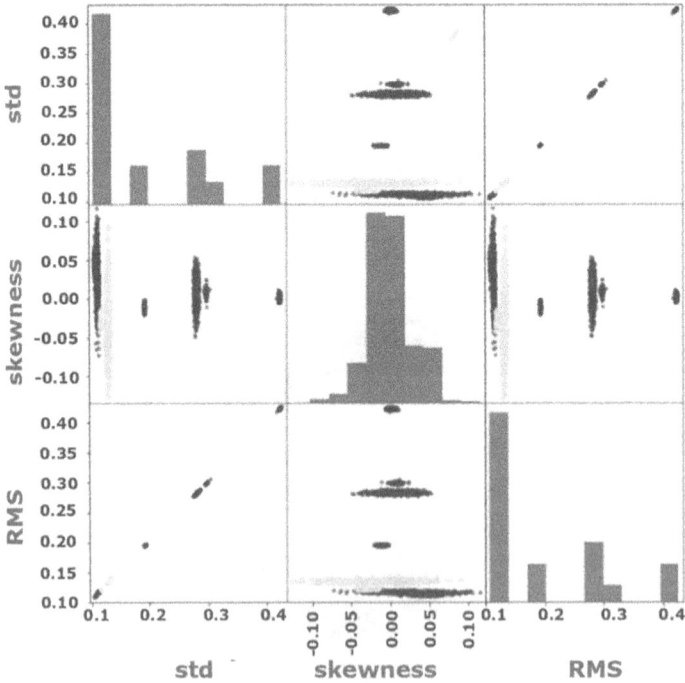

FIGURE 10.4   Pair plot for three-statistical features.

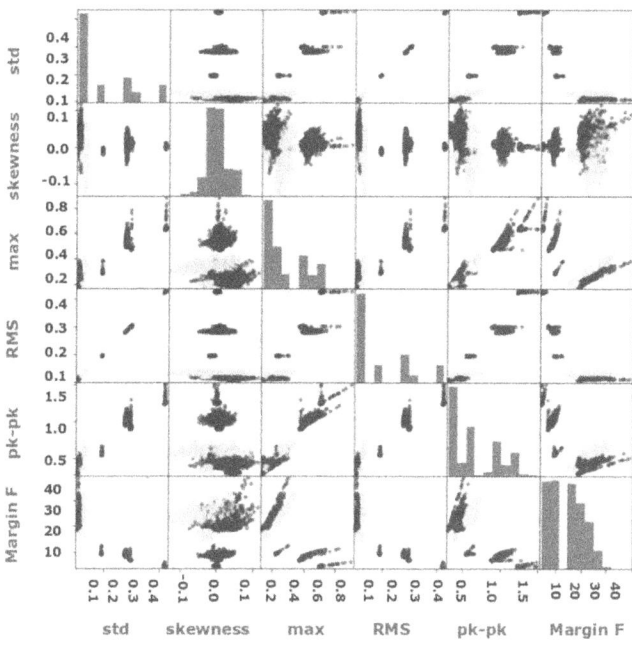

FIGURE 10.5 Pair plot for six-statistical features.

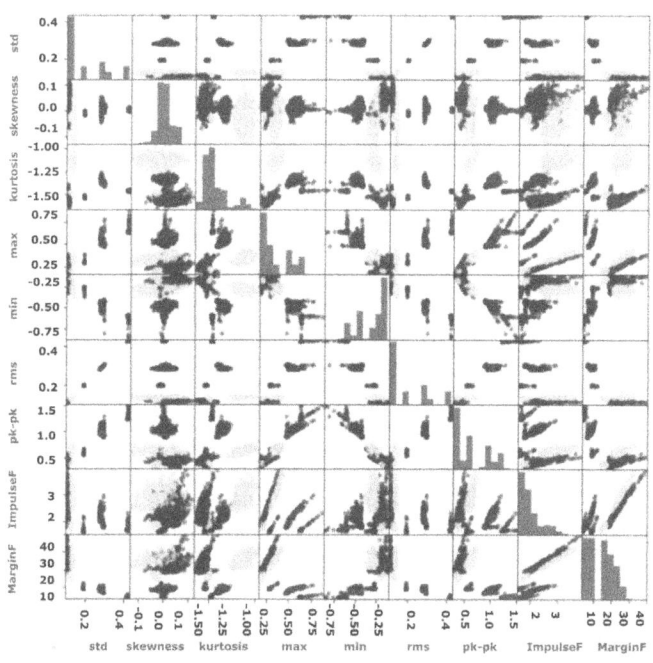

FIGURE 10.6 Pair plot for nine-statistical features.

10.4.4 NINE-STATISTICAL FEATURES

The best nine-features under the proposed model are standard deviation, skewness, RMS value, max, min, peak-peak, margin factor, impulse factor, and kurtosis, as per the ANOVA test.

10.5 CONCLUSIONS

The best possible model features were found through computer-aided advanced diagnostic systems with pattern recognition by using the ANOVA test. The extracted model features current data was then transmitted to generate fault diagnostic reports via the Zigbee-based Internet of Things (IoT) platform. The confusion matrix with the predicted label is extensively evaluated for different machine learning models and average class-wise precision, recall, and F-measure studied. Finally, the three best features model was selected since this accuracy obtained is high, and fewer data need to be transmitted via Zigbee at a transmission rate of about 180 Bps.

ACKNOWLEDGMENTS

The authors would like to express special thanks of gratitude to Birla Institute of Technology and Science, Pilani—Hyderabad for Additional Competitive Research Grant (BITS/GAU/ACRG/2019/H0595) support for a duration of 2 years.

REFERENCES

Ali, A. I., Partal, S. Z., Kepke, S., and Partal, H. P. 2019. Zigbee and LoRa based wireless sensors for smart environment and IOT applications, in 2019 1st Global Power, Energy and Communication Conference (GPECOM), IEEE, 19–23.

Basu, J. K, Bhattacharyya, D., and Kim, Th. 2010. Use of artificial neural network in pattern recognition. *International Journal of Software Engineering and its Applications* 4(2).

De Angelo, C. H., Bossio, G. R., Giaccone, S. J., Valla, M. I., Solsona, J. A., and García, G. O. 2009. Online model-based stator-fault detection and identification in induction motors. *IEEE Transactions on Industrial Electronics* 56(11): 4671–80.

Kuzminykh, I., Snihurov, A., and Carlsson, A. 2017. Testing of communication range in Zigbee technology, in 2017 14th International Conference the Experience of Designing and Application of CAD Systems in Microelectronics (CADSM), IEEE, 133–136.

Liu, R., Yang, B., Zio, E., and Chen, X. 2018. Artificial intelligence for fault diagnosis of rotating machinery: A review. *Mechanical Systems and Signal Processing* 108: 33–47.

Ndih, E. D. N., and Cherkaoui, S. 2016. On enhancing technology coexistence in the IoT era: Zigbee and 802.11 case. *IEEE Access* 4: 1835–44.

Radhika, S., Sabareesh, G., Jagadanand, G., and Sugumaran, V. 2010. Precise wavelet for current signature in 3ϕ im. *Expert Systems with Applications* 37(1): 450–55.

Ramya, C. M., Shanmugaraj, M., and Prabakaran, R. 2011. Study on Zigbee technology, in 2011 3rd International Conference on Electronics Computer Technology, IEEE, 6: 297–301.

Ranjan, G., Verma, A. K., and Radhika, S. 2019. K-nearest neighbors and grid search CV based real time fault monitoring system for industries, in 2019 IEEE 5th International Conference for Convergence in Technology (I2CT), IEEE, 1–5.

Setiawan, M. A., Shahnia, F., Rajakaruna, S., and Ghosh, A. 2015. Zigbee-based communication system for data transfer within future microgrids. *IEEE Transactions on Smart Grid* 6(5): 2343–55.

Vamsi, I. V., Abhinav, N., Verma, A. K., and Radhika, S. 2018. Random forest based real time fault monitoring system for industries, in 2018 4th International Conference on Computing Communication and Automation (ICCCA), IEEE, 1–6.

Verma, A. 2020. An efficient neural-network model for real-time fault detection in industrial machine. *Neural Computing and Applications*: 1–14. https://link.springer.com/article/10.1007/s00521-020-05033-z.

Verma, A. K., Radhika, S., and Padmanabhan, S. 2018. Wavelet based fault detection and diagnosis using online MCSA of stator winding faults due to insulation failure in industrial induction machine, in 2018 IEEE Recent Advances in Intelligent Computational Systems (RAICS), IEEE, 204–08.

Verma, A. K., Spandana, P., Padmanabhan, S., and Radhika, S. 2019. Quantitative modeling and simulation for stator inter-turn fault detection in industrial machine, in International Conference on Intelligent Computing and Communication, Springer, 87–97.

Verma, A. K., Akkulu, P., Padmanabhan, S., and Radhika, S. 2020. Automatic condition monitoring of industrial machines using FSA-based hall-effect transducer. *IEEE Sensors Journal* 21(2): 1072–81.

11 A Study on Time-Frequency Analysis of Phonocardiogram Signals

Samit Kumar Ghosh, Rajesh Kumar Tripathy, and R. N. Ponnalagu

CONTENTS

11.1 Introduction 189
11.2 Time-Frequency Analysis Approaches for PCG Signal 190
    11.2.1 Short Time Fourier Transform (STFT) 190
    11.2.2 Wavelet Transform (WT) 191
        11.2.2.1 Continuous Wavelet Transform 192
        11.2.2.2 Discrete Wavelet Transform 192
    11.2.3 Stockwell Transform 193
11.3 Results and Discussion 194
    11.3.1 STFT Analysis of PCG Signal 194
    11.3.2 CWT Analysis of PCG Signal 195
    11.3.3 ST Analysis of PCG Signal 197
11.4 Conclusion 199
References 200

11.1 INTRODUCTION

A phonocardiogram (PCG) is a graphically recorded heart sound signal, clearly depicting the mechanical activity of the heart. It acts as a noninvasive diagnostic tool for finding abnormalities in the functioning of heart valves. Careful study and analysis of the PCG signal help to detect heart valvular diseases at an early stage and thus prevent people from developing heart valve disorders (HVDs) [1]. PCG signals exhibit sudden frequency changes and transients and their statistical characteristics vary with time. Hence, they are called nonstationary or time varying signals [2]. In general, for analyzing and to extract useful information from a signal, the signal is processed either in a frequency or time domain and the accuracy of the analysis depends upon the technique used for processing and on the extracted features [3]. The time-domain analysis presents changes in the signal over a given period of time and the frequency domain analysis shows the variations in the signal over a given frequency range [4]. Fourier transform (FT) is widely used for processing signals because it explores the signal at various frequencies and provides the frequency response of the signal [5]. On the same lines, when FT is used for processing PCG signals, it provides its frequency response but it can not predict their location because it will not be able to follow the sudden changes in the signal amplitude, frequency, and phase because PCG is a nonstationary signal. Hence, either the time domain or the frequency domain description alone will not provide comprehensive information to analyze a nonstationary signal and to extract its attributes [6]. In order to analyze and evaluate such kind of a signal and to obtain valuable features such as specific time instants and their frequency shifted over time, it is necessary to integrate both time and frequency representations together and is known as time-frequency analysis (TFA) [7]. TFA includes various time-frequency (TF) representations that can simultaneously analyze a signal in both frequency and time domains [6, 8]. Some of those TFA approaches are short-time Fourier transform (STFT) [9], wavelet transform (WT) [10], and Stockwell transform (ST) [11]. The basic idea behind TFA is to separate the signal into small sections and analyze them individually to obtain a specific feature that will provide the characteristics of the signal on a TF plane [12]. The TF distributions provide a 2D representation of the time-varying signal. The primary objective of TFA is to obtain the energy concentration of the signal along the axis of frequency in a given time instant to acquire joint TF description of the signal [7]. STFT when used for analyzing nonstationary signals, yields better results compared to FT but its TF resolution is poor [13]. WT is better in terms of resolution but it uses a basis function that changes (expands and contracts) with change in frequency and does not include the phase information [14]. Also, the time scale plots produced by continuous wavelet transform (CWT) is complex. The lack of phase information and the complexity of plots obtained from WT leads to the development of ST. ST includes the useful attributes of both STFT and CWT. It can be seen either as a phase-corrected CWT or as a variable sliding window STFT [11, 15]. ST maintains complete phase information and good TF resolution over a wide range of frequencies [11]. In this work, we have analyzed the PCG signal using TFA approaches such as STFT, WT, and ST to detect the pathology from the PCG signal and classify the signal as normal or pathological. The four pathological cases considered in this work are aortic stenosis (AS), mitral regurgitation (MR), mitral stenosis (MS), and mitral valve prolapse (MVP). Section 11.2 details the different transform techniques and Section 11.3 presents the results obtained from the analysis using different TFA techniques. A discussion and comparison of results obtained are also presented in the same section. Section 11.4 draws the conclusions of this chapter.

11.2 TIME-FREQUENCY ANALYSIS APPROACHES FOR PCG SIGNAL

Various TFA approaches used for the preprocessing of a PCG signal are described in the following subsections.

11.2.1 SHORT TIME FOURIER TRANSFORM (STFT)

In order to analyze a continuous nonstationary signal, a large transform like FT is not suitable because it lacks to provide simultaneous frequency and time information. This leads to the development of STFT where the idea is to divide the continuous signal into small segments of equal length of window $w(t-\tau)$ and then compute the FT, which can be either discrete Fourier transform (DFT) or fast Fourier transform (FFT) [15]. It's temporal and spectral resolutions are proportional to each other because a fixed value bounds both products. Mathematically, the continuous-time STFT is written

$$\text{STFT}\{x(t)\} = X(\tau, w) = \int_{-\infty}^{\infty} x(t)w(t-\tau)exp(-jwt)dt \tag{11.1}$$

where, $X(\tau, w)$ is the FT of $x(t)$ and $w(t-\tau)$. $w(\tau)$ is the window function centered to zero and τ is the time index. The intensity plot of STFT magnitude over time is visualized by spectrogram and is represented by the following expression:

$$\text{spectrogram}\{x(t)\} = |X(\tau, w)|^2 \tag{11.2}$$

In case of a discrete domain, the data will be broken into small chunks and DFT is computed for each of them and the results are obtained in the form of a matrix. Eq. (11.3) shows this:

$$\text{STFT}\{x[n]\} = X(m, w) = \sum_{n=-\infty}^{\infty} x[n]w[n-m]exp(-jwn) \tag{11.3}$$

STFT overcomes the limitations of FT, but the time and frequency resolution depends upon the window size chosen. In case a small window is selected, resolution in time scale will be good, but frequency resolution will be poor. If a good frequency resolution is needed, a large window must be selected. There occurs a trade-off between frequency and time resolutions, and selecting a proper window size is the main drawback of STFT. Hence, STFT false back in analyzing nonstationary signals like PCG whose frequency is also frequently varying. WT overcomes the limitation of STFT, which is discussed in the next subsection.

11.2.2 WAVELET TRANSFORM (WT)

A wavelet, unlike a sinusoid, is a rapidly decaying wave-like oscillation that has zero mean and extends to infinity for an infinite duration. Wavelets are of different sizes and shapes such as morlet, daubechies, coiflets, bioorthogonal, Mexican hat, symlets [10]. Nowadays, wavelets are widely used for analyzing a transient, time-varying, or nonstationary phenomenon. The basic idea of WT is to identify a basis function and include a useful framework for computations. In STFT, due to fixed-size windows, very small frequency components (which are critical in a PCG signal) cannot be obtained from the spectral response. To overcome this, in WT, variable-sized regions are introduced, and it provides better frequency resolution in the low-frequency range and a good time resolution at higher frequencies [16, 17]. Hence, WT is more suitable for analyzing heart sound. WT is classified into two categories, namely continuous wavelet transform (CWT) [18, 19] and discrete wavelet transform (DWT) [20, 21].

11.2.2.1 Continuous Wavelet Transform

The CWT of a time-domain signal $x(t)$ is defined as the integral transform of $x(t)$ with a family of wavelet function $\psi_{a,b}(t)$ [18]

$$X(a,b) = \frac{1}{\sqrt{a}} \int_{-\infty}^{\infty} x(t)\psi^{*}\left(\frac{t-b}{a}\right) dt \qquad (11.4)$$

where, the function $\psi(t)$ represents the mother wavelet, whereas the function $\psi_{a,b}(t)$ is known as the daughter wavelet. ψ^{*} represents the complex conjugate of wavelet function ψ. The daughter wavelets are derived by applying scaling and shifting operation in the mother wavelet. The parameters a and b are known as the scaling factor and shifting factor, respectively. In a discrete domain, the CWT can be expressed as

$$x[n,s] = \frac{1}{\sqrt{|s|}} \sum_{i=n-\frac{M}{2}}^{n+\frac{M}{2}} x[i]\psi\left[\frac{i-n}{s}\right] \qquad (11.5)$$

where, x represents the signal to be transformed, ψ denotes the mother wavelet function, n indicates the dilation parameter, and s is the translation parameter. M is the duration of the wavelet at $s = 1$. CWT is different from other transforms in such a way that the graphical representation is presented as time versus scale instead of time versus frequency. The wavelet scale spectrum (WSS) can be expressed in a discrete domain as

$$\mathrm{WSS}(x[n,s]) = \sum_{n=1}^{N} |x[n,s]|^{2} \qquad (11.6)$$

where, N indicates the samples number of the signal. The CWT method will result in strongly repetitive wavelet coefficients and requires large processing time. In order to remove the redundancy and enable faster computation, DWT is used.

11.2.2.2 Discrete Wavelet Transform

DWT is the combination of low pass filter (LPF) and high pass filter (HPF). It decomposes the input signal into two coefficients: details coefficient and approximate coefficient. The high-frequency components are present in the details coefficient whereas the low-frequency components are present in approximation coefficients. One can obtain several levels of DWT of a signal, from one level to the next level, it is based upon the approximation coefficients. The DWT is different from CWT in such a way that it uses scale and position values based on the power of 2 and the values are $a = 2^{j}$ and $b = k * 2^{j}$, $(j,k) \in Z^{2}$ given as [20]

$$\psi_{a,b}(t) = \frac{1}{\sqrt{2^{j}}} \psi\left(\frac{t-k*2^{j}}{2^{j}}\right) \qquad (11.7)$$

DWT is used to decompose the signal into high- and low-frequency components for analyzing the signals better, and to obtain the original signal, inverse DWT is required. The wavelet decomposition is based on the number of hierarchically decomposition levels. Based on the desired cut-off frequency, one can choose the decomposition level. Also, it is important to determine the mother wavelet and DWT level for the extraction of specific features from PCG signals.

11.2.3 STOCKWELL TRANSFORM

Stockwell transform (ST) is a hybrid combination of STFT and WT and it overcomes the limitations of both [11]. ST is a time-frequency decomposition that offers absolutely referenced frequency and phase information. In addition to that, it provides phase information of the signal, which is more helpful for analysis of the signal. It makes use of a Gaussian function, whose height and width are decided based on the frequency. The Gaussian window is considered because (1) it reduces the quadratic time-frequency moment about a TF point, (2) side lobes are not present in a Gaussian function, and (3) FT of a Gaussian is Gaussian because of the symmetric nature. As ST does not have cross-term issues compared to other transforms, the contrast of the signal is better. ST offers amplitude-frequency-time spectrum and the phase-frequency-time spectrum, which is helpful for determining local spectral characteristics. The performance of ST is superior to others because here, the modulating sinusoids are fixed while the Gaussian window contracts and expands. This results in an absolute phase spectrum and is always referred to origin of time axis (fixed reference point) [22]. Also, the time frequency resolution of ST is better. The mathematical expression for the standard ST of a continuous 1D signal $x(t)$ is [23]

$$ST(\tau, f) = \int_{-\infty}^{\infty} x(t) w(\tau - t, \sigma) e^{-j2\pi ft} dt \tag{11.8}$$

where, $w(\tau - t, \sigma)$ is called as the Gaussian window function, which is centered at $t = \tau$. In ST, the $w(\tau - t, \sigma)$ is selected as

$$w(\tau - t, \sigma) = \frac{1}{\sqrt{2\pi}\sigma} e^{-\frac{(\tau-t)^2}{2\sigma^2}} \tag{11.9}$$

where, τ is the translation factor, and σ is the dilation factor and it has inverse relation with frequency as $\sigma = \frac{1}{|f|}$. Window width always changes in the inverse of the frequency f because the time-frequency resolution of ST is dictated by a relatively fixed window, resulting in an invariable time-frequency resolution for different signals. The discrete transform of a signal $x[kT]$, $k = 0, 1, 2....., N-1$ corresponding to $x(t)$, and is given as

$$ST\left(jT, \frac{n}{NT}\right) = \sum_{m=0}^{N-1} X\left[\frac{m+n}{NT}\right] e^{-\frac{2\pi m^2}{n^2}} e^{\frac{i2\pi mj}{N}} \tag{11.10}$$

TABLE 11.1

Basics of Transforms

Transform	Basic Functions	Numerical Expression
FT	The complex exponential function with different frequency	$e^{j2\pi ft}$
STFT	The windowed or truncated complex exponential function	$w(t-\tau)exp(-jwt)$
WT	The translated and scaled form of mother wavelet	$\dfrac{1}{\sqrt{a}}\psi\left(\dfrac{t-b}{a}\right)$
ST	The Gaussian windowed complex exponential function	$\dfrac{1}{\sqrt{2\pi}\sigma}e^{-\frac{(\tau-t)^2}{2\sigma^2}}e^{-j2\pi ft}$

where, $X\left[\dfrac{n}{NT}\right]$ is the FT of $x[kT]$. In Eq. (11.10), $m = 0,1,2......,N-1$ and $n = 0,1,2......,N-1$. In order to obtain the time domain signal inverse, ST is computed as [24, 25]

$$x[kT] = \sum_{n=0}^{N-1}\left\{\frac{1}{N}\sum_{j=0}^{N-1}ST\left[jT,\frac{n}{NT}\right]\right\}e^{\frac{i2\pi mj}{N}} \tag{11.11}$$

The basic functions and the numerical expression of the transforms discussed in this section are presented in Table 11.1 [15].

11.3 RESULTS AND DISCUSSION

The various TFA approaches discussed in Section 11.2 are applied to PCG signals, and the results obtained are presented in this section. PCG signals used in the analysis are captured using an electronic stethoscope, and the sampling rate is 8,000 samples per second. The pathological cardiac conditions considered are HVDs such as AS, MS, MR, and MVP.

11.3.1 STFT ANALYSIS OF PCG SIGNAL

Figure 11.1a–o displays the different PCG signals, their TF plot obtained by applying STFT, and the corresponding spectrograms. Figure 11.1a–e shows the PCG signal of normal, AS, MR, MS, and MVP cases. The contour plots and spectrograms of respective categories of the PCG signal are illustrated in Figure 11.1f–j and k–o. To obtain the spectrogram, a 1,024 point FFT and a hamming window are applied. From Figure 11.1a–o, it is observed that heart sound components are clearly localized in both time as well as the frequency domain. From the contour plot and spectrogram, it can be seen that in a normal PCG signal, the fundamental heart sounds (S1, S2) are localized in the frequency range of 50–150 Hz, and are uniformly distributed, but in pathological cases, the heart sounds lie above 150 Hz and have more peaks and are nonuniformly distributed.

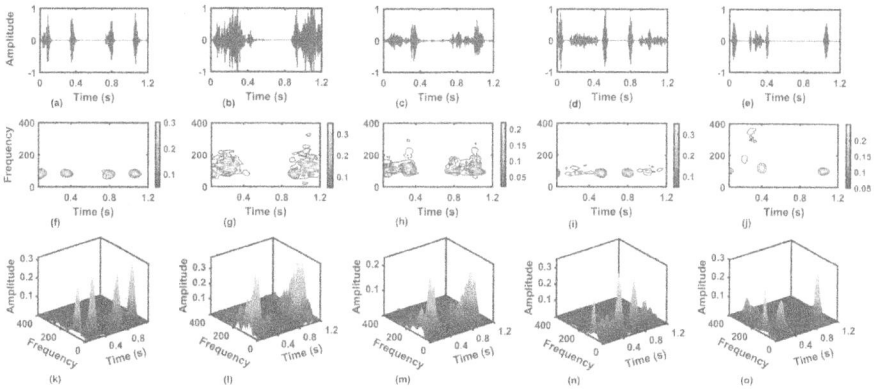

FIGURE 11.1 STFT analysis of a PCG signal: Time domain PCG signal **(a)** normal, **(b)** AS, **(c)** MR, **(d)** MS, and **(e)** MVP; Contour plots of corresponding PCG signal **(f)** normal, **(g)** AS, **(h)** MR, **(i)** MS, and **(j)** MVP; Spectrogram of corresponding PCG signal **(k)** normal, **(l)** AS, **(m)** MR, **(n)** MS, and **(o)** MVP.

11.3.2 CWT ANALYSIS OF PCG SIGNAL

Figure 11.2a–c displays a normal PCG signal, its TF contour plot is obtained by applying CWT and the corresponding scalogram (spectrum of time-scale versus amplitude). In a similar way, Figures 11.3–11.6 show the PCG signals of the four pathological cases, their respective TF contour plots, and scalograms.

CWT provides unique time domain data, which helps to translate the signal from time-domain (1D) to a time-frequency domain (2D). The scalogram presented in (c) in Figures 11.2–11.6 depicts the variation in the intensity of the color code. It can be seen that the intensity is dispersing smoothly in the case of a normal signal and disperses abruptly in the case of the pathological signal. Hence, careful study from

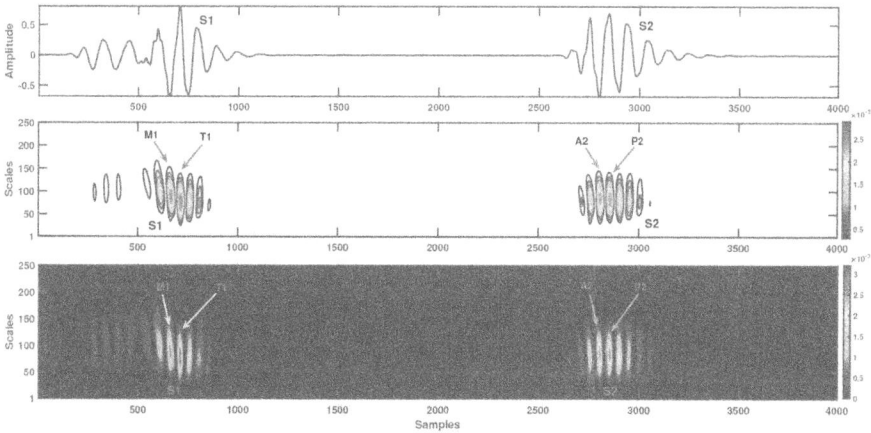

FIGURE 11.2 **(a)** Time domain PCG signal (normal); **(b)** contour plot of (a) after application of CWT; **(c)** scalogram of (a).

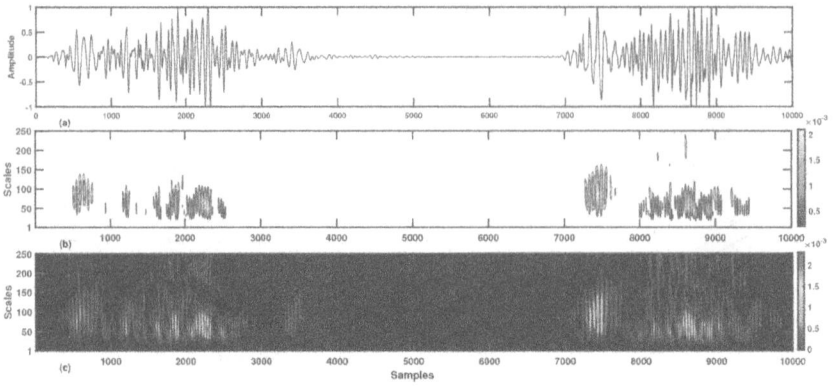

FIGURE 11.3   (a) Time domain PCG signal (AS); (b) contour plot of (a) after application of
CWT; (c) scalogram of (a).

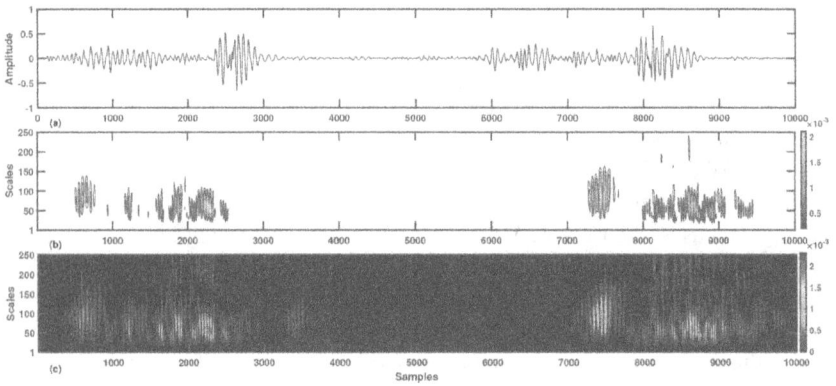

FIGURE 11.4   (a) Time domain PCG signal (MR); (b) contour plot of (a) after application
of CWT; (c) scalogram of (a).

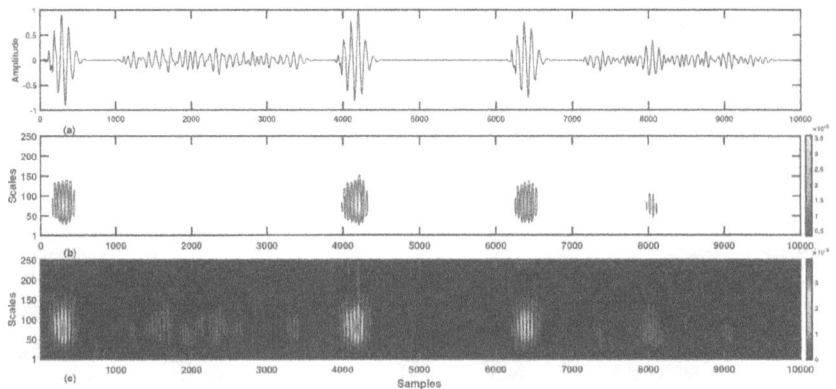

FIGURE 11.5   (a) Time domain PCG signal (MS); (b) contour plot of (a) after application of
CWT; (c) scalogram of (a).

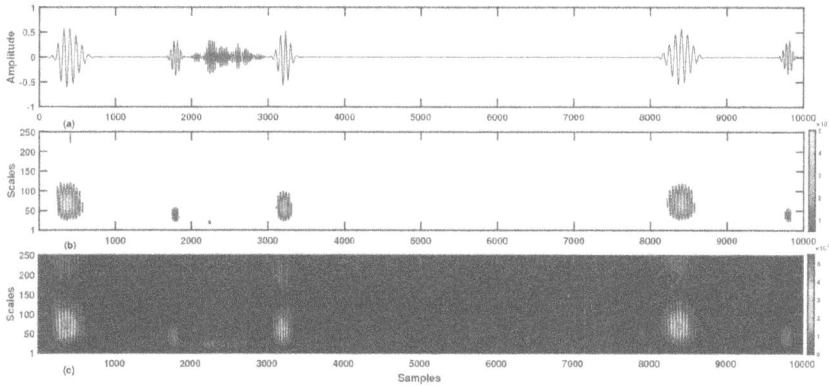

FIGURE 11.6   (a) Time domain PCG signal (MVP); (b) contour plot of (a) after application of CWT; (c) scalogram of (a).

the scalogram images reveals the difference between normal and pathological PCG signals. From the contour plot shown in Figure 11.2b, it can be observed that a space of 1,590 samples ($\approx 0.2s$) are present between heart sound S1 and S2. Also, it is evident from the figure that S1 sound has a lesser frequency component compared to S2 because the amount of blood present in the cardiac chamber is larger. It can also be noted that there are two major frequency components M1 and T1 [indicated in Figure 11.2b] present in the spectrum of heart sound S1 and the time delay between them is 6.3 ms. These two frequency components are produced due to the closure of mitral (M1) and tricuspid valve (T1). Similarly, the frequency components A2 and P2 [indicated in Figure 11.2b] generated due to the closure of the aortic (A2) and pulmonic valve (P2) are present in the spectrum of the second heart sound S2 and the delay between them is 5.8 ms. It can be noted that in the case of pathological conditions displayed in (b) in Figures 11.3–11.6, the above mentioned time delays are more compared to the normal case. This information is helpful in deducting the pathological cases.

11.3.3   ST ANALYSIS OF PCG SIGNAL

The time-domain PCG signal (normal and various pathological cases) shown in (a) in Figures 11.7–11.11 is converted into the TF domain by using a Gaussian window according to the principle of ST and the resulting contour plots, energy spectrums, and phase plots are shown in (b–d), respectively, in Figures 11.7–11.11. From the contour plot, it can be clearly observed that the frequency components of PCG signals are uniformly distributed in the normal case. In contrast, it is nonuniformly distributed in the pathological cases, and also the range of frequencies is higher than the normal case. The switching events between S1 to S2 and S2 to S1 is clearly visualized in the normal condition than in the abnormal cases. The energy spectrums (c) and phase spectrums (d) shown in Figures 11.7–11.11 portray the difference between normal and pathological cases. Hence, ST provides more precise TF localization compared to STFT and WT.

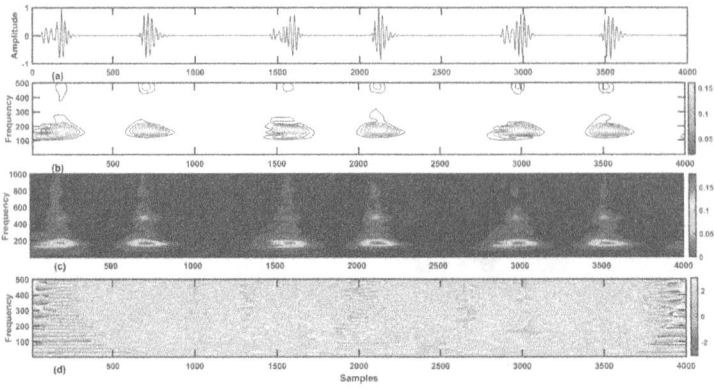

FIGURE 11.7   **(a)** Time domain PCG signal (normal); **(b)** contour plot of (a) after application of ST; **(c)** energy spectrum of (a); **(d)** phase spectrum of (a).

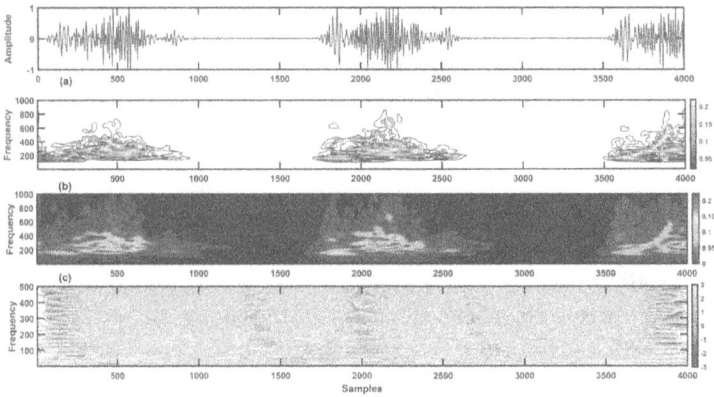

FIGURE 11.8   **(a)** Time domain PCG signal (AS); **(b)** contour plot of (a) after application of ST; **(c)** energy spectrum of (a); **(d)** phase spectrum of (a).

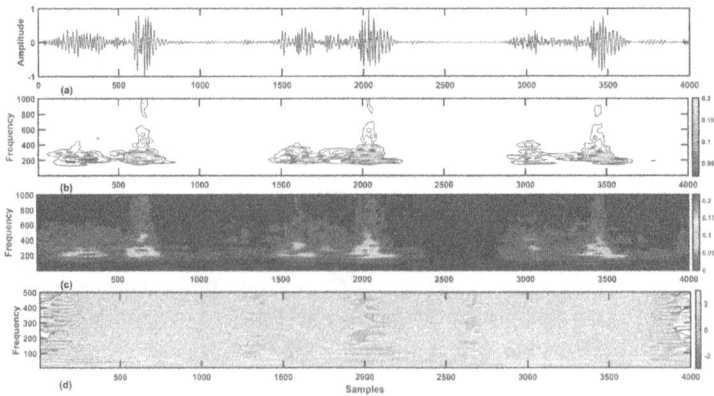

FIGURE 11.9   **(a)** Time domain PCG signal (MR); **(b)** contour plot of (a) after application of ST; **(c)** energy spectrum of (a); **(d)** phase spectrum of (a).

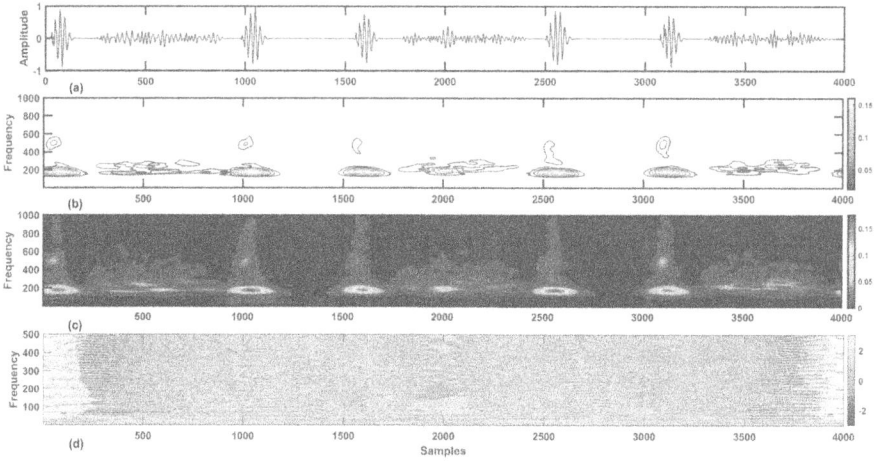

FIGURE 11.10 (**a**) Time domain PCG signal (MS); (**b**) contour plot of (a) after application of ST; (**c**) energy spectrum of (a); (**d**) phase spectrum of (a).

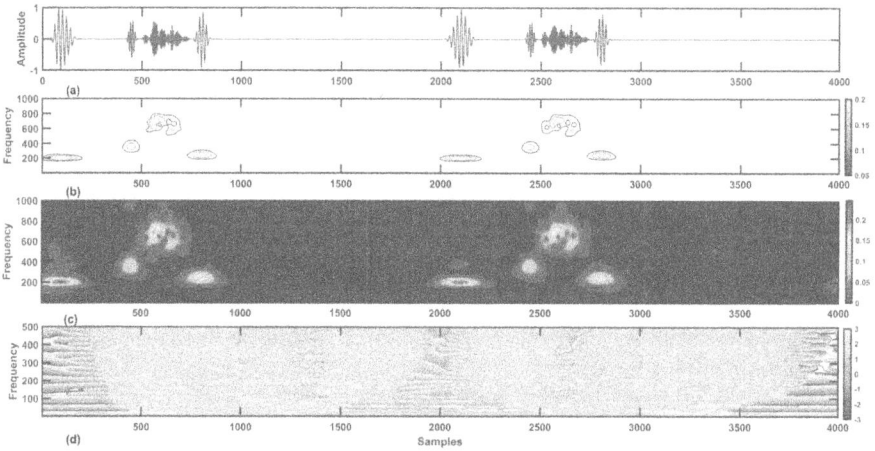

FIGURE 11.11 (**a**) Time domain PCG signal (MVP); (**b**) contour plot of (a) after application of ST; (**c**) energy spectrum of (a); (**d**) phase spectrum of (a).

11.4 CONCLUSION

In this work, three different types of TFA approaches, their advantages, drawbacks, and suitability for analyzing PCG signals are presented. STFT evaluates both the time duration and frequency of every heart sound component present in the PCG signal, but it is not possible to determine the time duration between valve closures, which is very crucial in the diagnosis of HVDs. WT overcomes these issues but lacks in providing phase information. Among the three TFA approaches, ST is more suitable for analyzing the PCG signal. TFA analysis of PCG signals provides plots and descriptions, which could be helpful in identifying pathological PCG signals. It also provides the necessary information to clinical experts for diagnosing HVDs.

REFERENCES

1. Kao, Wen-Chung, and Chih-Chao Wei. "Automatic phonocardiograph signal analysis for detecting heart valve disorders," *Expert Systems with Applications* 38, no. 6 (2011): 6458–68.
2. Obaidat, M. S. "Phonocardiogram signal analysis: Techniques and performance comparison," *Journal of Medical Engineering & Technology* 17, no. 6 (1993): 221–27.
3. Eisenstein, B., and John Fehlauer. "Signal processing for feature extraction and pattern recognition," in *ICASSP'76. IEEE International Conference on Acoustics, Speech, and Signal Processing*, 1, pp. 749–752. IEEE, 1976.
4. Cohen, Leon. *Time-frequency analysis*. Vol. 778. Prentice Hall, 1995.
5. Rabiner, Lawrence R., and Bernard Gold. "Theory and application of digital signal processing," Prentice-Hall, Englewood Cliffs, NJ, 1975.
6. Boashash, Boualem. Time-frequency signal analysis and processing: A comprehensive reference. Academic Press, 2015.
7. Qian, Shie, and Dapang Chen. "Joint time-frequency analysis," *IEEE Signal Processing Magazine* 16, no. 2 (1999): 52–67.
8. Molla, Md Khademul Islam, Mostafa Al Masum Shaikh, and Keikichi Hirose. "Time-frequency representation of audio signals using Hilbert spectrum with effective frequency scaling," in *2008 11th International Conference on Computer and Information Technology*, pp. 335–340. IEEE, 2008.
9. Griffin, Daniel, and Jae Lim. "Signal estimation from modified short-time Fourier transform," *IEEE Transactions on Acoustics, Speech, and Signal Processing* 32, no. 2 (1984): 236–43.
10. Holschneider, Matthias, Richard Kronland-Martinet, Jean Morlet, and Ph Tchamitchian. "A real-time algorithm for signal analysis with the help of the wavelet transform," n *Wavelets*, pp. 286–97. Springer, Berlin, Heidelberg, 1990.
11. Stockwell, R. G., L. Mansinha, and R. P. Lowe. "Localisation of the complex spectrum: The S transform," *Journal of Association of Exploration Geophysicists* 17, no. 3 (1996): 99–114.
12. Stankovic, LJubisa. "A method for time-frequency analysis," *IEEE Transactions on Signal Processing* 42, no. 1 (1994): 225–29.
13. Lee, Jung Jun, Sang Min Lee, In Young Kim, Hong Ki Min, and Seung Hong Hong. "Comparison between short time Fourier and wavelet transform for feature extraction of heart sound," in *Proceedings of IEEE. IEEE Region 10 Conference. TENCON 99.'Multimedia Technology for Asia-Pacific Information Infrastructure' (Cat. No. 99CH37030)*, vol. 2, pp. 1547–50. IEEE, 1999.
14. Subbarao, M. Venkata, and P. Samundiswary. "Time-frequency analysis of non-stationary signals using frequency slice wavelet transform," in *2016 10th International Conference on Intelligent Systems and Control (ISCO)*, pp. 1–6. IEEE, 2016.
15. Ranjan, Rajeev, Neeru Jindal, and A. K. Singh. "Fractional S-transform and its properties: A comprehensive survey." *Wireless Personal Communications* (2020): 1–23.
16. Chui, Charles K. Wavelets: A mathematical tool for signal analysis. Society for Industrial and Applied Mathematics, 1997.
17. Chui, Charles K. An introduction to wavelets. Elsevier, 2016.
18. Gao, Robert X., and Ruqiang Yan. "Continuous wavelet transform," in *Wavelets*, pp. 33–48. Springer, Boston, MA, 2011.

19. Khorrami, Hamid, and Majid Moavenian. "A comparative study of DWT, CWT and DCT transformations in ECG arrhythmias classification," *Expert Systems with Applications* 37, no. 8 (2010): 5751–57.

20. Rao, K. Deergha, and M. N. S. Swamy. "Discrete wavelet transforms," in *Digital Signal Processing*, pp. 619–91. Springer, Singapore, 2018.

21. Rioul, Olivier, and Pierre Duhamel. "Fast algorithms for discrete and continuous wavelet transforms," *IEEE Transactions on Information Theory* 38, no. 2 (1992): 569–86.

22. Dehghani, Mohammad Javad. "Comparison of S-transform and wavelet transform in power quality analysis," *World Academy of Science, Engineering and Technology* 50, no. 4 (2009): 395–98.

23. Stockwell, Robert Glenn. "A basis for efficient representation of the S-transform," *Digital Signal Processing* 17, no. 1 (2007): 371–93.

24. Du, Jingde, M. W. Wong, and Hongmei Zhu. "Continuous and discrete inversion formulas for the Stockwell transform," *Integral Transforms and Special Functions* 18, no. 8 (2007): 537–43.

25. Wang, Yanwei, and Jeff Orchard. "Fast discrete orthonormal Stockwell transform," *SIAM Journal on Scientific Computing* 31, no. 5 (2009): 4000–12.

12 A Study on the Performance of Solar Photovoltaic Systems in the Underwater Environment

Challa Santhi Durganjali, Sudha Radhika,
R. N. Ponnalagu, and Sanket Goel

CONTENTS

12.1 Introduction 204
12.2 A Review on Performance Analysis of a Solar PV Cell..............................204
   12.2.1 PV Cell Characteristics..205
      12.2.1.1 Irradiation Dependence on PV Cell Characteristics..........206
      12.2.1.2 Temperature Dependence on PV Cell Characteristics.......206
      12.2.1.3 Material Dependence on PV Cell Characteristics 208
   12.2.2 Factors Affecting Solar Cell Efficiency...208
12.3 Efficiency Improvement of Solar Cells...209
   12.3.1 Forced Cooling Systems 210
      12.3.1.1 Active Cooling Method..210
      12.3.1.2 Passive Cooling Method 210
      12.3.1.3 Heat Pipe Cooling Method 211
      12.3.1.4 Nanofluids Cooling Method...211
      12.3.1.5 Thermoelectric Cooling Method 211
   12.3.2 Floating Solar Photovoltaic Panels 213
      12.3.2.1 Construction of an FPV System...213
      12.3.2.2 Classification of an FPV System..214
      12.3.2.3 Advantages of an FPV System...216
12.4 Hydro-Optical Characteristics of Water and Its Suitability
   for Solar Cells 217
12.5 Solar Irradiation Inside Ocean Water at Different Depths..........................218
12.6 Recent Usage of Underwater Solar PV Concept 219
12.7 Conclusion 220
References..221

12.1   INTRODUCTION

Future energy and energy needs in the world are endangered due to the shortage of fossil fuels. Many researchers predict that the current reserves of natural gas and oil will diminish within 70 and 40 years, respectively. Also, the carbon emissions are increasing at more than 3% per year, which in turn is increasing the amount of fossil fuels such as carbon monoxide and carbon dioxide in the atmosphere (Energy Information Administration of US Department of Energy 2016). Hence, alternative energy resources available now, to meet demand, include solar energy, wind energy, tidal energy, and thermal energy, which are all renewable energy resources; solar energy plays a major role because it contributes to 70% of the total energy. The use of solar energy began in the 7th century BC in different applications, but the conversion of solar energy from light to electricity began after finding the photovoltaic (PV) effect.

In 1839, French scientist Edmond Becquerel while experimenting on an electrolytic cell, discovered the PV effect (Becquerel 1839). He found that solar energy can be captured by a semiconductor device through the PV effect and the device was thus named as a PV device. After continuous efforts on semiconductor materials with the PV effect, Bell Telephone Laboratories patented the "Light-Sensitive Electric Device Including Silicon" on June 15, 1948. This achievement initiated the usage of silicon solar cells. Silicon solar cells have higher conversion efficiencies ranging between 15% to 18% compared to other PV materials (Energy Informative 2013).

Essig et al. (2015) proved that the silicon cell efficiency can be improved to 30% by decreasing the surface temperature of a silicon PV cell using a cooling mechanism on the cell surface. As such, most commercialized solar cells are made up of silicon materials. Here, water is utilized as a cooling liquid to acquire higher efficiency from the solar cell. Water is preferred for a cooling agent here because of its nontoxicity, stability, high heat transfer capacity, low cost, and abundance. Muaddi and Jamal (1991) stated that the solar cell spectral response matches with the absorption coefficient of the sunlight in water. Since 71% of the total surface of Earth is covered with water and of which 96.5% is contributed by oceans (UNESCO 2012), the marine or ocean environment provides numerous renewable energy resources, in the form of tidal energy, thermal energy, wind energy, light energy, etc. These renewable energy resources, if efficiently trapped, can be used for many onshore and off-shore applications (Trapani and Redõn Santafé 2015). In this chapter, solar PV cells and submerged solar cells are explained in detail. The ocean environments and its irradiation variation with depth of water are detailed.

12.2   A REVIEW ON PERFORMANCE ANALYSIS
OF A SOLAR PV CELL

As mentioned in the Introduction, Becquerel (1839) found that solar energy can be captured by a semiconductor device through the PV effect forming solar cells. The collection of solar cells is arranged into a framework, known as a solar panel or PV panel. Würfel (2007) reported in his book that when photons fall on the PV cells' surface, the energy of the photon will undergo three main processes:

1. **The absorption of the incident radiation energy:** As photons fall on the solar cell, all photons are not absorbed by the cell. Photons that have sufficient energy to move from the valence band to the conduction band by excitation of an electron and cause the current to flow will be absorbed. Any photon with lesser energy than the energy of the bandgap will not be absorbed by the PV cells and cannot excite the electrons, and will be either transmitted or reflected.

2. **The hole-electron pair thermalization:** In thermalization, the heat energy from sun is converted to chemical energy through the heating process of electrons and holes causes bombarding and creating electron-hole pairs at the P-N junction.

3. **Energy conversion:** The conversion into electrical energy from chemical energy is known as energy conversion. Where holes are moved to P-type and electrons are moved to N-type and become segregated. When these two ends are connected through an external circuit, we can see movement of electrons and that obtained current is known as a PV current.

If irradiation falls on a solar cell surface, we can see that the PV current and the maximum current obtained from a solar cell depends on its type of construction. The solar cell construction depends on cell technology, type of top layer, and number of junctions in the solar cell. Based on all the above mentioned parameters, the solar PV cells are classified again as shown in Figure 12.1.

12.2.1 PV CELL CHARACTERISTICS

The performance of solar cells can be estimated from their electrical and thermal characteristics. The electrical characteristics include I-V and P-V [power (P), current (I),

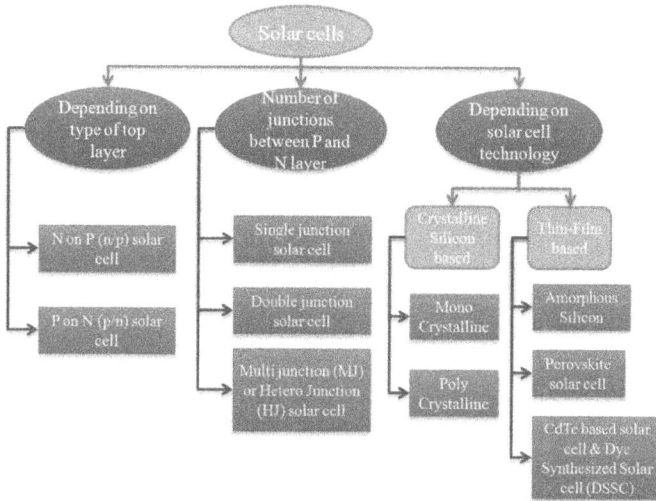

FIGURE 12.1 Classification of solar PV cells.

and voltage (V)] characteristics through which the PV cell efficiency can be deliberated. From the I-V characteristics, the open circuit voltage (V_{oc}) and short circuit current (I_{sc}) are noted. From P-V characteristics, the maximum output power (P_{max}) from the PV cell can be obtained, from which the PV cell efficiency can be evaluated. Irradiation from the sun is the source for a solar cell and temperature is an element that affects the performance of solar cells. Depending on irradiation, temperature, and type of cell materials, the analysis of solar cell performance is further carried out.

12.2.1.1   Irradiation Dependence on PV Cell Characteristics

When the intensity of radiation on a PV cell increases, I_{sc} will increase and vice versa. But V_{oc} of solar cells will mostly remain constant irrespective of radiation. Since intensity of radiation varies, depending on the rate of generation of electrons and holes, the current output from the PV cell varies, which in turn varies the power output. In general, with the surge in irradiation, the short circuit current of the solar cell escalates and by this, the output power also increases. At maximum irradiation point (1000W/m²) and 25°C temperature conditions, i.e., at standard test conditions (STC), it is possible to get maximum output power from the PV cell.

The V_{oc} of the cell is constant and depends on the material used. To have more open circuit voltage, the solar cells are joined in parallel and series. If the irradiation falling on a solar cell is continuous, then the temperature on the top of the solar cell increases even though the environmental temperature is less. Werner Luft (Luft 1970)experimentally compared the performance of 5-grid and 13-grid silicon (Si) and gallium arsenide (GaAs) solar cells at 30–150°C different temperature ranges and radiation intensities ranging from 0.07–2.8 W/cm² and concluded that, compared to GaAs solar cells, the 3-grid silicon solar cell showed a better performance.

12.2.1.2   Temperature Dependence on PV Cell Characteristics

The performance of PV cells under different temperature and irradiation conditions will vary depending on the cell type. For example, an anomalous behavior is observed at less temperatures in the "N on P" cells; namely, the V_{oc} becomes nearly self-reliant of temperature beneath a transition temperature depending on the sunlight intensity. Smith et al. (1963)explained the working of an "N-on-P" solar cell and its characteristics when it is used in a spacecraft application. Kennerud (1967) experimentally determined the values of V_{oc}, I_{sc}, and P_{max} for "N on P" and Prince and Wolf (1958) determined "P on N" silicon solar cells from −177°C to +50°C temperature ranges under equivalent sunlight intensity of 58 mW/cm²and the I-V characteristics are shown in Figure 12.2a and b. It is detected from the I-V characteristics that V_{oc} lessens in a linear fashion with temperature, while I_{sc} increases.

Durganjali and Sudha (2019) made a mathematical modeling of a solar cell that will work in both positive and negative temperature conditions. It is perceived that the performance of the PV cell with the varying temperature (−45°C to +50°C) levels and at different irradiations (600W/m², 1000W/m², and 200W/m²) is analyzed and at 600W/m², the I-V and P-V characteristics analysis is shown in Figure 12.3a and b.

The research is concentrated on the performance of PV cells under different temperature stages. The electrical parameters are measured at greater than the absolute

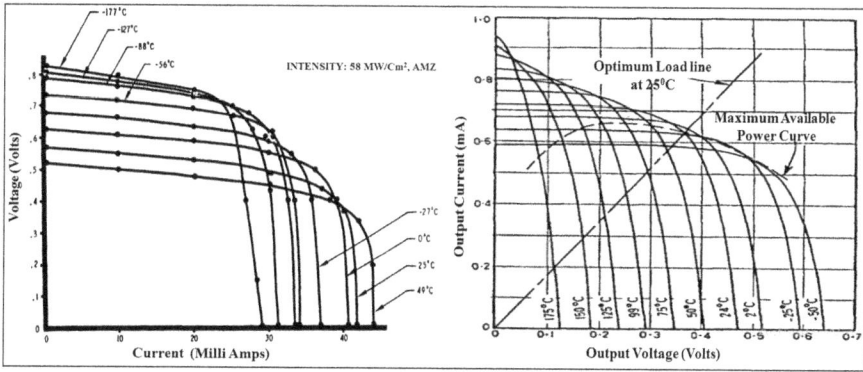

FIGURE 12.2   **(a)** I-V characteristics of a silicon "N on P" solar cell; **(b)** I-V characteristics of a "P on N" silicon solar cell at temperatures ranging from −177°C to +50°C. (From Kennerud 1967 and Prince and Wolf 1958.)

zero Kelvin temperature. The electrical parameters include P_{max}, V_{oc} and I_{sc}. The preferred test conditions (STC), top and truncated temperatures recorded in India, are taken into consideration for PV performance comparison. The electrical parameters

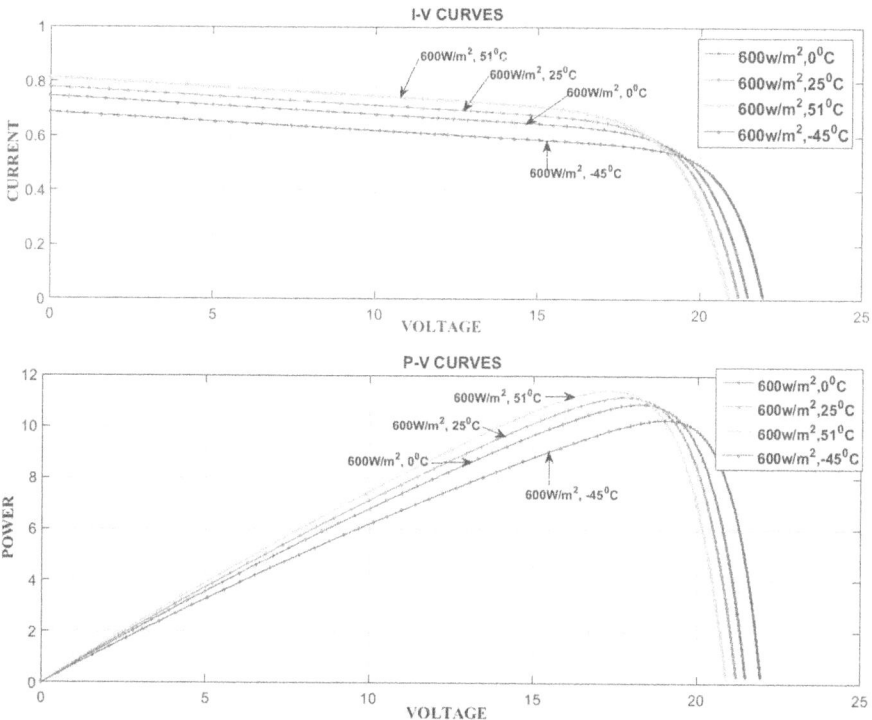

FIGURE 12.3   Characteristics of solar cells at 600W/m² and at varying temperature levels (−45, 0, 25 and 51⁰C) for: **(a)** I-V; **(b)** P-V. (From Durganjali and Sudha (2019).)

with different irradiation and temperature variations are simulated in MATLAB®/SIMULINK. The efficiency, maximum output power, V_{oc} falls with the increase in temperature and I_{sc} increases simultaneously.

12.2.1.3 Material Dependence on PV Cell Characteristics

The material utilized for the construction of solar cells plays a vital role. Different materials used for making solar cells include "Silicon (Si), Copper Indium Gallium Selenide (CIGS), Germanium (Ge), Gallium Arsenide (GaAs), thin-film solar cells, Cadmium Telluride (CdTe)" etc. The majority of solar cells are fabricated using silicon as the material and the efficiency of commercialized silicon solar cells is 15–21% and the experimental solar cells have more than 20% and reach up to 30% of conversion efficiency, which is described in the next section.

Solar cell conversion rate or efficiency refers to the rate of conversion of the incoming solar energy into the electrical power. Typically, the commercialized solar panels operate in the range 15–21% of efficiency (Energy Informative 2013). Experiments are being carried out to improve the cell efficiency with different types of fabrications. In Metal Wrap Through (MWT), Upgraded Metallurgical Grade (UMG), Passivated Emitter Rear and Totally diffused Rear-Junction (PERT_RJ), and Passivated Emitter and Rear Solar Cells (PERC), solar cells are reaching conversion or quantum efficiency of up to 30%, but these are not commercialized yet. Some of the fabrications and inventions made by different researchers are explained below with their respective efficiencies also listed in Table 12.1.

12.2.2 Factors Affecting Solar Cell Efficiency

Thus, it is noticed that the main disadvantage of solar PV cell efficiency lies in the sun's energy physical conversion. The fundamental principle of the solar PV industry is a study by H. Queisser and W. Shockley developed in 1961. The maximum efficiency of 33.7% is possible from the physical theory in which a PV cell can achieve

---

TABLE 12.1
List of Different Silicon Solar Cells and Their Efficiencies

Type of Silicon (Si) Solar Cell	Conversion Efficiency
Triple junction-high frequency	11–12% (Banerjee et al. 2012)
MWT	19–19.6% (Yin et al. 2013)
Bifacial N-type solar cell	>20% (Böscke et al. 2013)
PERC solar cell	20% (Muller et al. 2015)
Boron-doped P-type mono-crystalline Cz silicon wafers	20–21% (Zhang et al. 2016)
Selective FSF N-type rear-junction laser-doped solar cells	20–21% (Wang et al. 2015)
P-type multi-crystalline silicon	20–22% (Deng et al. 2016)
From 100% UMG silicon feedstock N-type Czochralski-grown silicon solar cell	>21% (Zheng et al. 2017)
N-PERT_RJ	22–23% (Peng et al. 2019)
Hetero Junction(HJ)	>24% (Taguchi et al. 2014)

from a light source to obtain electricity. This theory is called the Shockley-Queisser limit. This is related with the process of photon absorption to generate an electron (e⁻) and then pass it to the conduction band (Shockley and Queisser 1961). There is no specific technology or manufacturing process development that can change the limitation fact. This Shockley-Queisser limit is the energy conversion efficiency of a solar cell with the maximum theoretical boundary. Some of the factors affecting solar cell efficiency are:

1. Snow, ice, dust, and humidity
2. Insulation resistance
3. Temperature
4. Solar cell type (crystalline, amorphous, thin-film, and perovskite)
5. Design configuration

By observing the V-I characteristics of "P on N" and "N on P" solar cells and also with lithium doped solar cells, the surface effects in solar cells can be analyzed. Lithium doped solar cells performance with respect to variation in temperature indicates that lithium in the region of N-type "P on N" silicon solar cells interact by 1 MeV electrons (e⁻) or 16.8 MeV protons (Wysocki 1966) exposure with induced radiation damage. The minority-carrier life time cannot be degraded by the interaction of centers; these types of cells are most radiation-resistant ones. The movement of lithium interaction involves the cell radiation rate and temperatures are essential in establishing the degradation amount that is noticed. At low radiation rates and room temperature, the PV cell power output does not decreases and at high radiation rates and temperature, the cell output decreases but it retrieves after irradiation degradation, i.e., the Lithium doped PV cells acts as self-healing, summarized by Wysocki (1966). Proton-induced humiliation of the cell characteristics can be summarized before, after, and during proton exposure. Anomalous damage, including significant losses of P_{max}, and partially recoverable losses of V_{oc} can be identified within different regions of the PV cells with the penetration of proton into depths by correlation (Brown 1967).

Tallent and Oman (2013) tested silicon solar cell performance with the Archimedes array of mirrors, which were capable of concentrating the sun's energy. The V-I characteristics are drawn at a 40°C temperature and at different illumination levels. Concentrating the sun's energy increased the output, but with increase in temperature, the efficiency of the cell decreased and it increased heat on the surface of the solar cell causing damage to the cell. In practical terms, the energy conversion efficiency can be affected by one of the major factors, which is temperature, as discussed in Section 12.2.1.2. The temperature effect can be decreased with the use of cooling methods. The efficiency of solar cells improves with the decrease of the temperature effect possibly with the use of cooling techniques.

12.3 EFFICIENCY IMPROVEMENT OF SOLAR CELLS

The efficiency of PV cells decreases with an elevation in the temperature. The rate of temperature degradation varies from 0.25% to 0.5% per °C, with the dependability on the type of PV cell material utilized. Specifically, in the case of concentrated PV (CPV) cells, which utilize focused sunlight and gives large power with less expensive

PV equipment. It is also observed that the high temperature significantly reduces the working life of the CPV systems. The effect of temperature on PV cells can be reduced by adapting cooling techniques for PV systems. There are two kinds of cooling techniques: forced and natural cooling. The forced cooling systems are further subdivided as active cooling, passive cooling, thermo-electric cooling, and nanofluid cooling. The natural cooling system includes floating PV (FPV) systems that use natural water as a cooling liquid. Sections12.3.1 and 12.3.2 discuss the respective techniques and their advantages in detail.

The temperature effect can be decreased with the following technical methods:

1. Forced cooling systems
2. Natural cooling systems

12.3.1 FORCED COOLING SYSTEMS

The forced cooling system uses additional equipment like pipes, motor drives, etc., to force the cooling liquid to flow in the front side, backside, or on both sides of the PV panels and the coolant used may be air, water, metal oxides, and organic or inorganic liquids. The different types of forced cooling mechanisms are discussed in Krauter (2004), Royne, Dey, and Mills (2005), Kumar and Rosen (2011), Daghigh, Ruslan, and Sopian (2011), and Makki, Omer, and Sabir (2015). A 5% increase in output power is obtained by cooling techniques in Smith et al. (2014). Nevertheless, 87% of irradiated energy will convert into heat. The waste heat has been harnessed into useful thermal energy in recent developments. In general, hybrid elements that mobilize both the thermal and electrical power of solar are known as PV-thermal units (PV/T unit). These PV/T units generally have lower specific efficiency but higher overall efficiency when collated with solar collectors and stand-alone PV (He, Zhang, and Ji 2011). Early in PV exploitation, cooling techniques for heat applications were proposed (Tonui and Tripanagnostopoulos 2006). The commonly used forced cooling techniques are passive cooling and active cooling and this classification relies on whether power is consumed during the cooling process or not.

12.3.1.1 Active Cooling Method

Active cooling methods consume power continuously to force the coolant (air or water) through the panel sides. The primary power consumption unit is either a pump or fan, which is used for circulating the fluid. In Nižetić et al. (2016), the cooling methods from both back and front side were tried. The flow of water varied to 0.0625 kg/s, which is its maximum value, through a pump to apply force to the cooling liquid used. Water flow was applied through a jet, which increases the effect of cooling on solar panels. The pump used will itself consume more power than the generated power.

12.3.1.2 Passive Cooling Method

The passive cooling method utilizes conduction/convection to enable extraction of heat naturally. The passive methods of cooling are subdivided into three main groups: conductive cooling, water cooling, and air cooling. The conductive cooling is similar to air cooling but a significant difference is that the heat transfer mechanism from solar cells is conductive. Phase Change Material (PCM) cooling is a selected type of passive conductive cooling. The PCM in the strict sense can't be observed as cooling, but it helps in maintaining a less steady temperature. PCM cooling can be considered as a passive cooling technique because to take away the heat, no additional power is needed, and the heat dissipation is mainly conductive in nature.

The global potential of PCM was observed in Smith, Forster, and Crook (2014). Han, Wang, and Zhu (2013) used PV cell immersion in different types of cooling fluids. The three different immersion liquids are isolation liquid, organic liquids, and deionized water. The irradiance considered is 10, 20, and 30 suns, where 1 sun is equal to 1,000 W/m². The achieved efficiency is 15% under one sun but under 10, 20, and 30 suns, is quite higher than the 1 sun but thermal effects are again increased. In practice, more than 1 sun cannot act on the surface of a solar panel.

12.3.1.3   Heat Pipe Cooling Method

Heat pipe cooling is a mix of PCM in conjunction with the convection mechanism of cooling medium and it is one type of passive cooling. This type of cooling uses an additional setting of pipes. The cooling medium expands and evaporates (or rises, depending on the type) at one side, taking up the heat. The cooling medium on the other side releases and condensates the heat to the surroundings. The cooling medium via capillary tubes travels back as liquid and it evaporates, by which completing one cycle. Some of the references listed in Table 12.2 used heat pipes to flow the cooling liquid.

12.3.1.4   Nanofluids Cooling Method

Nanofluids are the combination of solid nanoparticles such as aluminum oxide (Al_2O_3) and copper oxide (CuO) and cooling fluid. The weight of these particles is between 0.1 to 2.0% and experiences Brownian motion throughout the cooling fluid. The main advantages of nanofluids are reasonably higher heat capacity and their excellent thermal conductivity (Al-Shamani et al. 2014). A 5% improvement in efficiency is obtained by utilizing nanofluids as the cooling mechanism in solar thermal collectors.

12.3.1.5   Thermoelectric Cooling Method

Thermoelectric cooling is dependent on the Peltier effect. This effect is observed at a junction where electrification occurs when heat flows in a particular direction (Webster, Gurevich, and Velazquez-Perez 2014). The cooling effect is on one side, and on the other side of the junction, it produces heating. The cooling/heating strength depends on the difference in temperature and current/voltage intensity. The thermoelectric cooling effect requires more electricity.

A review on different cooling methods is presented by Kalaiselvan et al. (2018). They have also compared active and passive cooling methods on various parameters. Odeh and Behnia (2009) used water as the cooling liquid and obtained increased efficiency. Through experimental results, the heat loss in between the PV panel's upper surface and water by convection causes an increase in efficiency of about 15% and is achieved as output at maximum radiation conditions. Table 12.2 provides

TABLE 12.2

List of Authors Who Researched Forced Cooling Methods

Cooling Method	Type of Solar Cell	Area	Coolant Used and Mechanism	Temperature Decrease	Efficiency Improvement	Author
Active	polycrystalline solar PV module	0.924 m²	Water and Aluminum casing on the backside to act as a flow channel	12°C	8.9%	Farhana et al. (2012)
Active	Mono-crystalline solar PV module	0.152 m²	Water Backside cooling via two aluminum pipes	10°C (peak temperature 60°C)	0.8%	Du, Hu, and Kolhe (2012)
Active	Mono-crystalline solar PV module	1.24 m²	Water Backside cooling via closed casing	10°C	2.8%	Bahaidarah et al. (2013)
Active	PV module	374 cm²	Water Both front and back side	15–26°C	Back side14.8%, front side19.1%, Both sides20.4%	Rahimi et al. (2014)
Passive	Mono-crystalline PV module	—	Thermosyphon effect is used with PCM	—	19%	El-Seesy, Khalil, and Ahmed (2012)
Passive	Mono-crystalline PV module	0.36 m²	Water At the rear side of the module cotton wick structures wrapped spirally	—	1.4%	Chandrasekar et al. (2013)
Passive	Mono-crystalline PV module	0.150 m²	Water Heat pipes are constructed	13°C	6%	Moradgholi, Nowee, and Abrishamchi (2014)
Nano fluid-cooling	Crystalline and thin film PV modules	—	Metal oxides and Brownian motion of nanoparticles	10–30°C	6–12% (minimum 5%)	Al-Shamani et al. (2014)

details of different types of forced cooling methods used for improving the efficiency of solar panels from selected references.

The active cooling techniques result in more accessible thermal energy and produced power, but the power produced itself is sufficient to drive the cooling equipment used to cool the PV cell. Whereas, when CPV cells are used, this type of cooling method can comfortably be availed because of the ability to use less cooling fluid and less fluid-to-cell mass ratio. Thus, very little power is needed to maintain the CPV system. The main disadvantage of nanofluid cooling is the overall change in flow regime, and pumping process, that is, at different speeds and geometries, natural turbulent flow occurs, when compared with regular fluids. Water is an economic liquid and has higher thermal capacity, hence we can ensure that the passive water cooling is more efficient compared to the other cooling methods.

Even though forced cooling systems improve efficiency, they have certain disadvantages like a requirement of a separate setup to take out heat from the solar cells. The construction and maintenance of such a setup is expensive and the cost of maintaining the system outweighs the advantages of the electrical output improvement. Hence, another way of improving efficiency by cooling is by natural cooling, which includes the use of solar panels on water surfaces. This can be in two ways, either FPV panels or immersed/submerged PV panels. The major disadvantages of forced cooling systems are overcome by the FPV system and FPV systems have been tested and studied in different environments and water types (Redón Santafé et al. 2014; Gozálvez et al. 2012; Ferrer-Gisbert et al. 2013; Lee, Joo, and Yoon 2014;Y.Choi 2014; Tina, Rosa-Clot, and Rosa-Clot 2011; Choi, Lee, and Kim 2013; Y.-K. Choi 2014; Ho, Chou, and Lai 2015; Ho, Jou, and Lai 2017).Environmental conditions such as high/low tide, wind speed, summer, winter, and rainy seasons involve variation of irradiation and temperature falling on solar panels. Water types such as ocean water, lake water, and distilled water, which have a variation in salinity, turbidity, and algae formation, are considered because of the variation in irradiation and temperature transmission into the water bodies.

12.3.2 FLOATING SOLAR PHOTOVOLTAIC PANELS

FPV systems install PV modules on water bodies; in general, on human-constructed water bodies such as irrigation, storage, retention lakes, or reservoirs and ponds, and the capacity of plants varies from 4 kW to 20 MW (Holm 2017).By placing PV modules on water bodies, on one hand, the power output of the PV module increases by 5.9% due to the backwater cooling of modules (Majid et al. 2014) and on the other hand, water is conserved as the evaporation of water reduces up to 70% from water bodies (Sharma, Muni, and Sen 2015). The first FPV system was installed in California, in 2007,and the other FPV systems existing in different regions of the world were mostly established after 2014. Worldwide, the installed capacity of FPVs is almost 94 MW, and the plants installed in Japan contribute the most.

12.3.2.1 Construction of an FPV System

The basic components/parts of an FPV system, as shown in Figure 12.4, are the mooring system, the floating system, the PV panel, and the connecting cables. The floating system is a mixer of floater and structure and the PV panel is placed above it. The mooring system is a construction to which a vessel may be attached by means of anchor or cables. It prevents the installed PV modules from floating away or turning. The PV system has power conditioning devices and solar modules that converts into electrical energy from solar energy. In general, crystalline PV cells are utilized in FPVs, but in PV modules, fabrication research is needed that will adapt to the water

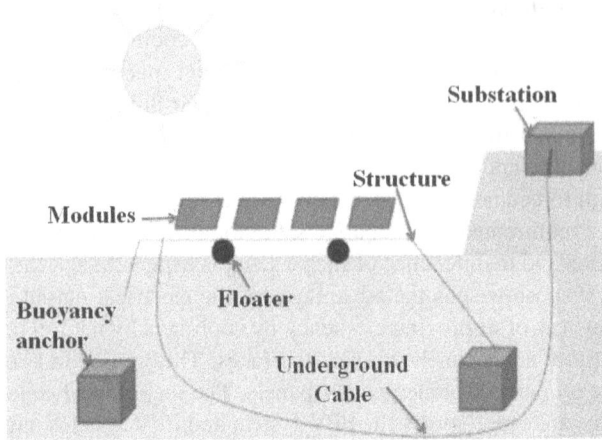

FIGURE 12.4   FPV system model.

bodies' or reservoir atmosphere. Underground cables are required to transmit the generated electrical energy from the FPVs to land. Later, the power can be stored in batteries or can be fed to the grid (Sahu, Yadav, and Sudhakar 2016).

12.3.2.2   Classification of an FPV System

FPV systems are classified into different types based on the module tracking system and the floating system used and the classification is shown in Figure 12.5. In a fixed-type FPV system, at a certain angle, PV modules are fixed. The fixed system has a normal design but to avoid turning away of the PV module, the mooring

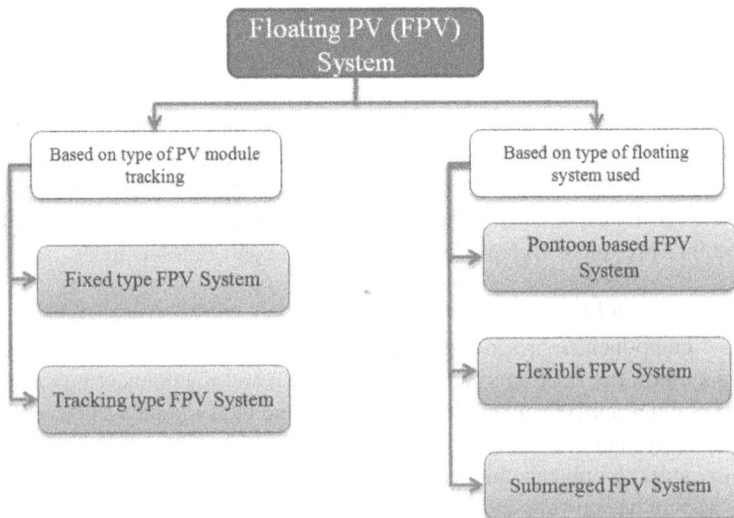

FIGURE 12.5   Classification of an FPV system.

should be constructed precisely. The main advantage with fixed FPV systems is that the PV system weight used can be less, which eases in selecting a less mechanical strength flotation structure (Y-K.Choi et al. 2014). In tracking-type FPV, to track the altitude of the sun and azimuth angle, a tracking system is installed. (Y-K.Choi et al. (2014) have proposed an algorithm for a 100 kW floating plant and by using both passive and active tracking systems, the azimuth angle tracking is attained. A fiber-reinforced plastic polymer membrane is used as a round rotary material and it was detected to be more stable and durable than aluminum and steel (Choi and Lee 2014). A 25% increase in the efficiency is achieved by utilizing a vertical axis system tracking stated by (Ueda et al. 2008). In general, a tracking type floating system has a power output 60–70% more than a fixed plant (Cazzaniga et al. 2012).

Based on the type of floating, FPVs are classified as pontoon, flexible, and submerged PV systems. A pontoon has good buoyancy enough to float on itself with high load and is referred to a floating device. Most of the FPV systems that are pre-installed are pontoon based. Pontoons are manufactured by rotational molding and are usually made from polyethylene, which has medium density. The main drawback of pontoons is that they cannot withstand drastic environmental conditions and they also create a limitation on the size of the plant.

Flexible FPV is a thin-film concept stated to have more reliability, with the significant performance of the system. The performance of a flexible FPV system is compared against a ground-mounted PV system by Trapani and Millar (2014). They reported an increase in electrical power an average of 5% due to the effect of cooling the water. Flexible FPVs can simply distort with the wave motion and the infrastructure requirement is also less. The radiation will fall with several angles of incident on the surface of solar cells and to maintain the close contact of PV array with the water surface, the surface tension is used (Trapani and Millar 2014).

12.3.2.2.1 Submerged PV System

In the submerged PV system, the solar panels are immersed in shallow water. The solar panels are more vulnerable to thermal degradation at more than the critical temperature value. For efficiency improvement, one of the options is temperature reduction. An attempt can be made to study the effect of cooling on temperature decrease. At various water flow rates and different depths in an underwater environment, the performance can be achieved better with the irradiation presence in depth of water, which is also called a submerged PV system. The performance of PV solar panels in submerged systems is affected due to the change of the radiation spectrum and by the decrease in utilizing temperature of PV modules. The performance-affecting factors of submerged systems depend on the solar cell technology, environmental conditions, and on the depth of water.

Tina et al. (2012) explored the submerged PV panels' energy advantages by studying their thermal and optical behavior for a depth of 1 cm, 5 cm, 10 cm, and 15 cm water. From submerged PV panels, there is a sizable increment in the power output due to two main reasons: absence of thermal drift and reduction of light reflection. Lanzafame et al. (2010) examined the electrical and thermal ways of a mono-crystalline module that is submerged by changing the water depth from 1 cm to 15 cm.

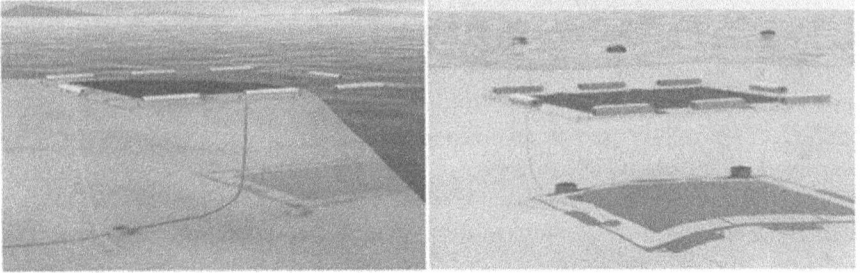

FIGURE 12.6   SCINTEC submergible PV concept. (From Mittal, Saxena, and Rao 2017.)

A 10–20% increase in efficiency is obtained at a depth of 8–10 cm water. They analyzed a best depth of water that exists until the module efficiency rises. Scienza Industria Technologia (SCINTEC) designed a solar cell that will work under 0–2 m of water as shown in Figure 12.6, also known as the submergible PV concept (Mittal, Saxena, and Rao 2017).

12.3.2.3   Advantages of an FPV System

There are several advantages of the FPV system and a few are listed here. As the FPV system does not require any foundation work, it can be easily deployed and maintained. Placement of PV modules on the water surface and in the water body reduces the evaporation of water, algae growth on water bodies, and also conserves land space. Dust accumulation and its impact on PV panels will be reduced, which will improve the efficiency and another major reason for increase in efficiency of PV systems is because of the cooling effect of water present in the water bodies. But as discussed in Sections 12.3.1 and 12.3.2, the liquid used for cooling the PV panels needs to satisfy the following requirements such as good heat conduction, nontoxic, good chemical stability, and be economical and easily available. Also, the absorption of sunlight by the liquid should match the spectral response of solar cells. Considering all these factors, water is the most suitable liquid because it has a high thermal capacity and ensures passive cooling, which will improve the efficiency of the PV panels.

The above specified data indicates that the solar panel improved performance in different water environmental conditions. Different designs and structural materials specified above improve the performance of a PV system. The issues encountered in an off-shore environment, and design complexity, prevents implementation of the large-scale FPV system, even though the natural evaporative cooling and reflection of light from the water can uphold the temperatures lesser than a land-based PV panel and therefore increases efficiency. The FPV system contributes shading on the surface of water and decreases evaporation. The reduced photosynthesis and algae growth lead to better water quality. Generally, areas with more potential of solar energy tend to be arid and dusty, so in the counterparts of a ground-mounted system, FPV systems can work in a less dust ambience. It saves precious area for tourism, mining, agricultural, and other land-impulsive actions and turns nonprofits generating and unexploited surfaces of water into profit-oriented PV power plants.

12.4   HYDRO-OPTICAL CHARACTERISTICS OF WATER AND ITS SUITABILITY FOR SOLAR CELLS

Natural waters, both saline and fresh, are of diffused and impure matter. The solutes and particles are both highly variable and optically significant in concentration and type. The hydro-optical properties of water show more spatial and temporal variations and resemble those of pure water. The water large-scale optical properties are divided into two unique classes, mutually: one is inherent and the other is apparent (Pozdnyakov et al. 1999; Mackenzie 2018; Morris 2009; Mobley 1995).

Inherent optical properties (IOPs) rely solely on the medium, and are not dependent on the field of "ambient light" inside the medium. The main IOPs are the volume scattering function and the absorption coefficient. Other IOPs contain the beam attenuation coefficient, the single-scattering albedo, and the index of refraction. Apparent optical properties (AOPs) depend both on the "geometric (directional) structure" and on the medium of the "ambient light field," and that shows ample stability and regular characteristics to be useful descriptors of a water body. Generally used AOPs are the various diffuse attenuation coefficients, the average cosines, and the irradiance reflectance.

Radiative transfer theory came up with the connection between the AOPs and the IOPs. The water body's physical territory—the incident radiance from the sky, waves on its surface, the character of its bottom—via the boundary conditions enters the theory to have a solution for the equations arising within the theory. The IOPs specify the need of radiative transfer theory through the optical properties of natural waters.

The IOPs include:

1. index of refraction
2. volume scattering function
3. scattering phase function
4. absorption coefficient
5. beam attenuation coefficient
6. single-scattering albedo
7. scattering coefficient
   - forward scattering coefficient
   - backward scattering coefficient

The AOPs include:

1. distribution function
2. remote sensing reflectance
3. average cosine of light field of
   - downwelling light dimensionless
   - upwelling light dimensionless
4. irradiance reflectance
5. diffuse attenuation (vertical) coefficients of
   - downward irradiance

- upward irradiance
- downward scalar irradiance
- upward scalar irradiance
- total scalar irradiance
- PAR

The seawater created by ions resulting from the dissolved salts is a better electricity conductor than pure water. The seawater conductivity is more than one million times when compared with pure water. The absorption at wavelengths cannot be affected by ions. However, the conductor behavior of sea water gives it a higher absorption than pure water at very long wavelengths. The electromagnetic (EM) radiation wavelength is in the order of thousands of kilometers and the sea water has an equitable absorption at visible wavelengths to the low values.

12.5 SOLAR IRRADIATION INSIDE OCEAN WATER AT DIFFERENT DEPTHS

If we take the planet as a whole, we will be able to notice that all water bodies are not a uniform blue and preferably vary in shade based on the depth of water. The deeper waters are more darker blue and lighter blue at shallower waters. In natural light without any artificial source of light, if we take a photograph, everything takes on a bluish hue, and the further that if we go down—30 meters, 100 meters, 200 meters, and more—everything appears blue.

When light passes through the atmosphere, it primarily absorbs instead of scattering. Water is like the atmosphere, it is made out of finite-size molecules: smaller than visible light wavelengths. Water molecules have a preference that they can absorb the wavelengths. Rather than a straightforward wavelength dependence, the water absorbs ultraviolet light, infrared light, and red visible light. When we go down to a moderate depth, the Sun's warmth we won't experience, things will start to turn blue, and we will be protected from ultraviolet radiation as the red light is taken away. Heading down a little further, the orange then the yellows, violets, and greens start to disappear. As we go to depths of multiple kilometers, the blue light finally disappears as well. That's why the deepest ocean appears a dark blue: as all the wavelengths other than blue get absorbed.

In water, among all the wavelengths of light, the deepest blue is unique and has a higher probability of getting reemitted and reflected. The reflectivity or global average albedo of our planet is 0.30, which means the incident light reflected back into the space is 30%. But if we consider that Earth is a totally deep-water ocean, the albedo would be just 0.11. The ocean is actually good at absorbing sunlight (Siegel, 2019). About 2% of the incident solar energy at the air-water interface is reflected back to the atmosphere, and the remaining is transmitted. The reflected solar energy fraction at the interface is more in the area of the solar spectrum short-wavelength part. The deletion of scattering centers due to dust particles and colloid materials makes seawater similar in its properties of light transmission in the pure water. In seawater, the dissolved salts of 3.5% by weight make no contribution for the process of solar energy absorption, but they enhance the seawater scattering coefficient by an insignificant amount. Solar energy inside water depth does not show a trend of an exponential decrease with depth.

The solar radiation beam, at a depth more than 3 meters in water, behaves as a mono-energetic radiation beam. The 3 m of water depth acts as a filter and separates the lower energy part of the spectrum. At the lower energy part, the extinction coefficient is high. A considerable fraction of absorption was noticed by Muaddi and Jamal (1990)in the first centimeter of water and is about 27% of the energy transmitted by solar, and in the first 3 m, about 70% and predicted absorption at a depth of 100 m is about 0.25% of the solar energy is transmitted.

The solar spectrum transmitted in pure water exhibits different interesting characteristics. At the first centimeter, above 1,300nm wavelength photons are absorbed by the water and fully detached from the spectrum. The spectral distribution becomes limited at a depth more than 2 m (i.e., to the visible region only). The solar energy beam steadily loses its heterogeneity with depth of water. In the solar spectrum infrared region, selective absorption grabs place because of the active vibration modes of the water molecules inside the water. In the spectral range of 300–2,500 nm, ionization cannot take place in pure water because photons' energies are less than 12.6eV which is the initial potential ionization of water. The seawater transparency is less than pure water delinquent to the particulate substances scattering at which it acts as soothe agents. The seawater will have optical properties that are similar as pure water, by removing the scattering centers, as explained by Muaddi and Jamal (1991).

12.6   RECENT USAGE OF UNDERWATER SOLAR PV CONCEPT

Remote marine sensing systems such as telemetry tags or underwater autonomous vehicles are limited in deployment duration and collection of data because of the finite energy available from the battery placed onboard. With limited power available, maintaining a high data resolution, and migrating to large distances over deep-rooted deployments is difficult and such systems often yield non ideal data sets. In telemetry tags, the electronic systems used are often potted in epoxy making the battery recharging or replacing inappropriate. The application of solar PV panels on such marine systems and harvesting energy could improve the tag longevity and/or collected data fidelity. Hahn et al. (2018)presented model assessment, which evaluates the output energy of a migrating or stationary solar cell below or above the surface of the ocean. The assumptions and theory beyond the model are explained in detail, which includes review concepts established for the purpose of variable consistency and consolidation.

Jenkins et al. (2014) demonstrated that maximum utilized power is procured at a water depth of 9.1 m utilizing high-bandgap InGaP solar cells. These solar cells absorb the "blue" segment of the spectrum and the light persisting is transmitted. The light absorbed by the N-contact of the solar cell has a decreased bandgap. At the greatest depth of 9.1 m, the power output obtained was 7 W/m^2 of solar cells, which will be sufficient for sensor systems used in modern electronics.

Sheeba, Rao, and Jaisankar (2015) conducted experiments at the Center for Energy and Environmental Science and Technology (CEESAT), an India-UK Renewable Energy Corporation project at the National Institute of Technology-Trichy, India, in both continuous and batch mode. Most of the investigations aimed at assessing the efficiency deviation of an amorphous-silicon PV panel by changing the depth of water in the range of 2–20 cm in submersed condition. The efficiency deviation with various water flow rates of 20, 30, and 40 ml/sec is also considered for the optimal depth.

Walters et al. (2015) presented the pattern design of a novel organic PV device tailored for underwater operation. The organic PV cell is multi-junction design with same spectral response of two absorber layers. The UW environment analysis is also presented at which the advantages of organic PV highlight. An organic PV enabled efficient conversion of the narrow underwater spectrum, resulting in high voltage.

Mol'kov and Dolin (2016) determined the inherent hydro-optical performance of water utilizing the vision means underwater. The analytical methods of the solar underwater path, framed by single and multiple scattered light, and direct light, are suggested. Using numerical simulation, the optical depths at which the water-scattered light contribution is predominant are estimated. Algorithms for attenuation coefficients and rejuvenating the water scattering from the underwater sculpture of solar paths are also suggested.

Tina et al. (2019) designed the feasibility to utilize PV modules under a layer of water. They considered a PV cell under different conditions: submerged, under a translucent box that carries water, and covered by a water layer. All the above test conditions have the benefits of water as a filter for the solar radiation spectrum and to decrease the heating of the PV cells. Highly depending on PV cell technology, the effect of the radiation spectrum varies on the PV cells.

Rosa-Clot et al. (2010) used a technique in the submerged water to cool the mono-crystalline module. The temperature at 30°Cyields an increase in efficiency of 20%, but irradiation intensity decreases with water depth. The relative efficiency is increased by 11% at a depth of 4 cm.

Abdulgafar, Omar, and Yousif (2007) compared 0.12 watts and 15 cm^2 polycrystalline solar cells efficiencies drenched in deionized depths of water. At the lowest depth of 1 cm, the overall power gained is high. However, at a depth of 6 cm, the greatest efficiency of 22% was achieved.

12.7 CONCLUSION

From the extensive literature survey done, it has been observed that by decreasing the temperature on the surface of PV cells, efficiency of the cells can be improved tremendously. The temperature can be decreased with the use of two types of cooling techniques: forced cooling and natural cooling. Forced cooling systems have additional electrical equipment that consumes power more than generated power, and the natural cooling technique has different construction for installation of PV panels. It is also observed that while comparing the literature based on forced and natural cooling techniques, the natural cooling techniques were mostly preferred with the PV systems. This is mainly due to the effective operation of PV cells working with natural cooling techniques as well as the cost factor when taken into account.

The natural cooling system involves FPV systems. In FPV systems, the submerged PV system type is mostly considered. The factors affecting the PV cell efficiency are majorly eliminated with the submerged PV systems. Even water has two types of optical properties—inherent and apparent—which shows the presence of irradiation at different water depths. The spectrum has different wavelengths and the distribution of wavelengths inside water also varies with the presence of particulate matter and type of water (normal, ocean, lake, organic, etc.). At different depths, the optical properties of solar cells vary with the environmental conditions but are suitable for submerged PV systems to operate and produce electricity.

Submerged PV systems save valuable land for tourism, mining, agricultural, and other land-impulse actions and turns non-profit generating and unexploited surface of water into profit-oriented PV power plants. The hydro-optical characteristics of solar cells prove that the irradiation presence in water and the physical properties of water makes a decrease in temperature by increasing the efficiency of submerged PV cells. On the whole, the above advantages will allow submerged PV systems that are suitable for energy generation in the present scenario.

REFERENCES

Abdulgafar, Sayran A, Omar S Omar, and Kamil M Yousif. 2007. "Improving the efficiency of polycrystalline solar panel via water immersion method," *International Journal of Innovative Research in Science, Engineering and Technology (An ISO Certified Organization)*.

Al-Shamani, Ali Najah, Mohammad H. Yazdi, M. A. Alghoul, Azher M. Abed, M. H. Ruslan, Sohif Mat, and K.Sopian. 2014. "Nanofluids for improved efficiency in cooling solar collectors–A review," *Renewable and Sustainable Energy Reviews*. https://doi.org/10.1016/j.rser.2014.05.041.

Bahaidarah, H., Abdul Subhan, P. Gandhidasan, and S. Rehman.2013. "Performance evaluation of a PV (Photovoltaic) module by back surface water cooling for hot climatic conditions." https://doi.org/10.1016/j.energy.2013.07.050.

Banerjee, Arindam, Frank Shengzhong Liu, Dave Beglau, Tining Su, Ginger Pietka, Jeff Yang, and Subhendu Guha.2012. "12.0% efficiency on large-area, encapsulated, multijunction Nc-Si:H-based solar cells," *IEEE Journal of Photovoltaics* 2 (2): 104–08. https://doi.org/10.1109/JPHOTOV.2011.2181823.

Becquerel, E. 1839. "Mémoire sur les effets électriques produits sous l'influence des rayons solaires," *Comptes Rendus Hebdomadaires Des Séances de l'Académie Des Sciences* 9: 561–67.

Böscke, T. S., D. Kania, A. Helbig, C. Schöllhorn, M. Dupke, P. Sadler, M. Braun, et al. 2013. "Bifacial N-type cells with >20% front-side efficiency for industrial production," *IEEE Journal of Photovoltaics* 3 (2): 674–77. https://doi.org/10.1109/JPHOTOV.2012.2236145.

Brown, R. 1967. "Surface effects in silicon solar cells," *IEEE Transactions on Nuclear Science* NS-14 (6): 260–65. https://doi.org/10.1109/TNS.1967.4324805.

Cazzaniga, R., Marco Rosa-Clot, Paolo Rosa-Clot, and Giuseppe M. Tina. 2012. "Floating tracking cooling concentrating (FTCC) systems," in *Conference Record of the IEEE Photovoltaic Specialists Conference*. https://doi.org/10.1109/PVSC.2012.6317668.

Chandrasekar, M., S. Suresh, T. Senthilkumar, and M. Ganesh Karthikeyan. 2013. "Passive cooling of standalone flat PV module with cotton wick structures," *Energy Conversion and Management*. https://doi.org/10.1016/j.enconman.2013.03.012.

Choi, Y. 2014. "A case study on suitable area and resource for development of floating photovoltaic system," *International Journal of Electrical, Computer, Energetic, Electronic and Communication Engineering* 8 (5): 828–32.

Choi, Young Kwan. 2014. "A study on power generation analysis of floating PV system considering environmental impact," *International Journal of Software Engineering and Its Applications*, 18. https://doi.org/10.14257/ijseia.2014.8.1.07.

Choi, Young-Kwan, Nam-Hyung Lee, and Kern-Joong Kim. 2013. "Empirical research on the efficiency of floating PV systems compared with overland PV systems," *Ces-Cube 2013, Astl.*

Choi, Young-Kwan, Nam-Hyung Lee, An-Kyu Lee, and Kern-Joong Kim. 2014. "A study on major design elements of tracking-type floating photovoltaic systems," *International Journal of Smart Grid and Clean Energy.* https://doi.org/10.12720/sgce.3.1.70-74.

Choi, Young Kwan, and Young Geun Lee. 2014. "A study on development of rotary structure for tracking-type floating photovoltaic system," *International Journal of Precision Engineering and Manufacturing.* https://doi.org/10.1007/s12541-014-0613-5.

Daghigh, R., M. H. Ruslan, and K. Sopian. 2011. "Advances in liquid based photovoltaic/ thermal (PV/T) collectors," *Renewable and Sustainable Energy Reviews.* https://doi. org/10.1016/j.rser.2011.07.028.

Deng, Weiwei, Daming Chen, Zhen Xiong, Pierre Jacques Verlinden, Jianwen Dong, Feng Ye, Hui Li, et al. 2016. "20.8% PERC solar cell on 156 mm × 156 mm P-type multicrystalline silicon substrate," *IEEE Journal of Photovoltaics* 6 (1): 3–9. https://doi. org/10.1109/JPHOTOV.2015.2489881.

Du, Bin, Eric Hu, and Mohan Kolhe. 2012. "Performance analysis of water cooled concentrated photovoltaic (CPV) system," *Renewable and Sustainable Energy Reviews* 16 (9): 6732–36. https://doi.org/10.1016/j.rser.2012.09.007.

Durganjali, C. Santhi, and Radhika Sudha. 2019. "PV cell performance with varying temperature levels," in *2019 Global Conference for Advancement in Technology, GCAT 2019.* https://doi.org/10.1109/GCAT47503.2019.8978302.

El-Seesy, Ibrahim E., T. Khalil, and Mohamed H. Ahmed.2012. "Experimental investigations and developing of photovoltaic/thermal system," *World Applied Sciences Journal.* https://doi.org/10.5829/idosi.wasj.2012.19.09.2794.

Energy Information Administration of US Department of Energy. 2016. "International energy outlook 2016-natural gas," *International Energy Outlook 2016*, 37–60.

Energy Informative. 2013. "What factors determine solar panel efficiency?" https://energyinformative.org/solar-panel-efficiency.

Essig, Stephanie, Jan Benick, Michael Schachtner, Alexander Wekkeli, Martin Hermle, and Frank Dimroth. 2015. "Wafer-bonded GaInP/GaAs//Si solar cells with 30% efficiency under concentrated sunlight," *IEEE Journal of Photovoltaics* 5 (3): 977–81. https://doi. org/10.1109/JPHOTOV.2015.2400212.

Farhana, Z., Y. M. Irwan, R. M. N. Azimmi, A. R. N. Razliana, and N. Gomesh. 2012. "Experimental investigation of photovoltaic modules cooling system," in *2012 IEEE Symposium on Computers and Informatics, ISCI 2012.* https://doi.org/10.1109/ ISCI.2012.6222687.

Ferrer-Gisbert, Carlos, José J. Ferrán-Gozálvez, Miguel Redón-Santafé, Pablo Ferrer-Gisbert, Francisco J. Sánchez-Romero, and Juan Bautista Torregrosa-Soler. 2013. "A new photovoltaic floating cover system for water reservoirs," *Renewable Energy.* https://doi. org/10.1016/j.renene.2013.04.007.

Gozálvez, José Javier Ferrán, Pablo S. Ferrer Gisbert, Carlos M. Ferrer Gisbert, Miguel Redón Santafé, Juan Bautista Torregrosa Soler, and Emili Pons Puig. 2012. "Covering reservoirs with a system of floating solar panels: Technical and financial analysis," in *International Congress on Project Engineering, Valencia.*

Hahn, Gregory G., Lauren Andrea Adoram-Kershner, Heather P. Cantin, and Michael W. Shafer. 2018. "Assessing solar power for globally migrating marine and submarine systems," *IEEE Journal of Oceanic Engineering*, 1–14. https://doi.org/10.1109/ JOE.2018.2835178.

Han, Xinyue, Yiping Wang, and Li Zhu. 2013. "The performance and long-term stability of silicon concentrator solar cells immersed in dielectric liquids," *Energy Conversion and Management*. https://doi.org/10.1016/j.enconman.2012.10.009.

He, Wei, Yang Zhang, and Jie Ji. 2011. "Comparative experiment study on photovoltaic and thermal solar system under natural circulation of water," *Applied Thermal Engineering*. https://doi.org/10.1016/j.applthermaleng.2011.06.021.

Ho, C. J., Wei Len Chou, and Chi Ming Lai. 2015. "Thermal and electrical performance of a water-surface floating PV integrated with a water-saturated MEPCM layer," *Energy Conversion and Management*. https://doi.org/10.1016/j.enconman.2014.10.039.

Ho, C. J., Bor-Tyng Jou, and Chi-Ming Lai. 2017. "Thermal and electrical performance of a PV module integrated with double layers of water-saturated MEPCM," *Applied Thermal Engineering* 123(August): 1120–33. https://doi.org/10.1016/j.applthermaleng.2017.05.166.

Holm, A. 2017. "Floating solar photovoltaics gaining ground." National Renewable Energy Lab, blog. https://www.nrel.gov. https://www.nrel.gov/state-local-tribal/blog/posts/floating-solar-photovoltaics-gaining-ground.html#:~:text=The%20system%20consists%20of%20994,acres%20of%20productive%20vineyard%20land.

Jenkins, Phillip P., Scott Messenger, Kelly M. Trautz, Sergey I. Maximenko, David Goldstein, David Scheiman, Raymond Hoheisel, and Robert J. Walters. 2014. "High-bandgap solar cells for underwater photovoltaic applications," *IEEE Journal of Photovoltaics* 4 (1): 202–06. https://doi.org/10.1109/JPHOTOV.2013.2283578.

Kalaiselvan, S., V. Karthikeyan, G. Rajesh, A. Sethu Kumaran, B. Ramkiran, and P. Neelamegam. 2018. "Solar PV active and passive cooling technologies-A review," *7th IEEE International Conference on Computation of Power, Energy, Information and Communication, ICCPEIC 2018*, 166–69. https://doi.org/10.1109/ICCPEIC.2018.8525185.

Kennerud, Kenneth L. 1967. "Electrical characteristics of silicon solar cells at low temperatures," *IEEE Transactions on Aerospace and Electronic Systems* AES-3(4): 586–90. https://doi.org/10.1109/TAES.1967.5408834.

Krauter, Stefan. 2004. "Increased electrical yield via water flow over the front of photovoltaic panels," in *Solar Energy Materials and Solar Cells*. https://doi.org/10.1016/j.solmat.2004.01.011.

Kumar, Rakesh, and Marc A. Rosen. 2011. "A critical review of photovoltaic-thermal solar collectors for air heating," *Applied Energy*. https://doi.org/10.1016/j.apenergy.2011.04.044.

Lanzafame, Rosario, Silvia Nachtmann, Marco Rosa-Clot, Paolo Rosa-Clot, Pier Francesco Scandura, Stefano Taddei, and Giuseppe M. Tina. 2010. "Field experience with performances evaluation of a single-crystalline photovoltaic panel in an underwater environment," *IEEE Transactions on Industrial Electronics*. https://doi.org/10.1109/TIE.2009.2035489.

Lee, Young Geun, Hyung Joong Joo, and Soon Jong Yoon. 2014. "Design and installation of floating type photovoltaic energy generation system using FRP members," *Solar Energy*. https://doi.org/10.1016/j.solener.2014.06.033.

Luft, Werner. 1970. "Silicon solar cell performance at high intensities," *IEEE Transactions on Aerospace and Electronic Systems* AES-6 (6): 797–803. https://doi.org/10.1109/TAES.1970.310161.

Mackenzie, Fred T. 2018. "Chemical and physical properties of seawater," *Seawater*, 1–24.

Majid, Z. A. A., M. H. Ruslan, K. Sopian, M. Y. Othman, and M. S. M. Azmi. 2014. "Study on performance of 80 watt floating photovoltaic panel," *Journal of Mechanical Engineering and Sciences*. https://doi.org/10.15282/jmes.7.2014.14.0112.

Makki, Adham, Siddig Omer, and Hisham Sabir. 2015. "Advancements in hybrid photovoltaic systems for enhanced solar cells performance," *Renewable and Sustainable Energy Reviews*. https://doi.org/10.1016/j.rser.2014.08.069.

Mittal, Divya, Bharat Kumar Saxena, and K. V. S. Rao. 2017. "Floating solar photovoltaic systems: An overview and their feasibility at Kota in Rajasthan," *Proceedings of IEEE International Conference on Circuit, Power and Computing Technologies, ICCPCT 2017.* 2017. https://doi.org/10.1109/ICCPCT.2017.8074182.

Mobley, Curtis D. 1995. "Chapter 12: Terrestrial optics." *Handbook of Optics V.1*, 1664.

Mol'kov, A., and L. S. Dolin. 2016. "The possibility of determining optical properties of water from the image of the underwater solar path," *Radiophysics and Quantum Electronics* 58 (8): 586–97. https://doi.org/10.1007/s11141-016-9630-9.

Moradgholi, Meysam, Seyed Mostafa Nowee, and Iman Abrishamchi. 2014. "Application of heat pipe in an experimental investigation on a novel photovoltaic/thermal (PV/T) system," *Solar Energy.* https://doi.org/10.1016/j.solener.2014.05.018.

Morris, D. P. 2009. "Optical properties of water," *Encyclopedia of Inland Waters*, 682–89. https://doi.org/10.1016/B978-012370626-3.00069-7.

Muaddi, J. A., and M. A. Jamal. 1990. "Solar energy at various depths below," *International Journal of Energy Research* 14 (February): 859–67.

Muaddi, J. A., and M. A. Jamal. 1991. "Solar spectrum at depth in water." *Renewable Energy* 1 (1): 31–35. https://doi.org/10.1016/0960-1481(91)90100-4.

Muller, Jens, David Hinken, Susanne Blankemeyer, Heike Kohlenberg, Ulrike Sonntag, Karsten Bothe, Thorsten Dullweber, Marc Kontges, and Rolf Brendel.2015. "Resistive power loss analysis of PV modules made from halved 15.6×15.6 cm^2 silicon PERC solar cells with efficiencies up to 20.0%," *IEEE Journal of Photovoltaics* 5 (1): 189–94. https://doi.org/10.1109/JPHOTOV.2014.2367868.

Nižetić, S., D. Čoko, A. Yadav, and F. Grubišić-Čabo. 2016. "Water spray cooling technique applied on a photovoltaic panel: The performance response," *Energy Conversion and Management.* https://doi.org/10.1016/j.enconman.2015.10.079.

Odeh, Saad, and Masud Behnia. 2009. "Improving photovoltaic module efficiency using water cooling," *Heat Transfer Engineering* 30 (6): 499–505. https://doi.org/10.1080/01457630802529214.

Peng, Zih Wei, Thomas Buck, Lejo J. Koduvelikulathu, Valentin D. Mihailetchi, and Radovan Kopecek. 2019. "Industrial screen-printed n-PERT-RJ solar cells: Efficiencies beyond 22% and open-circuit voltages approaching 700 MV," *IEEE Journal of Photovoltaics* 9 (5): 1166–74. https://doi.org/10.1109/JPHOTOV.2019.2919117.

Pozdnyakov, Dmitry V., Anton V. Lyaskovsky, Fred J. Tanis, and David R. Lyzenga. 1999. "Modeling of apparent hydro-optical properties and retrievals of water quality in the Great Lakes for SeaWiFS: A comparison with in situ measurements," *International Geoscience and Remote Sensing Symposium (IGARSS)* 5: 2742–44. https://doi.org/10.1109/igarss.1999.771637.

Prince, M. B., and M. Wolf. 1958. "New developments in silicon photovoltaic devices," *Journal of the British Institution of Radio Engineers* 18 (10): 583–94. https://doi.org/10.1049/jbire.1958.0062.

Rahimi, Masoud, Peyvand Valeh-E-Sheyda, Mohammad Amin Parsamoghadam, Mohammad Moein Masahi, and Ammar Abdulaziz Alsairafi. 2014. "Design of a self-adjusted jet impingement system for cooling of photovoltaic cells," *Energy Conversion and Management.* https://doi.org/10.1016/j.enconman.2014.03.053.

Redón Santafé, Miguel, Juan Bautista Torregrosa Soler, Francisco Javier Sánchez Romero, Pablo S. Ferrer Gisbert, José Javier Ferrán Gozálvez, and Carlos M. Ferrer Gisbert. 2014. "Theoretical and experimental analysis of a floating photovoltaic cover for water irrigation reservoirs," *Energy* 67: 246–55. https://doi.org/10.1016/j.energy.2014.01.083.

Rosa-Clot, M., P. Rosa-Clot, G. M. Tina, and P. F. Scandura. 2010. "Submerged photovoltaic solar panel: SP2," *Renewable Energy.* https://doi.org/10.1016/j.renene.2009.10.023.

Royne, Anja, Christopher J. Dey, and David R. Mills. 2005. "Cooling of photovoltaic cells under concentrated illumination: A critical review," *Solar Energy Materials and Solar Cells.* https://doi.org/10.1016/j.solmat.2004.09.003.

Sahu, Alok, Neha Yadav, and K. Sudhakar. 2016. "Floating photovoltaic power plant: A review," *Renewable and Sustainable Energy Reviews*. https://doi.org/10.1016/j. rser.2016.08.051.

Sharma, P., B. Muni, and D. Sen. 2015. "Design parameters of 10kw floating solar power plant." International Advanced Research Journal in Science, Engineering and Technology (IARJSET) National Conference on Renewable Energy and Environment (NCREE-2015), Vol. 2, Special Issue 1, 85-89. https://doi.org/10.17148/ IARJSET.

Sheeba, K. N., R. Madhusudhana Rao, and S. Jaisankar. 2015. "A study on the underwater performance of a solar photovoltaic panel," *Energy Sources, Part A: Recovery, Utilization and Environmental Effects* 37 (14): 1505–12. https://doi.org/10.1080/1556 7036.2011.619632.

Shockley, William, and Hans J.Queisser.1961. "Detailed balance limit of efficiency of P-Njunction solar cells," *Journal of Applied Physics* 32 (3): 510–19. https://doi. org/10.1063/1.1736034.

Siegel, Ethan. 2019. "This is why Earth's oceans and skies are blue–Starts with a bang!" *Medium*, May 24. https://medium.com/starts-with-a-bang/this-is-why-earths-oceans-and-skies-are-blue-c408ce7d0d45.

Smith, Christopher J., Piers M. Forster, and Rolf Crook. 2014. "Global analysis of photovoltaic energy output enhanced by phase change material cooling," *Applied Energy*. https://doi.org/10.1016/j.apenergy.2014.03.083.

Smith, K. D., H. K. Gummel, J. D. Bode, D. B. Cuttriss, R. J. Nielsen, and W. Rosenzweig. 1963. "The solar cells and their mounting," *Bell System Technical Journal* 42 (4): 1765–1816. https://doi.org/10.1002/j.1538-7305.1963.tb04050.x.

Smith, Matthew K., Hanny Selbak, Carl C. Wamser, Nicholas U. Day, Mathew Krieske, David J. Sailor, and Todd N. Rosenstiel. 2014. "Water cooling method to improve the performance of field-mounted, insulated, and concentrating photovoltaic modules," *Journal of Solar Energy Engineering, Transactions of the ASME*. https://doi. org/10.1115/1.4026466.

Taguchi, Mikio, Ayumu Yano, Satoshi Tohoda, Kenta Matsuyama, Yuya Nakamura, Takeshi Nishiwaki, Kazunori Fujita, and Eiji Maruyama. 2014. "24.7% record efficiency HIT solar cell on thin silicon wafer," *IEEE Journal of Photovoltaics* 4 (1): 96–99. https://doi. org/10.1109/JPHOTOV.2013.2282737.

Tallent, R. J., and Henry Oman. 2013. "Solar-cell performance with concentrated sunlight," *Transactions of the American Institute of Electrical Engineers, Part II: Applications and Industry* 81 (1): 30–33. https://doi.org/10.1109/tai.1962.6371786.

Tina, G. M., M. Rosa-Clot, and P. Rosa-Clot. 2011. "Electrical behavior and optimization of panels and reflector of a photovoltaic floating plant," in *Proceedings of the 26th European Photovoltaic Solar Energy Conference and Exhibition (EU PVSEC'11)*, 4371–75.

Tina, G. M., M. Rosa-Clot, P. Rosa-Clot, and P. F. Scandura. 2012. "Optical and thermal behavior of submerged photovoltaic solar panel: SP2," *Energy*. https://doi.org/10.1016/j. energy.2011.08.053.

Tina, Giuseppe Marco, Marco Rosa-Clot, Vesna Lojpur, and Ivana Lj Validzic. 2019. "Numerical and experimental analysis of photovoltaic cells under a water layer and natural and artificial light," *IEEE Journal of Photovoltaics* 9 (3): 733–40. https://doi. org/10.1109/JPHOTOV.2019.2896669.

Tonui, J. K., and Y. Tripanagnostopoulos. 2006. "Improved PV/T solar collectors with heat extraction by forced or natural air circulation," *International Journal of Hydrogen Energy*. https://doi.org/10.1016/j.ijhydene.2006.02.009.

Trapani, Kim, and Dean L. Millar. 2014. "The thin film flexible floating PV (T3F-PV) array: The concept and development of the prototype," *Renewable Energy*. https://doi. org/10.1016/j.renene.2014.05.007.

Trapani, Kim, and Miguel Redõn Santafé. 2015. "A review of floating photovoltaic installations: 2007–2013," *Progress in Photovoltaics: Research and Applications.* https://doi.org/10.1002/pip.2466.

Ueda, Yuzuru, Tsurugi Sakurai, Shinya Tatebe, Akihiro Itoh, and Kosuke Kurokawa. 2008. "Performance analysis of PV systems on the water," in *23rd European Photovoltaic Solar Energy Conference,* 2670–73.

United Nations Educational Scientific and Cultural Organization (UNESCO). 2012. "The properties and availability of water: A fundamental consideration for life." Water Civilization International Centre, 1. http://www.unesco.org/new/fileadmin/MULTIMEDIA/FIELD/Venice/pdf/special_events/bozza_scheda_DOW02_1.0.pdf.

Walters, W. Yoon, D. Placencia, D. Scheiman, M. P. Lumb, A. Strang, Stavrinou, and P. P. Jenkins. 2015. "Multijunction organic photovoltaic cells for underwater solar power," *IEEE 42nd Photovoltaic Specialist Conference (PVSC),* New Orleans, LA, 2015, pp. 1–3, https://doi:10.1109/PVSC.2015.7355644.

Wang, Wei, Jian Sheng, Shengzhao Yuan, Yun Sheng, Wenhao Cai, Yifeng Chen, Chun Zhang, Zhiqiang Feng, and Pierre J. Verlinden. 2015. "Industrial screen-printed n-type rear-junction solar cells with 20.6% efficiency," *IEEE Journal of Photovoltaics* 5 (4): 1245–49. https://doi.org/10.1109/JPHOTOV.2015.2416919.

Webster, John G., Yu. G. Gurevich, and J. E. Velazquez-Perez. 2014. "Peltier effect in semiconductors," *Wiley Encyclopedia of Electrical and Electronics Engineering,* no. November 2017: 1–21. https://doi.org/10.1002/047134608x.w8206.

Würfel, Peter. 2007. *Physics of solar cells: From principles to new concepts.* Wiley-VCH Verlag. https://doi.org/10.1002/9783527618545.

Wysocki, J. J. 1966. "Lithium-doped radiation resistant silicon solar cells," in *IEEE RANSACTIONS on NUCLEAR SCIENCE,* 13: 168–73. https://doi.org/10.1109/iedm.1966.187649.

Yin, Weiwei, Xusheng Wang, Feng Zhang, and Lingjun Zhang. 2013. "19.6% cast mono-MWT solar cells and 268 W modules," *IEEE Journal of Photovoltaics* 3 (2): 697–701. https://doi.org/10.1109/JPHOTOV.2013.2239357.

Zhang, Shu, Xiujuan Pan, Haijun Jiao, Weiwei Deng, Jianmei Xu, Yifeng Chen, Pietro P. Altermatt, Zhiqiang Feng, and Pierre Jacques Verlinden. 2016. "335-W world-record p-type monocrystalline module with 20.6% efficient PERC solar cells," *IEEE Journal of Photovoltaics* 6 (1): 145–52. https://doi.org/10.1109/JPHOTOV.2015.2498039.

Zheng, Peiting, Fiacre Emile Rougieux, Xinyu Zhang, Julien Degoulange, Roland Einhaus, Pascal Rivat, and Daniel H. MacDonald. 2017. "21.1% UMG silicon solar cells," *IEEE Journal of Photovoltaics* 7 (1): 58–61. https://doi.org/10.1109/JPHOTOV.2016.2616192.

13 A Review on Brain Tumor Segmentation Algorithms Using Recent Deep Neural Network Architectures and a Gentle Introduction to Deep Neural Network Concepts

B. Dheerendranath, B. V. V. S. N. Prabhakar Rao, P. Yogeeswari, C. Kesavadas, and Venkateswaran Rajagopalan

CONTENTS

13.1 About Brain Tumors 228
13.2 Brain Tumor Segmentation Importance and Challenges.............................. 228
13.3 What Is a CNN?.. 230
13.4 Common DNN and CNN Terminologies 233
    13.4.1 Receptive Field 233
    13.4.2 Dilated Convolution... 234
    13.4.3 Batch and Group Normalization.. 235
    13.4.4 Stride.. 236
13.5 CNN-Based Brain Tumor Segmentation Algorithms.................................... 237
    13.5.1 3D Autoencoder-Decoder Architecture... 237
    13.5.2 Cascaded Convolutional Networks... 238
    13.5.3 3D Convolutional Neural Networks 239
    13.5.4 U-Net-Based Fully Convolutional Network....................................... 240
13.6 Conclusion 242
Acknowledgment 242
References... 242

13.1 ABOUT BRAIN TUMORS

Tumors are lesions, a mass of abnormal cells derived from elevated proliferation of the affected tissues (Garman 2011). Tumors of the central nervous system (brain) can be comprehensively classified according to their subtype and grade using the diagnostic criteria published by the World Health Organization (WHO) (Louis et al. 2016). A brain tumor is the most common cause for the increase in mortality in children (Ostrom et al. 2015) and adults worldwide. According to the American Cancer society, in 2020 it is estimated that 18,020 people will die of brain cancer in United States and 23,890 new cases will be reported (Siegel, Miller, and Jemal 2020). According to WHO reports, 296,851 new cancer cases of the brain and nervous system were recorded in 2018.

Brain tumors are typically classified into two groups: primary and secondary brain tumors. A primary brain tumor is an abnormal growth that originates in the brain and usually does not spread to other organs in the body. Secondary (metastatic) brain tumors begin as cancer elsewhere in the body and spread to the brain (Fox et al. 2011). Primary brain tumors are classified as benign (noncancerous) or malignant (cancerous). Benign brain tumors have sluggish growth rate, usually have a clear border demarcating the surrounding tissues, and rarely spread to proximal tissues. Malignant tumors provoke life-threatening conditions because they grow rapidly, have irregular boundaries, and spread to neighboring brain areas. In contrast to other types of malignant tumors (lung, liver, breast, etc.), brain cancers are mostly restricted to its site of origin and rarely spread to other body parts, outside of the brain (El-Sayed et al. 2014). Despite whether a brain tumor is benign or malignant, all are potentially life-threatening. The tumor compresses and displaces normal brain tissue, which may affect vital body functions controlled by that part of the brain. Some brain tumors cause a blockage of cerebrospinal fluid that flows around and through the brain, which can increase intracranial pressure and enlarge the ventricles (Lin and Avila 2017).

The WHO classification of tumors affecting the brain is based on histologic features under microscopic appearance of the tumor specimen (Cha 2006). This is subjective and is prone to operator bias. Despite this drawback, this classification scheme is used for the scientific study, therapy, and prognosis of brain tumors (Cha 2006). Therefore, there is a need for robust classification of brain tumors that can provide better diagnosis, aid in prognosis, and in designing therapeutic paradigms.

13.2 BRAIN TUMOR SEGMENTATION IMPORTANCE AND CHALLENGES

Brain biopsy, usually under stereotactic guidance, is the standard clinical procedure to establish diagnosis and decide therapeutic choices in brain tumors located in non-eloquent brain areas. However, there are serious limitations associated with brain biopsy when it comes to tumors located in eloquent cortex or in deep brain regions. Moreover, due to its invasive nature, all biopsy procedures carry potential complications such as infection, risk for neurological deficits, and surgical site complications such as hematoma. In addition, the biopsy can be inconclusive in many conditions

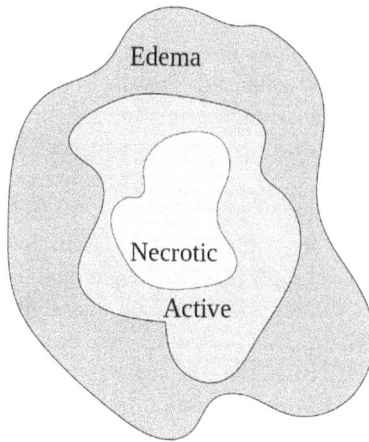

FIGURE 13.1   Pictorial representation of edema, necrotic core, and active region in a brain tumor.

(Heper et al. 2005, Koszewski, Kroh, and Kunert 2002). Therefore, developing a noninvasive quantitative assessment of brain tumors is of vital importance and can change the current management of brain tumors.

In this chapter, we focus on gliomas, a very common type of brain tumor. Early diagnosis of gliomas is essential to improve treatment strategies (Iúõna, Direkoglu, and Sùah 2016). The therapeutic strategies require a detailed and precise analysis of the affected region within the brain of patients. The tumor segmentation is a pathway to outline and segregate healthy brain tissue from tumorous regions in radiologic images (which include active tumor region, necrotic core, and edema (Iúõna, Direkoglu, and Sùah 2016) as shown in Figure 13.1.

In clinical practice, manual segmentation of a brain tumor (by brain tumor, we refer to gliomas in this chapter) from radiological images is the common procedure. Accurate tumor segmentation is critical in deciding surgery and other treatment strategies such as radiation and chemotherapy (Liu et al. 2014). However, manual segmentation is very time consuming, not reproducible, and prone to operator bias (Angulakshmi and Lakshmi Priya 2017). Hence, automated tumor segmentation methods are required in order to reduce time and can provide reproducible results. But automated brain tumor segmentation still remains a challenging task because tumors are heterogeneous in shape and in location within the brain.

Neuroimaging modalities, especially magnetic resonance imaging (MRI), have shown promise in brain tumor diagnosis and prognostication by assessing tumor type, position, and size noninvasively. MRI is routinely employed in clinical/radiological diagnosis of brain tumors because it does not use any ionizing radiations like computed tomography or does not require any invasive procedures like positron emission tomography (Wu and Shu 2018). However, MRI is qualitative (for example, gliomas appear as hyperintense in T2-weighted images) and has difficulty in distinguishing tumor recurrence from old tumors, tumors from non-tumoral lesions such as ischemia, and tumor grading (Verma et al. 2013).

Further, MRI has different imaging sequences that can provide different tissue contrast of the tumor and the healthy brain tissue. Some common clinical sequences include T1-weighted, contrast enhanced T1-weighted, T2-weighted, and FLAIR. Several brain tumor segmentation algorithms for MR images have been developed from a simple image thresholding approach to advanced approaches such as fuzzy c-means and Markov random field; a detailed review of these algorithms is given in Liu et al. (2014). Although these methods were successful in detecting the tumor region, their accuracy levels are still suboptimal. With the availability of advanced computing facilities such as a graphics processing unit and the development of novel convolutional neural networks (CNNs) and deep neural networks (DNNs) in recent years, several researchers have developed automatic brain tumor segmentation algorithms using CNNs and DNNs. In the past few years, in different areas of medical imaging, DNNs and CNNs have produced promising results, especially in diagnosis, detection, and treatment of vulnerable diseases (Bakator and Radosav 2018). Similarly, several robust CNN- and DNN-based algorithms have been proposed for automatic brain tumor segmentation. These algorithms have demonstrated good accuracy in detecting different tumor regions in the brain such as edema, core tumor, and enhancing tumor (Wang et al. 2019).

13.3   WHAT IS A CNN?

A CNN was proposed back in 1987, but gained popularity by the end of the 20th century and the beginning of the 21st century. The conventional artificial neural network consists of an input layer followed by one or more hidden layers (choosing the number of hidden layers depends on the complexity of problem) and an output layer. Commonly in conventional artificial neural networks, handcrafted features are given as input (by the user) to the network. Based on these input features, the network performs classification on any given input data. Figure 13.2, for example, shows a conventional fully connected (meaning every neuron from a layer is connected to

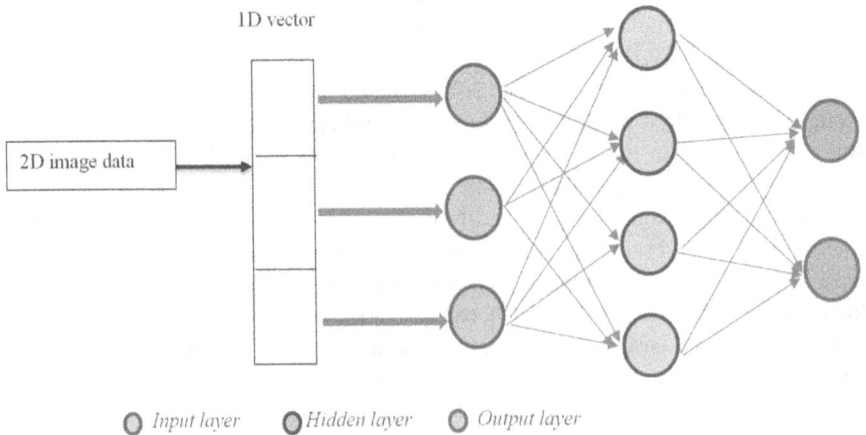

FIGURE 13.2   Schematic representation of conventional artificial neural network architecture.

a   b

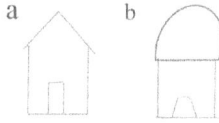

FIGURE 13.3 Schematic of two different types of house images that are to be classified. (a) House with roof having sharp edge (b) House with roof having curvy edge

every other neuron in the subsequent layer) neural network for a three inputs and two outputs network.

The main difference between the CNN and the conventional artificial neural network is that the former extracts (by itself) the relevant features required for the classification of the given input data, whereas the latter does not. This is a significant achievement in the field of artificial intelligence because the accuracy of classification in the conventional artificial neural networks was bounded by the handcrafted features (Sultana, Sufian, and Dutta 2018). We illustrate below the concept of CNN using an example. Consider the problem of classifying two different types of house images shown in Figure 13.3a and b. The network architecture of a simple CNN model is shown in Figure 13.4. The convolution layer takes the given input image and convolves using different types of filters to extract the desired features from the given image. For example, when the network is trained by giving several images of the house type shown in Figure 13.3a (the same way it will also work for Figure 13.3b from the training data set), the convolution filters could extract features like horizontal line, vertical line, line inclined at +45° and line inclined at −45°. We can see why this is the case because for the house type shown in Figure 13.3a, the horizontal and vertical lines denote the walls and door of the house whereas the roof of the house has +45° and −45° lines. So, these horizontal, vertical, and +45°, −45° lines form the feature set that can then be used to classify the house type shown in Figure 13.3a from Figure 13.3b. Similarly, the network will learn the feature set for the house type shown in Figure 13.3b. These features are then used to classify given any input image of these two house models. Next, how does the network learn the values of convolution filters that will detect the horizontal, vertical, +45°, −45° angle lines?

To understand this, let us consider a 3 × 3 convolution kernel for identifying the four features: horizontal, vertical, +45°, −45° angle lines. The user will randomly initialize the values for these convolution filters, and an example of this is shown in Figure 13.5.

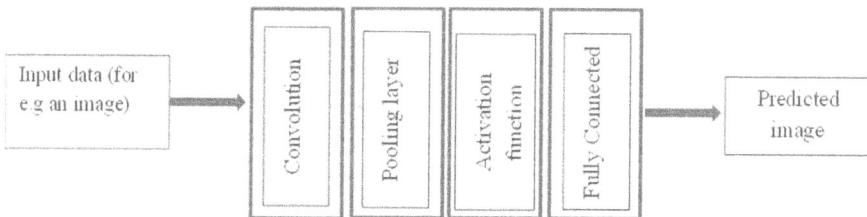

FIGURE 13.4 Schematic of a typical CNN architecture for image classification.

0.5	0.5	0.5	0.2	0.05	-0.2	0.7	0.7	0.01	0.02	0.4	0.4
0.1	0.1	0.1	0.2	0.05	-0.2	0.7	0.01	-0.7	-0.4	0.02	0.4
-0.5	-0.5	-0.5	0.2	0.05	-0.2	0.01	-0.7	-0.7	-0.4	-0.4	0.02

FIGURE 13.5   Randomly initialized convolution filter kernel values to detect horizontal, vertical, +45°, −45° lines (from left to right) in a given image.

Once the network is trained, that is, after giving several input images like the house type shown in Figure 13.3a, the randomly initialized values of the convolution filter will get modified so as to detect a particular feature. A modified/learned convolutional filter for detecting horizontal, vertical, +45°, −45° lines most probably will be like the one shown in Figure 13.6.

These filter masks are then convolved over the given input image. The horizontal mask after convolution with the input image will yield high values in regions where there are horizontal edges and low values otherwise; similarly, the vertical and other filter masks. The next layer, which is the pooling layer, will take this convolved output/features and will scale down the information by considering only the maximum of the convolved values (in such case, it is called as max pooling), or average the convolved values over a particular kernel size (called as average pooling) etc. The output of the pooling layer is then flattened (that is, a 2D image will be converted into a 1D row or column vector by concatenating the second row of pixels to the first row so on and so forth as depicted on the input side of Figure 13.2). After, flattening a fully connected layer using the activation function provides classification of the given input image. A DNN uses several such CNNs to build architectures that can learn complex patterns (features) in the image data set and predict/classify any given image. (Note in CNN, the neurons in hidden layers are not fully connected to every single neuron in the preceding layer unlike a multilayer perceptron network.) The hidden layers include convolution, pooling, and activation functions, which are stacked together to perform a sequential operation so as to classify the final image. The inspiration for the CNN came from the visual cortex in the brain where each set of neurons analyze a small region or feature in the given image. Also, the visual

1	1	1	1	0	-1	1	1	0	0	1	1
0	0	0	1	0	-1	1	0	-1	-1	0	1
-1	-1	-1	1	0	-1	0	-1	-1	-1	-1	0

FIGURE 13.6   From left to right filter kernel that detects horizontal, vertical, +45°, −45° lines in a given image.

cortex processes information sequentially over several layers starting with detecting simple features in its first layers to more complex features in its deeper layers. Overall, the main goal of this chapter is to present a review of the DNN and CNN architectures that have been developed for MRI-based brain tumor segmentation.

13.4 COMMON DNN AND CNN TERMINOLOGIES

13.4.1 RECEPTIVE FIELD

In simple terms a receptive field is the area/region in the input image that the convolution kernel looks/affects it. Figure 13.7 shows the receptive fields for two different convolution filter kernels.

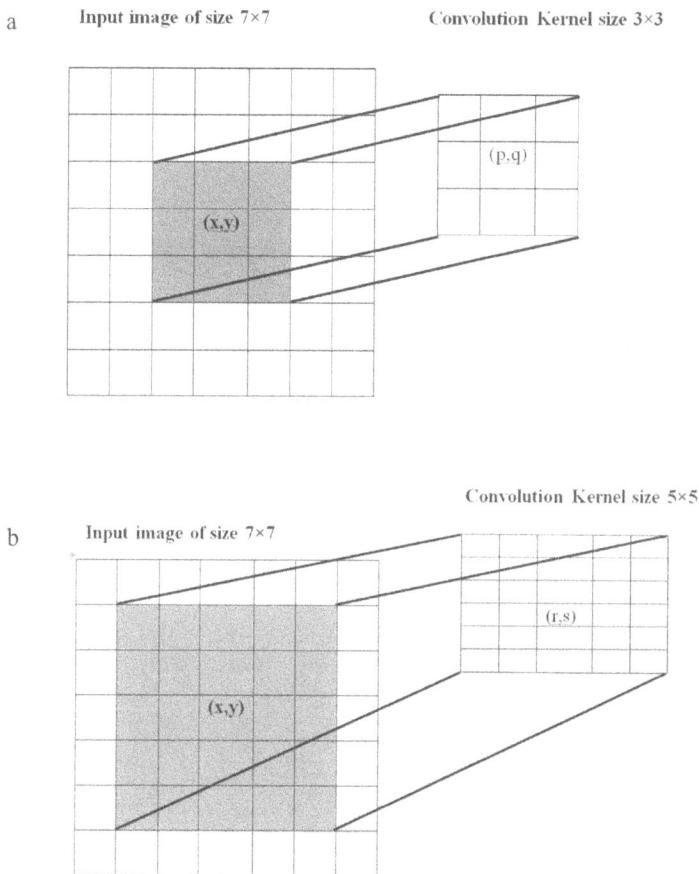

FIGURE 13.7 (a) The blue color area/region in the input image is the receptive field seen by the filter kernel of size 3×3 shown on the right side while calculating the feature map; (b) the green area/region in the input image is the receptive field seen by the filter kernel of size 5×5 shown on the right side while calculating the feature map. The (x,y), (p,q), (r,s) denote the center coordinates of the input image, convolution masks of size 3×3 and 5×5, respectively.

13.4.2 DILATED CONVOLUTION

The dilated convolution concept is used when a large receptive field has to be considered during convolution operations but without any additional/increase in the computational cost. For example, when one is interested in getting the global features from an image using a convolution kernel of size 3 × 3, a dilated convolution is used. We can understand this from Figure 13.8a and b. In Figure 13.8a, a common convolution mask of size 3 × 3 is used. The receptive field for this is shown in the

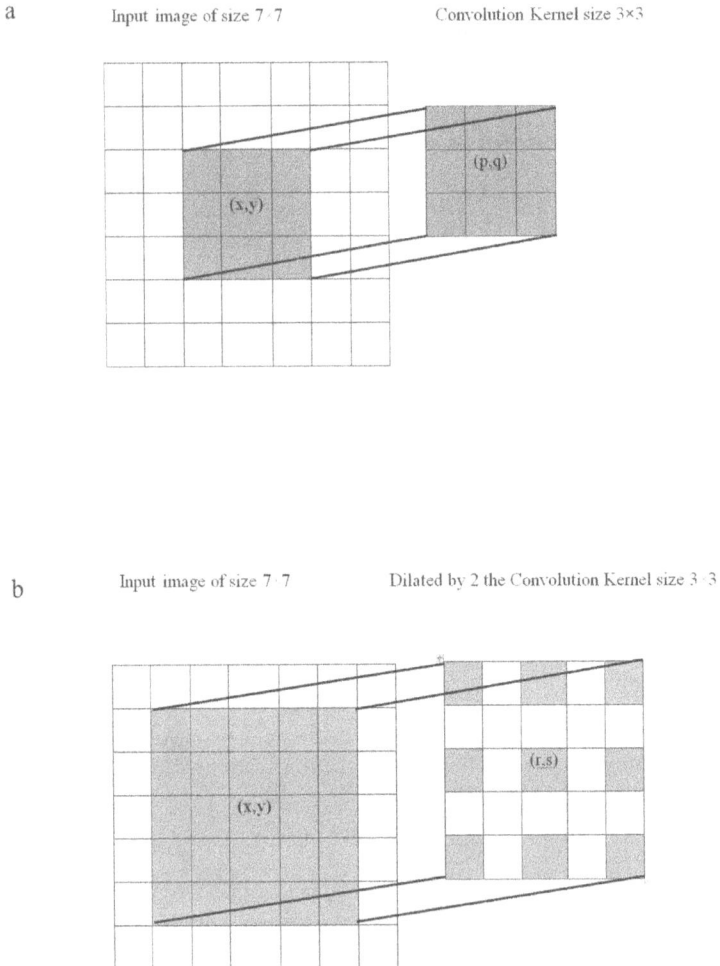

a Input image of size 7×7   Convolution Kernel size 3×3

b Input image of size 7×7   Dilated by 2 the Convolution Kernel size 3×3

FIGURE 13.8   (a) The blue color area/region of the input image is the receptive field seen by the filter kernel of size 3 × 3 shown on the right side while calculating the feature map; (b) the green area/region of the input image is the receptive field seen by the dilated convolution process, the white cells in the dilated convolution mask are 0s. The (x,y), (p,q), (r,s) denote the center coordinates of the input image, convolution masks of size 3 × 3 and convolution mask dilated by a factor of 2.

corresponding input image. For the dilated convolution mask shown in Figure 13.8b, the number of features for this dilated mask is the same as that of 3×3, such as nine features/convolution values (green-colored cells in the convolution mask). But we can see that the receptive field (green color in the input image) for this dilated convolution mask is increased when compared to the normal convolution mask of size 3×3 in Figure 13.8a. Hence, the dilated mask will enable us to obtain more global features as opposed to the normal convolution mask given in Figure 13.8a.

13.4.3 BATCH AND GROUP NORMALIZATION

In a basic multilayered neural network, changes to weight values in one layer will propagate and sequentially affect each succeeding layer. This weight modification for example in the case of sigmoid activation functions can lead to exploding gradients. Also, when the probability distribution of the input data changes (meaning for example during the trainning process I may use the image data set obtained by using different image capturing systems such as a digital camera, smarphone camera etc.), the network activation function has to continuously adapt based on this changing input data distribution. This variation in data distribution can be minimized by normalizing the input data to zero mean and unit variance. For this purpose, batch and group normalization approaches were developed. From Figure 13.9a and b, let us understand the difference between batch and group normalization. Here, we have considered four color image data sets: I1, I2, I3, and I4. Each color image is given in terms of three channels. I1R, I2R, I3R, and I4R denote the first pixel in each of the four images (Figure 13.9a) and I1R, I1B, and I1G denote the first pixel across the RGB channels in the first image. In the case of batch normalization, the mean and standard

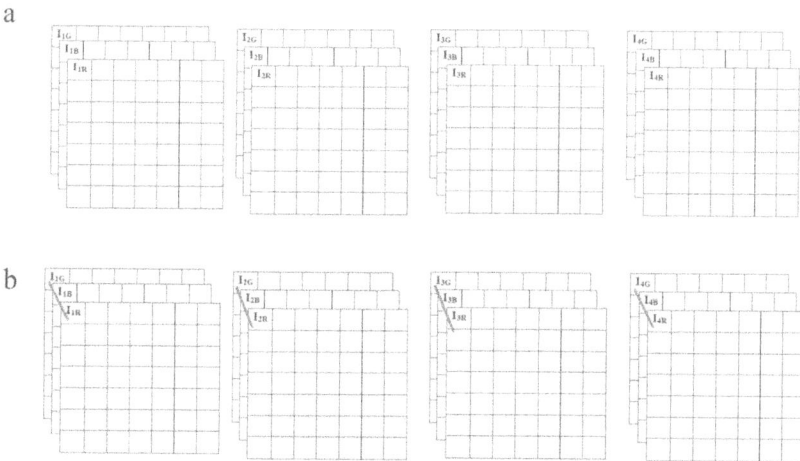

FIGURE 13.9 (a) Batch normalization shown across four different training data with each training set having three channels; (b) group normalization shown across four different training data with each training set having three channels. The mean and standard deviation are calculated along the green line shown.

deviation values will be computed across a channel considering all four images; that is, if we consider the red channel (R), then the mean and standard deviation values will be calculated by considering the pixel intensity values of I1R, I2R, I3R, and I4R. Similarly, for the channels blue (B) and green (G), mean and standard deviation values are computed across all four images. Hence in batch normalization, the mean and standard deviation values are calculated for each channel. In the case of group normalization for example, if we consider a group to comprise of three channels (R, G, and B), then we can see that we have four groups (I1, I2, I3, and I4) each of three channels. The mean and standard deviation will be calculated across the three channels and for each of the four images as shown by blue lines in Figure 13 9b. Batch normalization is used when a sufficient number of batches are available (Wu and He 2019). When a lower number of batches is available, then group normalization is best suited.

13.4.4 STRIDE

Stride controls the way that a filter is convolved with an image. This is illustrated using the following example. Consider an image of size 7×7 that is convolved using a filter of size 3×3 (blue rectangle shown in Figure 13.10a). For a stride value of 1, the filter kernel moves one step either in the horizontal or vertical direction. For a stride value of 2, the kernel moves two steps/pixels (this is shown in Figure 13.10b).

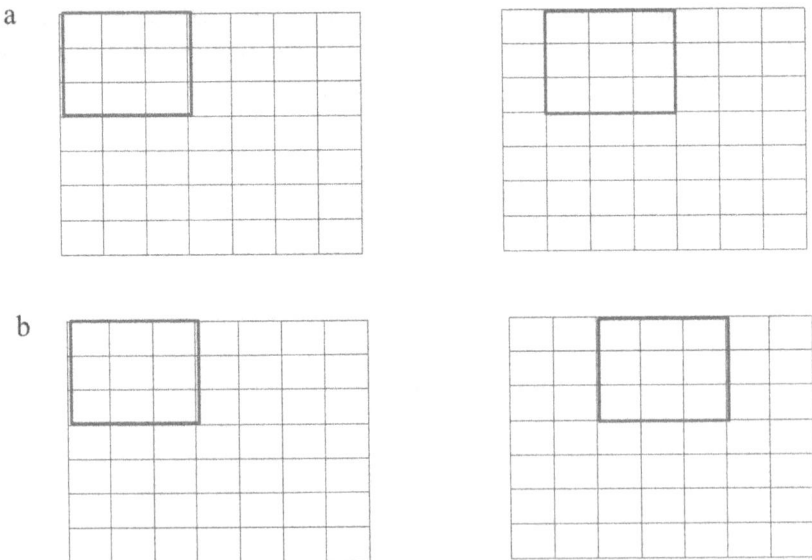

FIGURE 13.10   (a) Left image shows the image with filter kernel (blue color) and the right image shows when the filter kernel moved one step in the horizontal direction for the next calculation, which we call stride one; (b) left image shows the image with filter kernel (blue color) and the right image shows when the filter kernel moved by two steps in the horizontal direction for the next calculation, which we call stride two.

13.5 CNN-BASED BRAIN TUMOR SEGMENTATION ALGORITHMS

Typical flow diagram of a CNN-based brain tumor segmentation algorithm is shown in Figure 13.11. For comparing the efficacy of various brain tumor segmentation models using a CNN, a common unbiased data is required. All the algorithms discussed in this chapter were tested using the common Brain Tumor Segmentation Challenge (BraTS) (http://braintumorsegmentation.org/) database, which has a collection of multimodal MRI images [(T1-weighted, T1-Gd (gadolinium)].

13.5.1 3D AUTOENCODER-DECODER ARCHITECTURE

The aim of a BraTS challenge is to evaluate novel cutting-edge methods for brain tumor segmentation. One such method was proposed by Myronenko (2018) based on an autoencoder-decoder network. This architecture was built using the BraTS 2018 training data set, which includes T1-w, T1-Gd, T2-w, and FLAIR MR images for each of the 210 HGG and 75 LGG patients. They used a large image crop size of $160 \times 192 \times 128$ followed by a large encoder to extract features and a small decoder to reconstruct the segmented images. In addition to encoders, a variational auto-encoder (VAE) was added to the architecture to improve the tumor segmentation accuracy by providing further guidance and regularization to the encoder path. This is shown in Figure 13.12. The encoder part consists of a 3D convolution, rectified linear unit (ReLU) activation function with group normalization. Images were upsampled using 3D bilinear up-sampling in the decoder. Image preprocessing was a simple zero mean and unit standard deviation shift with data augmentation by random image axis shift and intensity shift. No other image preprocessing steps such as bias field correction or motion correction were performed. The network yielded an accuracy of 0.7664, 0.8839, and 0.8154 for enhanced tumor (ET), whole tumor (WT), and tumor core (TC) regions, respectively.

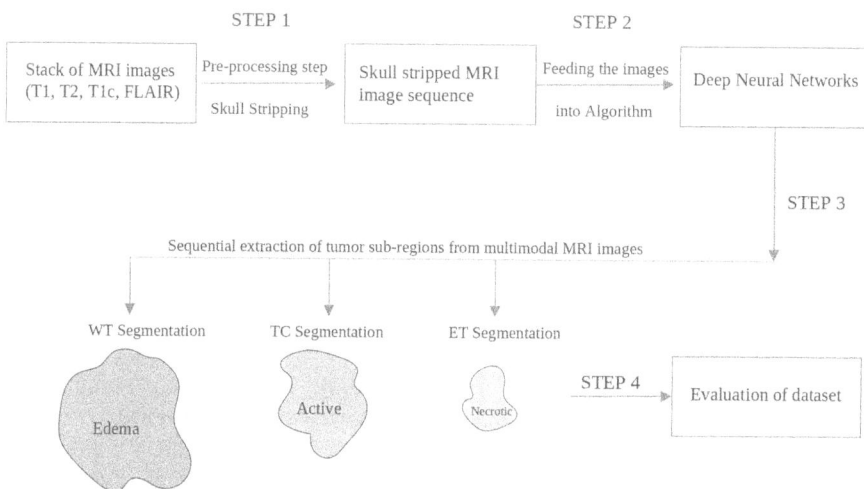

FIGURE 13.11 Typical flow diagram of a CNN-based brain tumor segmentation algorithm.

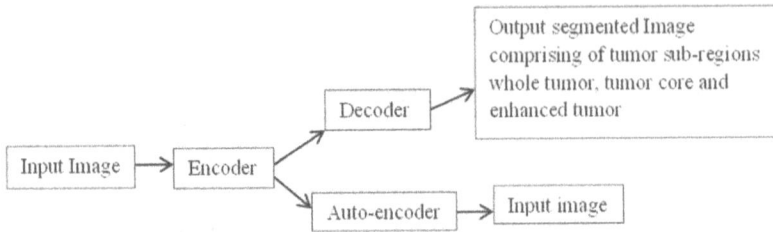

FIGURE 13.12   Schematic of the architecture proposed by Myronenko.

It was reported that no other data augmentation procedure or post processing technique improved the performance of the network. The auto-encoder not only improved the performance of the network but also significantly improved the accuracy for any random initialization.

13.5.2   CASCADED CONVOLUTIONAL NETWORKS

Havaei et al.'s (2017) work using cascaded DNNs with convolution layers for brain tumor segmentation was quite successful. Their architecture consisted of two main components: (1) two-pathway architecture to extract local and global features and (2) the spatial dependencies between adjacent labels were taken into account by inputting to the second CNN the output probability maps from the first CNN. Further, they performed training in two phases: (1) by constructing data patches with equiprobable healthy and tumorous pixels and (2) considering a data patch that is unbalanced (usually tumorous tissue occupies only 2% of the pixels in an image when compared to 98% of pixels with healthy tissue). They had only 30 training data sets (BraTS 2013); hence, to avoid overfitting, they used regularization methods such as L1, L2, and dropout, stride of 1. They cascaded the network in three different ways and studied their performance. For this, they considered a CNN network before the two-pathway architecture. The output of this first CNN network is then cascaded in three different ways: (1) output of the first CNN is given as additional image channels along with the T1-w, T1-Gd, T2-w, and FLAIR images as input to the two pathway architecture, and this network is called Input Cascaded CNN; (2) the output of the first CNN is cascaded (given as input) in the local pathway of the two-pathway architecture, and this is called Local Cascade CNN; and (3) the output of first CNN is concatenated just before the output of the two pathway architecture. Input Cascaded CNN architecture outperformed the other two architectures and found that the high performance was due to the local and global details captured by the two-pathway architecture.

A triple cascaded network was proposed by Wang et al. (2017). They have designed the architecture with three networks, to perform a binary segmentation in each network. As a result, the model has sequentially segmented the tumor subregions. Among the three networks, WNet was the first and it was designed to segment a whole tumor from the multimodal (T1-w, T1-Gd, T2-w, and FLAIR) input data. Cropped WT images were the input to the second stage TNet, which segments the tumor core. A bounding box image of this TC is then given as input to the third stage ENet to segment the enhanced tumor region. An important aspect is that they proposed an anisotropic network with a large receptive field in 2D and a small receptive field in the third direction. The isotropic kernel $3 \times 3 \times 3$ was decomposed into $3 \times 3 \times 1$ for intraslice and $1 \times 1 \times 3$ for interslice convolution. This was done to improve the speed, but at the same time to consider the 3D input images. They also performed multiview fusion to account for 3D segmentation. First, they obtained segmentation results independently in all the three orthogonal views and fused them by averaging their softmax outputs to obtain the final segmented images. The dice similarity coefficient (DSC) was 0.7859, 0.9050, and 0.8378 for ET, WT, and TC, respectively. This model gives comparatively high accuracy rate in segmenting tumor subregions. Overall, triple cascade networks architecture is simple, easy to implement, and reduces the parameter overfitting problem

Inspired by Havaei et al.'s architecture mentioned above, a multistep cascaded network was developed by Li et al. (2020). The brain tumor was segmented from coarse to fine using 3D U-Net with three auxiliary outputs in the expanding pathway to decrease the problem of vanishing gradient. The cascaded network was designed in such a way at first the WT region was obtained; this WT and T1-Gd was given as input to the next U-Net stage which resulted in TC, the TC and T1-Gd was then given to the final stage U-Net, which resulted in ET regions. Finally, WT, TC, and ET were combined into a segmented image. They also introduced focal loss function that down weighted easy examples in the training data set. They used an image crop size of $96 \times 96 \times 96$ data augmentation that consisted of rotating the images, flipping, and adding Gaussian noise to the input data. The DSC values were 0.886, 0.813, and 0.771 for WT, TC, and ET regions, respectively.

13.5.3 3D Convolutional Neural Networks

Kamnitsas et al. (2017b) developed a novel multiscale 3D CNN model with a fully connected conditional random field network to segment brain tumor subregions. They were influenced by Ronneberger and Fischer (2015) and Long, Shelhamer, and Darrell (2015) and proposed a twofold pathway architecture with 11 layers in each path. Although the parallel pathways share an identical number of layers, their functionalities are distinct with each pathway being assigned a specific task. The second path was designed to capture spatial information on down sampled images whereas, the first path apprehends local information from the 3D patches of multimodal MRI images. Further, they have used small 3^3 kernels instead of 5^3 kernels in both of the pathways to convolve faster with the input 3D patches and to maintain less weight for minimal memory usage. Finally, the soft segmented feature maps were fed into fully connected CRF to remove the false positive predictions by the proposed pipeline. They have trained and tested their model with the BraTS 2015 data set. The DSC was 0.901, 0.754, and 0.728 for WT, TC, and ET, respectively. This algorithm was ranked first in the BraTS 2016 challenge.

Kamnitsas et al. (2017) went further and proposed a state-of-the-art model in 2017. They created an ensemble algorithm by combining the ideas and architectures of multiple DNN models. Their main objective was to integrate multiple models that are robust, have high performance, and are optimized to withstand data variance and biased errors. The common approach to the ensemble method is averaging the variance of output data from different models. In their model, they integrated the Deep Medic, fully convolutional network and U-Net and built an ensemble model that they claimed as novel (because they have used different models instead of using the same model by changing the hyperparameters). Their training data include 210 cases with HGG and 75 cases with LGG. Later, the models were tested on a test data set of 46 cases (that includes both HGG and LGG) and the output was combined to give a predicted class label. In this process, they have used the BraTS 2017 data set for validation and testing of the model and it has achieved a DSC of 0.886, 0.785, and 0.729 for WT, TC, and ET, respectively. This model won first prize in the 2017 BraTS challenge.

13.5.4 U-Net-Based Fully Convolutional Network

U-Net is a comprehensive and high-throughput DNN created by Ronneberger and Fischer (2015) specifically for biomedical image segmentation tasks. The key aspect of U-Net is that it gives a good accuracy even with a limited data set unlike other DNN networks. The functionalities of U-Net and encoder-decoder models share similarity, but U-Net implements an autoencoder network with skip connections in its architecture to concatenate the encoder and decoder feature maps (in contraction and expansion path) as shown in Figure 13.13. Basically, the U-Net architecture has

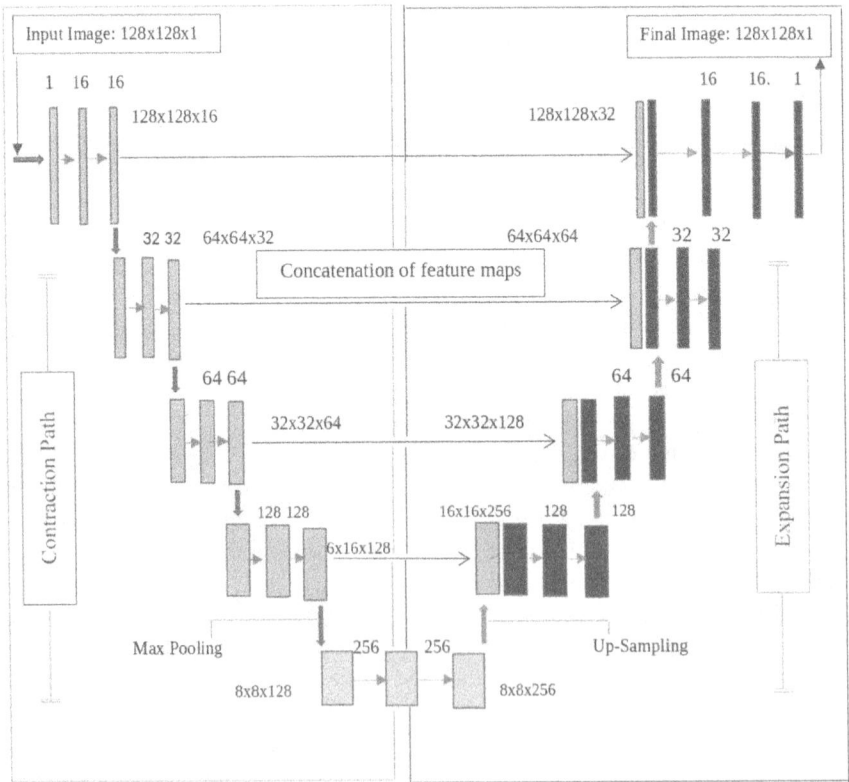

FIGURE 13.13   Schematic of the U-Net architecture.

two paths, which are known as contraction/encoding path and expansion/decoding and has connections between the two paths to retain the localization information. Briefly, the input images are given as input to the convolutional layers, of 3×3 or 5×5 kernel (selection of kernel matrix size depends upon the user) to extract the feature maps. Then, a down sampling is performed by passing the feature maps through a max pooling layer (2×2 matrix). Similar to other DNN models, U-Net also uses the ReLU activation function to normalize the feature maps. In every subsequent convolution layer, the number of feature maps increases by two when compared to its previous layer and the image dimension is halved; hence, it depicts the contraction phase. At the end of the contraction path, the feature maps in a bottleneck layer form as the input for the expansion path, where a deconvolution/upsampling was performed to increase the image dimension by subsequently reducing the number of feature maps into half in each expansion layer. At the end of the expansion path, a 1×1 convolution kernel was used to obtain the output image. Dong et al. (2017) proposed a variant U-Net architecture for brain tumor segmentation following the architecture of Ronneberger and Fischer (2015). They trained and tested this network using the BraTS 2015 data set and a DSC of 0.86, 0.86, and 0.65 for WT, TC, and ET (for both HGG and LGG cases) was obtained. They also showed that this outperformed the fuzzy c-means framework, Markov random field, and even the deep CNN model (Pereira et al. 2016). However, it showed poor performance in segmenting the tumor-enhanced region of an LGG tumor. Significant research has been carried out in U-Net architecture to improve its overall performance in the last few years until recently when Feng et al. (2020) extended the architecture that was applied to 2D images to 3D images and has proposed an efficient way to segment brain tumor subregions from 3D MR image sequences without increasing the computational demand. Feng et al. extracted 3D patches from the multimodal MRI sequences, which were given as input to the U-Net architecture to generate feature maps. To avoid overfitting problem, data augmentation and optimization of the model was performed. The BraTS 2018 data set was used to train and test the model. A very good DSC score of 0.9114, 0.8304, and 0.7946 was obtained for WT, TC, and ET, respectively. This architecture showed top-notch performance and was robust when compared to other deep CNN models (Isensee et al. 2018). Overall, U-Net has great potential to segment the tumor subregions with high accuracy and is worth exploring.

Further, Jiang et al. (2020) proposed a novel two-stage cascaded U-Net model for tumor segmentation. In their network, the first stage was used to train the coarse predictions and the second stage was used for refinement of prediction maps (by concatenating the initial prediction map with the original input image). Briefly, in the first stage, the extracted 3D patches ($4 \times 128 \times 128 \times 128$) from the multimodal MRI sequence were fed into the encoder path. It was followed by a convolution layer of $3 \times 3 \times 3$ filters using stride 2. Then, group normalization and the ReLU activation function was used to avoid random errors. By the end of the decoder path, a 1^3 convolution filter was applied to reduce the number of output channels. In the second stage of network, two decoders were used. Where one decoder was specifically used to train the data (trilinear interpolation decoder) and the other decoder path with deconvolution layers was used to test the data. Finally, they trained and tested the model with the BraTS 2019 data set and achieved a DSC score of 0.8879, 0.8369, and 0.8326 for the WT, TC, and ET, respectively. This algorithm was ranked first in the BraTS 2019 challenge. This architecture showed top-notch performance and is robust.

13.6   CONCLUSION

This chapter provides an overview of diverse DNN architectures used in automatic brain tumor segmentation from MRI. Overall, DNN algorithms have demonstrated satisfactory results. But still, there are certain challenges like distinguishing the enhanced tumor region from healthy tissues in LGG patients. However, from the above models (in terms of model complexity, training time, memory usage, and accuracy rate), it is evident that the 3D autoencoder regularization model, cascaded network architecture, and advanced implementations of U-Net models have great potential to achieve high accuracy in lesion segmentation.

ACKNOWLEDGMENT

We thank the Department of Biotechnology (project number BT PR 30007/ MED/32/660/2018) for providing funding support for this study.

REFERENCES

Angulakshmi, M., and G. G. Lakshmi Priya. 2017. "Automated brain tumour segmentation techniques–A review," *Imaging Systems and Technology*, 27: 66–77. doi: https://doi.org/10.1002/ima.22211.

Bakator, M., and D Radosav. 2018. "Deep learning and medical diagnosis: A review literature. Multimodal technologies and interaction," *Multimodal Technologies Interaction* 2 (3).

Cha, S. 2006. "Update on brain tumor imaging: from anatomy to physiology," *AJNR American Journal of Neuroradiology* 27 (3): 475–87.

Dong H., Yang G., Liu F., Mo Y., and Guo Y. 2017. "Automatic brain tumor detection and segmentation using U-Net based fully convolutional networks," in Valdés Hernández M., González-Castro V. (eds) *Medical Image Understanding and Analysis. MIUA* 723: 506–17.

El-Sayed, A., H.M.M. El-Dahshan, K. Revett, and M. Salem. 2014. "Computer-aided diagnosis of human brain tumor through MRI: A survey and a new algorithm." *Expert Systems with Applications* 41(11): 5526–45.

Feng, X., N. J. Tustison, S. H. Patel, and C. H. Meyer. 2020. "Brain tumor segmentation using an ensemble of 3D U-Nets and overall survival prediction using radiomic features," *Frontiers in Computational Neuroscience* 14: 25. doi: 10.3389/fncom.2020.00025.

Fox, B. D., V. J. Cheung, A. J. Patel, D. Suki, and G. Rao. 2011. "Epidemiology of metastatic brain tumors," *Neurosurgery Clinic North America* 22 (1): 1–6. doi: 10.1016/j.nec.2010.08.007.

Garman, R. H. 2011. "Histology of the central nervous system," *Toxicology Pathology* 39 (1): 22–35. doi: 10.1177/0192623310389621.

Havaei, M., A. Davy, D. Warde-Farley, A. Biard, A. Courville, Y. Bengio, C. Pal, P. M. Jodoin, and H. Larochelle. 2017. "Brain tumor segmentation with deep neural networks," *Medical Image Analysis* 35: 18–31. doi: 10.1016/j.media.2016.05.004.

Heper, A. O., E. Erden, A. Savas, K. Ceyhan, I. Erden, S. Akyar, and Y. Kanpolat. 2005. "An analysis of stereotactic biopsy of brain tumors and nonneoplastic lesions: A prospective clinicopathologic study," *Surgical Neurology* 64 Suppl 2: S82–8. doi: 10.1016/j.surneu.2005.07.055.

Isensee, F, P. Kickingereder, W. Wick, M. Bendszus, and K.H. Maier-Hein. 2018. "Brain tumor segmentation and radiomics survival prediction: Contribution to the BraTS 2017 challenge," in Crimi A., Bakas S., Kuijf H., Menze B., Reyes M. (eds) Brainlesion: Glioma, Multiple Sclerosis, Stroke and Traumatic Brain Injuries. *BrainLes 2017. Lecture Notes in Computer Science* 10670: 287–97.

IúÕna, A., C. Direkoglu, and M. Sùah. 2016. "Review of MRI-based brain tumor image segmentation using deep learning methods," *Procedia Computer Science* 102: 317–24.

Jiang Z., Ding C., Liu M., and Tao D. 2020. "Two-stage cascaded U-Net: 1st place solution to BraTS challenge 2019 segmentation task," in Crimi A., Bakas S. (eds) *Brainlesion: Glioma, Multiple Sclerosis, Stroke and Traumatic Brain Injuries. BrainLes 2019. Lecture Notes in Computer Science* 11992: 231–41.

Kamnitsas, K., W. Bai, E. Ferrante, S. McDonagh, M. Sinclair, N. Pawlowski, M. Rajchl, M. Lee, B. Kainz, D. Rueckert, and B. Glocker. 2017a "Ensembles of multiple models and architectures for robust brain tumour segmentation," in Crimi A., Bakas S., Kuijf H., Menze B., Reyes M. (eds) *Brainlesion: Glioma, Multiple Sclerosis, Stroke and Traumatic Brain Injuries. BrainLes 2017. Lecture Notes in Computer Science* 450–62.

Kamnitsas, K., C. Ledig, V. F. J. Newcombe, J. P. Simpson, A. D. Kane, D. K. Menon, D. Rueckert, and B. Glocker. 2017b. "Efficient multi-scale 3D CNN with fully connected CRF for accurate brain lesion segmentation," *Medical Image Analysis* 36: 61–78. doi: 10.1016/j.media.2016.10.004.

Koszewski, W., H. Kroh, and P. Kunert. 2002. "Difficulties in stereotactic biopsies of brain tumors," *Neurologia i Neurochirurgia Polska* 36 (3): 481–88.

Li X., Luo G., Wang K. 2020. "Multi-step cascaded networks for brain tumor segmentation," in Crimi A., Bakas S. (eds) *Brainlesion: Glioma, Multiple Sclerosis, Stroke and Traumatic Brain Injuries. BrainLes 2019. Lecture Notes in Computer Science* 11992: 163–73.

Lin, A. L., and E. K. Avila. 2017. "Neurologic emergencies in the patients with cancer," *Journal of Intensive Care Medicine* 32 (2): 99–115. doi: 10.1177/0885066615619582.

Liu, J., M. Li, J. Wang, F. Wu, T. Liu, and Y Pan. 2014. "A survey of MRI-based brain tumor segmentation methods," *Tsinghua Science and Technology* 19 (6): 578–95.

Long, J., E. Shelhamer, and T. Darrell. 2015. "Fully convolutional networks for semantic segmentation," *2015 IEEE Conference on Computer Vision and Pattern Recognition (CVPR)*, 7–12.

Louis, D. N., A. Perry, G. Reifenberger, A. von Deimling, D. Figarella-Branger, W. K. Cavenee, H. Ohgaki, O. D. Wiestler, P. Kleihues, and D. W. Ellison. 2016. "The 2016 World Health Organization classification of tumors of the central nervous system: A summary," *Acta Neuropathology* 131 (6): 803–20. doi: 10.1007/s00401-016-1545-1.

Myronenko, Andriy. 2018. "3D MRI brain tumor segmentation using autoencoder regularization." BrainLes@MICCAI.

Ostrom, Q. T., P. M. de Blank, C. Kruchko, C. M. Petersen, P. Liao, J. L. Finlay, D. S. Stearns, J. E. Wolff, Y. Wolinsky, J. J. Letterio, and J. S. Barnholtz-Sloan. 2015. "Alex's Lemonade Stand Foundation infant and childhood primary brain and central nervous system tumors diagnosed in the United States in 2007–2011," *Neuro-Oncology* 16 Suppl 10: x1–x36. doi: 10.1093/neuonc/nou327.

Pereira, S., A. Pinto, V. Alves, and C. A. Silva. 2016. "Brain tumor segmentation using convolutional neural networks in MRI images," *IEEE Transactions on Medical Imaging* 35 (5): 1240–51.

Ronneberger, O., T. Brox, and P. Fischer. 2015. "U-Net: Convolutional networks for biomedical image segmentation." *International Conference on Medical Image Computing and computer Assisted Intervention,* 234–241.

Siegel, R. L., K. D. Miller, and A. Jemal. 2020. "Cancer statistics, 2020," *CA: A Cancer Journal for Clinicans* 70 (1): 7–30. doi: 10.3322/caac.21590.

Sultana, F., A. Sufian, and P. Dutta. 2018. "Advancements in image classification using convolutional neural network," *2018 Fourth International Conference on Research in Computational Intelligence and Communication Networks (ICRCICN),* 22–23 Nov. 2018.

Verma, N., M. C. Cowperthwaite, M. G. Burnett, and M. K. Markey. 2013. "Differentiating tumor recurrence from treatment necrosis: a review of neuro-oncologic imaging strategies," *Neuro-Oncology* 15 (5): 515–34. doi: 10.1093/neuonc/nos307.

Wang G., W. Li, S. Ourselin, and T. Vercauteren. 2017. "Automatic brain tumor segmentation using cascaded anisotropic convolutional neural networks," in Crimi A., Bakas S., Kuijf H., Menze B., Reyes M. (eds) *Brainlesion: Glioma, Multiple Sclerosis, Stroke and Traumatic Brain Injuries. BrainLes 2017. Lecture Notes in Computer Science* 10670: 178–90.

Wang, G., W. Li, S. Ourselin, and T. Vercauteren. 2019. "Automatic brain tumor segmentation based on cascaded convolutional neural networks with uncertainty estimation," *Frontiers in Computational Neuroscience* 13: 56. doi: 10.3389/fncom.2019.00056.

World Health Organization (WHO). N.d. "International Agency for Research on Cancer." https://gco.iarc.fr/.

Wu, M., and J. Shu. 2018. "Multimodal molecular imaging: Current status and future directions," *Contrast Media Mol Imaging* 2018. doi: 10.1155/2018/1382183.

Wu, Yuxin, and Kaiming He. 2019. "Group normalization," *International Journal of Computer Vision* 128 (3): 742–755. doi: 10.1007/s11263-019-01198-w.

14 LabVIEW-based Simulation Modeling of Building Load Management for Peak Load Reduction

A. Ajitha, Sudha Radhika, and Sanket Goel

CONTENTS

14.1 Introduction 245
14.2 Methodology..246
14.3 Proposed DSM-DR Architecture..247
14.4 LabVIEW..247
14.5 Residential Load Management 249
14.6 Commercial (Office Building) Load Management......................................251
14.7 Results and Discussion 251
    14.7.1 Residential Load Management 251
    14.7.2 Commercial Load Management 251
14.8 Conclusion 254
References..255

14.1 INTRODUCTION

With rapid progress in both economic and industrial sectors, the need for electricity has been increasing worldwide for the last few decades with an average of 2.8% annually (Li et al. 2015, Isnen et al. 2020). Globally, building energy shares around 20.1% of total energy, which is delivered with an increase of 1.5% per year (Jamil and Mittal 2017). It has been reported that the residential sector energy demand will increase by 24% in the near decade (Li et al. 2015) due to the rise in living standards and increased usage of electrical appliances (Wang et al. 2018). This will certainly increase the burden on the existing grids, especially during peak load hours when demand exceeds supply. To fulfill the peak energy demand, utilities were forced to build excess capacity. Managing the power consumption during peak hours will yield positive results for power utilities rather than investing in new capacity generation (Isnen et al. 2020). Technological developments for smart grids can assist in reducing peak loads (Ertugrul et al. 2017). A smart grid is an intelligent electrical network that integrates power generation with consumers to deliver energy in a well-controlled way (Gelazanskas and Gamage 2014). Furthermore, with demand side management (DSM) and its techniques, the load profile can be improved and energy consumption management can be achieved (Samadi et al. 2010, Guo et al. 2013, Safdarian et al. 2014).

DSM is a technology with initiative programs through which utilities can motivate and control consumers to optimally utilize energy (X. Chen et al. 2013). Any measure taken to manage load on the demand side ranging from replacement of inefficient appliances with energy efficient ones to installing advanced systems is considered as DSM (Palensky and Dietrich 2011). Past literature provides an insight that DSM includes various traditional techniques such as peak clipping, load shifting, and strategic load conservation for energy management in buildings. Load shedding is also one of the load management methods for peak clipping. Influencing the consumer to change the energy usage pattern in accordance with the time dependent pricing of the utility is termed as demand response (DR) (Gatsis and Giannakis 2012, Paterakis et al. 2015), a short-term load management method under DSM. Information and communication technologies will make a responsive and efficient network in smart grids to achieve better consumer load flexibility with DR (Salehi et al. 2011).

Building loads that includes both residential and commercial are potential contributors for global energy usage (Lazos et al. 2014) and hence are befitting for the implementation of new DR strategies that influence the way of energy consumption and conservation (Missaoui et al. 2014) generally called building energy management systems (BEMS). However, the methods of energy consumption and management are different for residential and commercial loads as summarized in Lazos et al. (2014) depending on the nature of loads and occurrence of peaks. There exists significant literature on energy management in both residential and commercial sectors separately (Samadi et al. 2010; Ma et al. 2015; Hu et al. 2016; Ertugrul et al. 2017; Wang et al. 2018) and also under BEMS (Park et al. 2011; C-N. Chen et al. 2016).

This paper presents a LabVIEW-based software simulation model for implementing load management in both residential and commercial office buildings. The ability of interfacing with hardware employing real-time signals has made LabVIEW more competent with other simulation software (Salehi et al. 2011). The developed simulation detects the peak load on the grid and controls or manages the operation of loads in residential by shifting and in commercial building through programmable loads and designing of smart cabins.

14.2   METHODOLOGY

Understanding the nature of loads and their operating patterns is necessary for the implementation of DSM. Residential loads includes various electrical appliances that are necessary to meet consumer requirements like cooking, lighting, entertainment, cooling, etc., and commercial building loads were dominated by heating, ventilation, and cooling (HVAC) systems (Lazos et al. 2014).

In this work, residential loads are categorized into interruptible and noninterruptible based on the consumer comfort. Loads whose time of operation can be changed or shifted are called interruptible loads like washing machines, water pumps, water heaters, etc., and the loads whose operation cannot be interrupted because it causes high consumer discomfort are noninterruptible loads like lights, fans, TVs, air

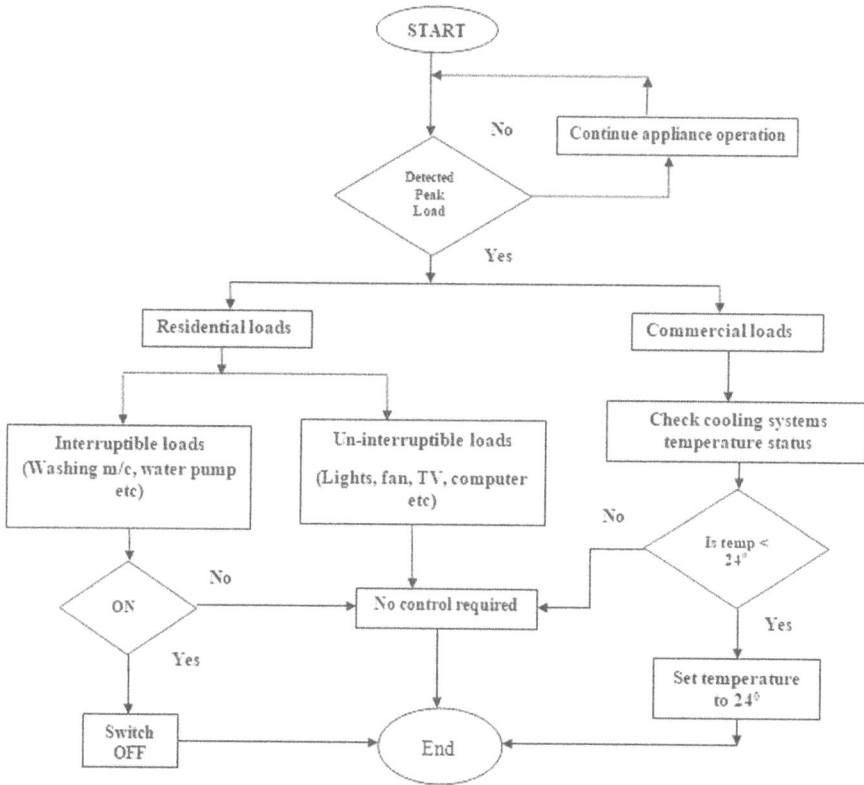

FIGURE 14.1   Flowchart for building load management.

conditioners, etc. In commercial loads, power consumption variation can be achieved for loads like air conditioners. The work is focused on these to reduce energy consumption in office buildings and designing smart cabins with sensor-based lights and fans. Figure 14.1 shows the flowchart of the proposed load management for both residential and commercial loads.

14.3   PROPOSED DSM-DR ARCHITECTURE

Load management during peak hours can possibly result in normalizing the load curve, which in turn improves system reliability. The peak clipping technique is a traditional DSM technique employed to reduce peak by load shifting. In the proposed architecture, the system responds to a peak occurrence on the grid. To realize the technique, the LabVIEW platform is adopted.

14.4   LabVIEW

Laboratory virtual instrument engineering workbench (LabVIEW) is a software based on graphical designing for testing, measuring, and automation (Salehi et al. 2011, Haribabu et al. 2018). It offers a simple design, flexible graphical, and hardware

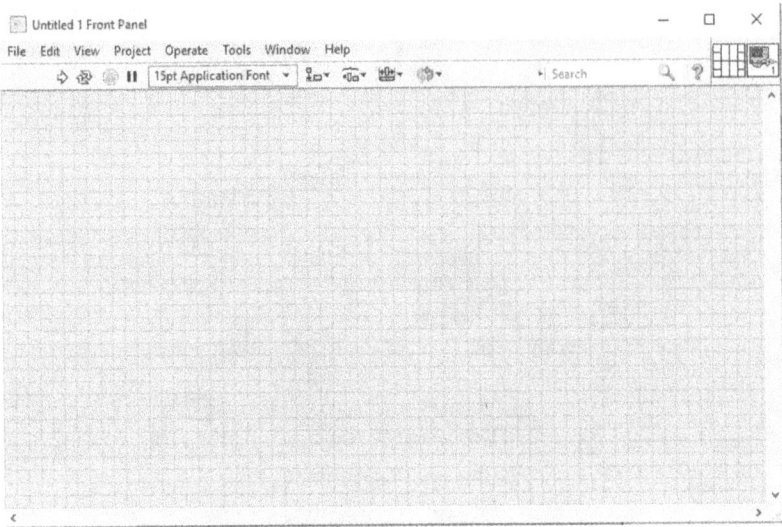

FIGURE 14.2   Front panel of LabVIEW.

interface with a wide variety of tools for numerous applications. In the current research, LabVIEW designs and runs the work as a simulation. But this can also be used as an application for data acquisition based needs with custom-built hardware from National Instruments for further experimental setup. LabVIEW contains two windows:

- Front panel: Used to view the output and as a user interface (Figure 14.2)
- Block diagram: The window where graphical code is written (Figure 14.3)

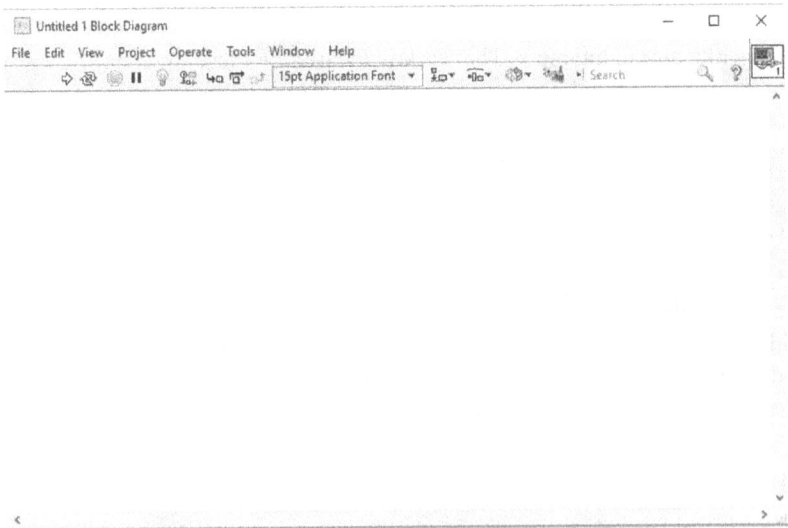

FIGURE 14.3   Block Diagram of LabVIEW.

14.5   RESIDENTIAL LOAD MANAGEMENT

Figure 14.4 shows the flowchart of load management for residential loads. The controller in a consumer house is designed to detect the peak load on the grid with the predefined peak set value. When the peak demand on the grid is sensed, the controller will automatically manage the loads as defined by the consumer based on comfort. By checking the operation status of interruptible loads during peak load condition, the controller switches their operation to nonpeak hours to achieve energy conservation.

Figure 14.5 shows the front panel design of residential load management on LabVIEW software. The DSM-DR method has both interruptible and uninterruptible loads designed using different user interface elements. Figure 14.6 shows the graphical programming on a block diagram panel.

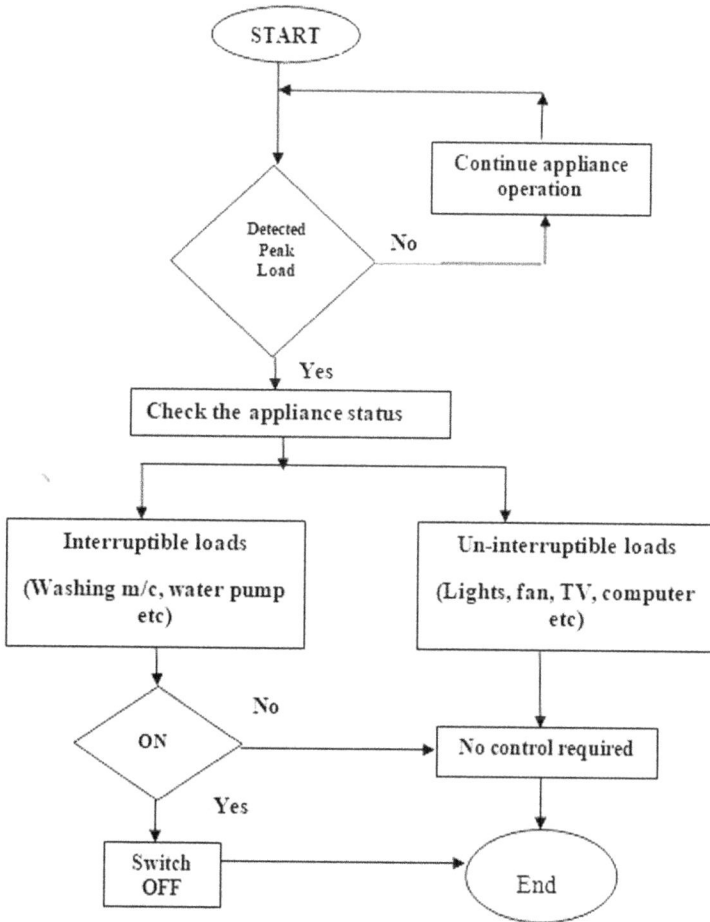

FIGURE 14.4   Flowchart for residential load management.

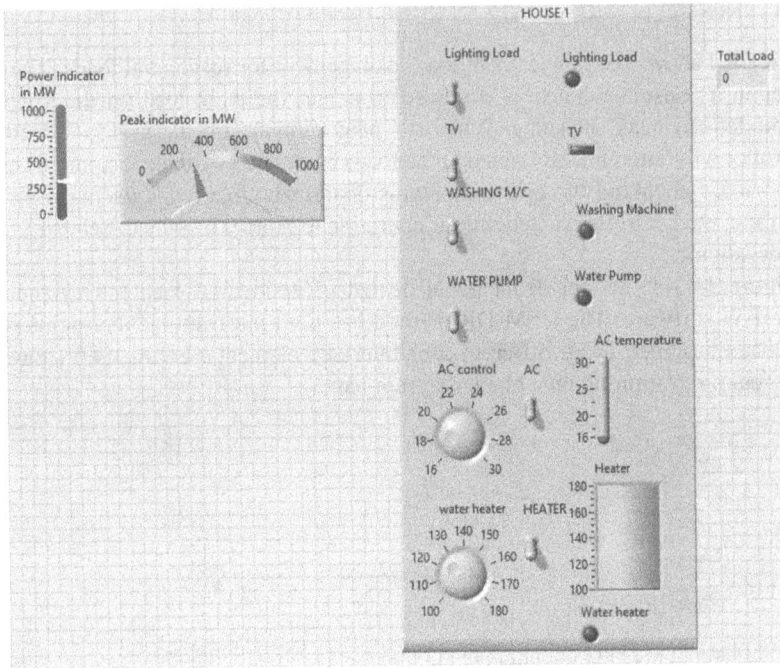

FIGURE 14.5   Residential setup (front panel).

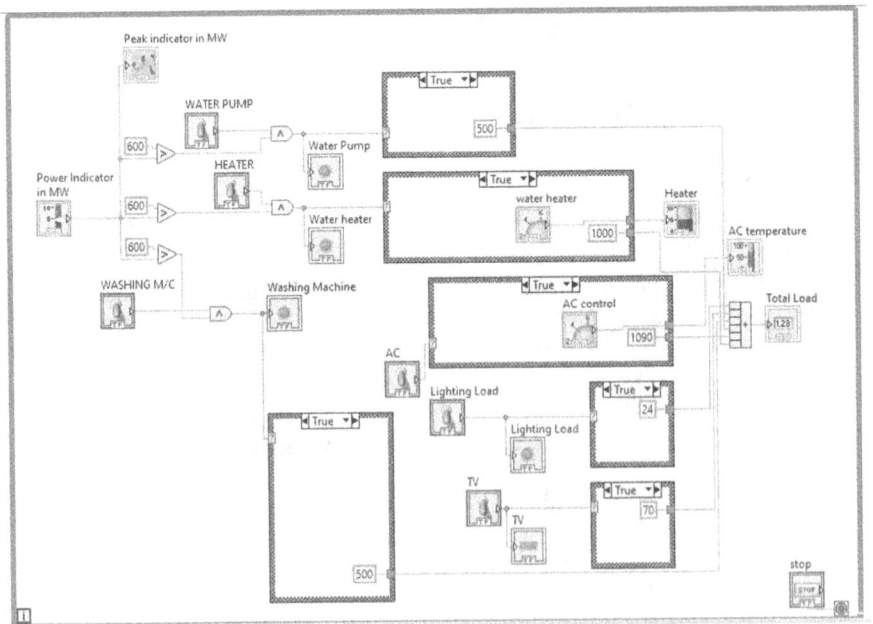

FIGURE 14.6   Residential set-up (block diagram: G programming).

14.6 COMMERCIAL (OFFICE BUILDING) LOAD MANAGEMENT

In commercial office buildings, the nature of loads is different compared to residential and also the time of peak occurrence, which normally happens to be during afternoons. The cooling system dominates the other loads like lightning. Therefore, energy conservation is achieved by managing the operation of cooling systems based on the occurrence of the peak. The controller in an office setup senses the peak condition and reduces the cooling systems' energy consumption level by setting the temperature to an optimal predefined value. Moreover, the lightning requirement in individual cabins is controlled based on occupancy.

A LabVIEW front panel design of a commercial office building with n number of cabins and a centralized air conditioning system is shown in Figure 14.7 and a respective block diagram is shown in Figure 14.8.

14.7 RESULTS AND DISCUSSION

14.7.1 RESIDENTIAL LOAD MANAGEMENT

The peak load detector indicated on the front panel design was given a preset value of 600 MW peak demand condition. When it senses exceeding the preset peak value, then it automatically turns off the interruptible loads without disturbing uninterruptible loads (Figure 14.9) and under nonpeak hours, all the loads are operated normally (Figure 14.10). It is observed that 62.8% of energy has been saved through the load management during peak hours with the designed residential setup.

14.7.2 COMMERCIAL LOAD MANAGEMENT

In a commercial building environment, load management and energy conservation are targeted with the main focus on cooling systems. Figure 14.11 shows the normal operation of cooling systems as per consumer requirements. When peak condition is

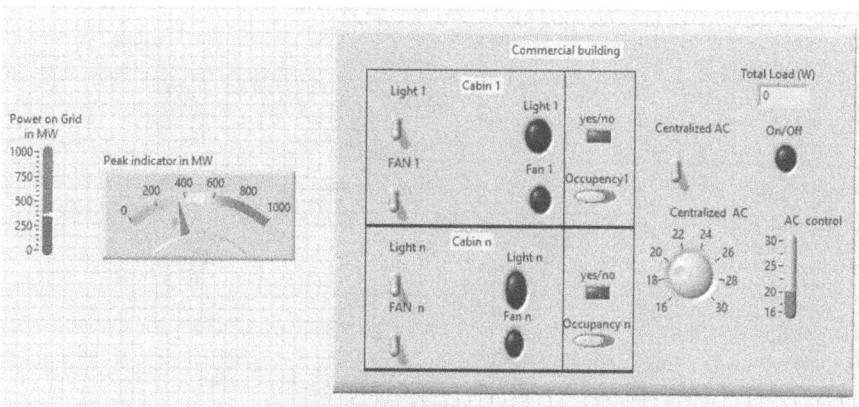

FIGURE 14.7 Commercial office design (front panel).

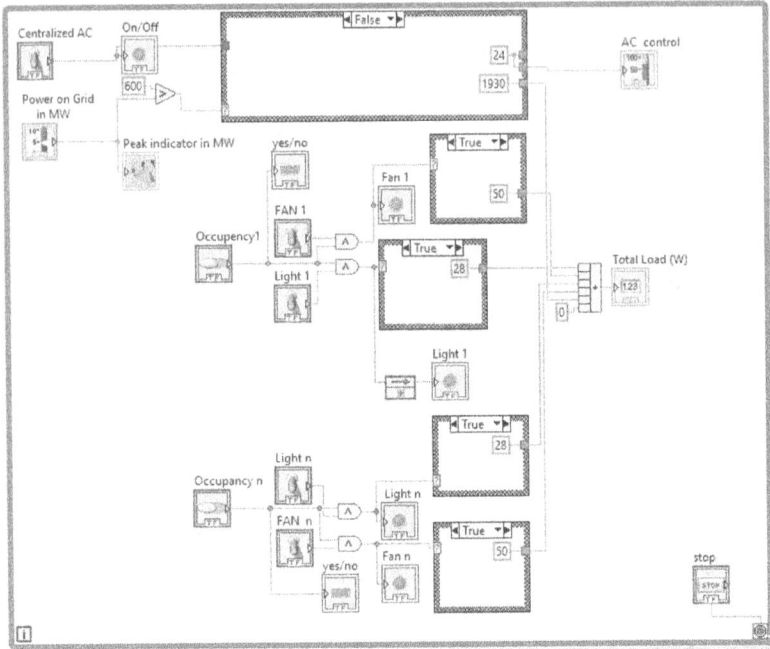

FIGURE 14.8   Commercial office design (block diagram).

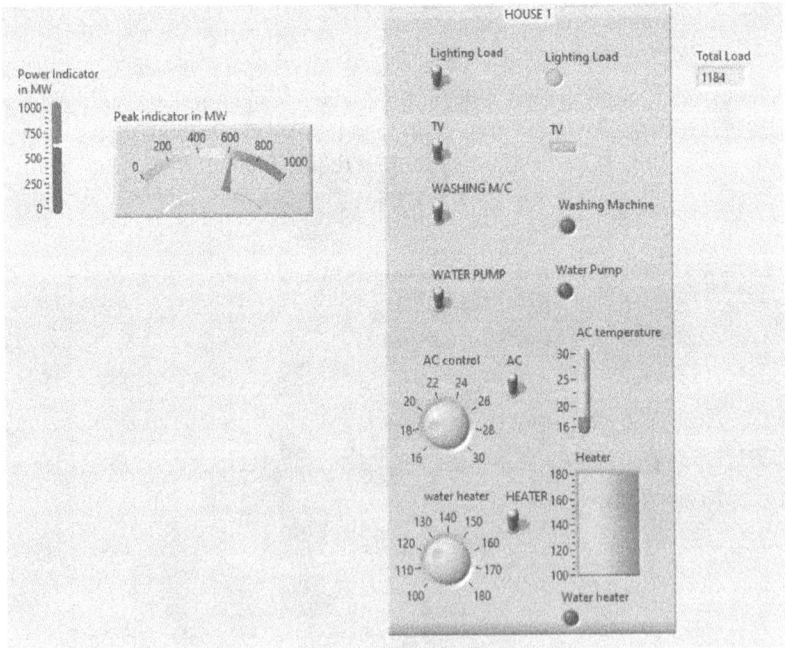

FIGURE 14.9   Operating condition of residential loads under peak conditions.

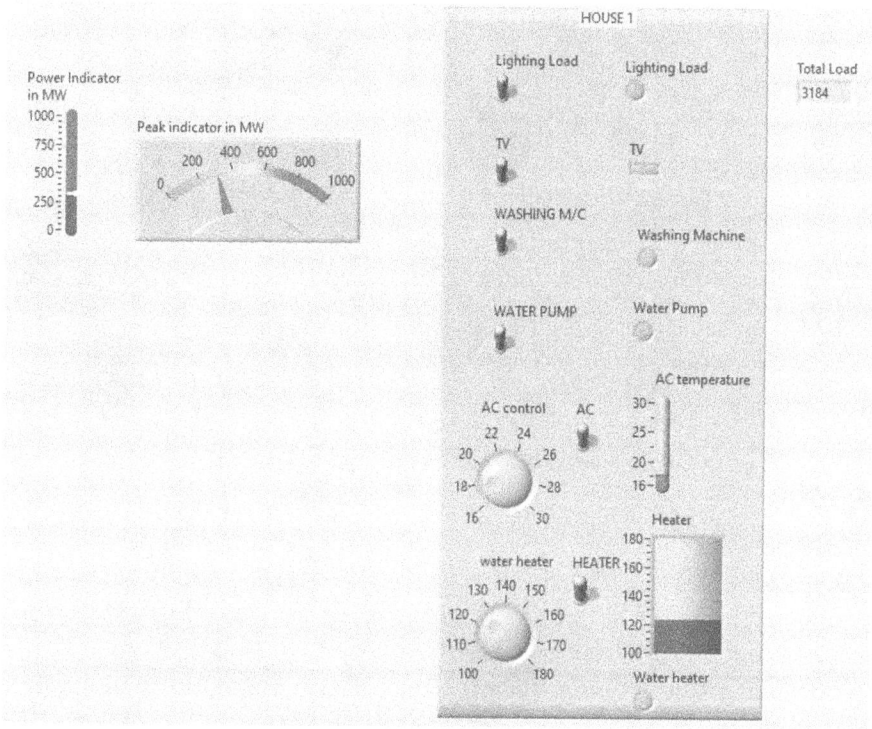

FIGURE 14.10 Operating condition of residential loads under nonpeak conditions.

detected, then the energy consumption of the cooling system is automatically changed to a predefined value (24°C) without affecting the working ambiance (Figure 14.12). A significant amount of energy saving can be achieved with smart cabins where the respective light and fan is operated based on place occupancy (Figure 14.13).

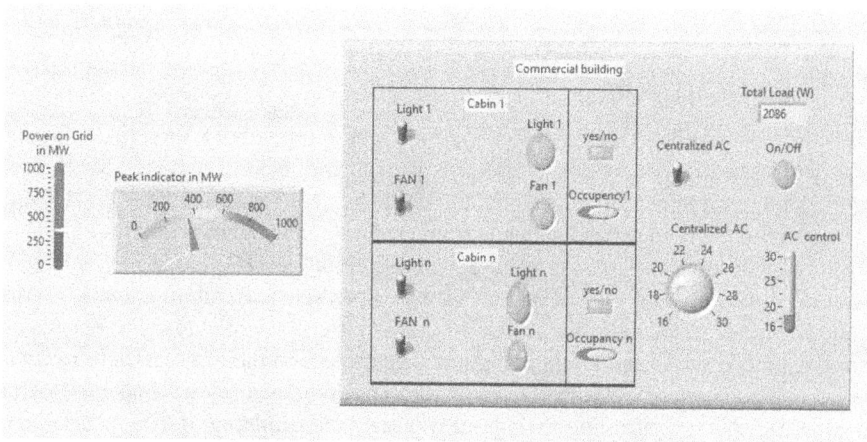

FIGURE 14.11 Cooling system load condition under nonpeak conditions.

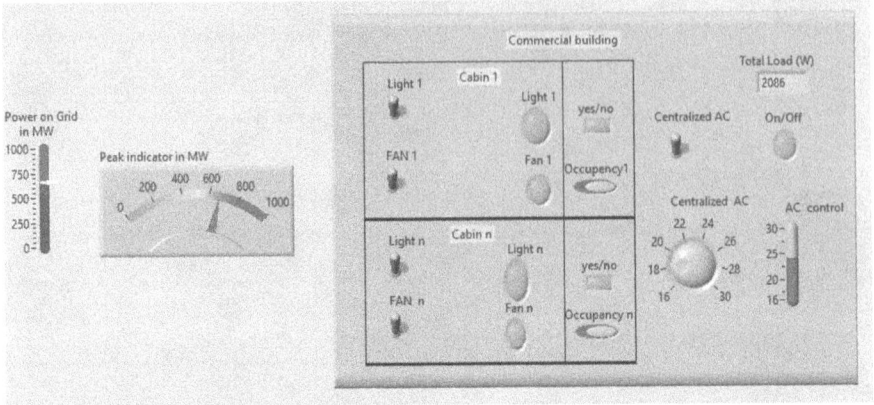

FIGURE 14.12 Cooling system load condition under peak conditions.

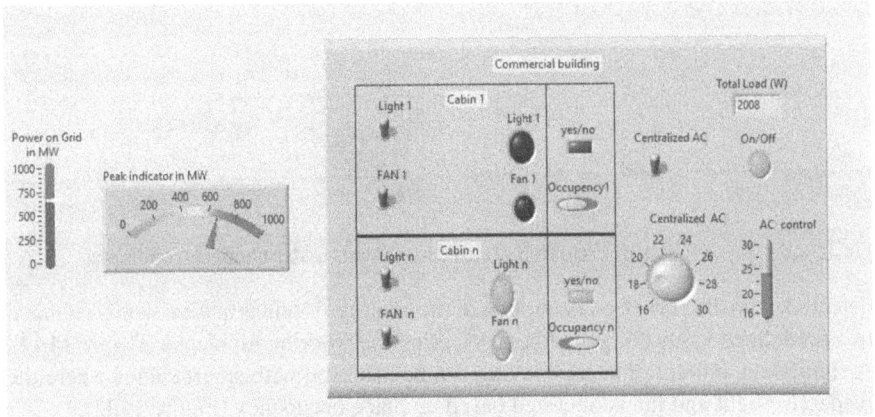

FIGURE 14.13 Smart cabin operating condition.

14.8 CONCLUSION

The rapid increase for electricity demand in the building sector with a rise in living standards can be successfully addressed through the techniques of DSM. In this work, load management for residential loads during peak load conditions is achieved through shifting of interruptible loads from peak to nonpeak hours and for commercial office buildings by managing energy consumption levels of cooling systems. A simulated model using LabVIEW software was proposed. To attain automatic load control during peak conditions sensed on the grid, the residential loads were categorized as interruptible and noninterruptible loads based on their operation flexibility. Load management of commercial loads that are dominated by cooling systems is achieved through controlling their level of energy consumption during peak hours and also through the designing of smart cabins. The aim of saving a significant amount of energy is realized with the proposed model.

REFERENCES

Chen, C.-N., M.-Y. Cho and C.-H. Lee (2016). "Design and implementation of building energy management system," *2016 3rd International Conference on Green Technology and Sustainable Development (GTSD)*, IEEE.

Chen, X., T. Wei and S. Hu (2013). "Uncertainty-aware household appliance scheduling considering dynamic electricity pricing in smart home," *IEEE Transactions on Smart Grid* **4**(2): 932–41.

Ertugrul, N., C. E. McDonald and J. Makestas (2017). "Home energy management system for demand-based tariff towards smart appliances in smart grids," *2017 IEEE 12th International Conference on Power Electronics and Drive Systems (PEDS)*, IEEE.

Gatsis, N. and G. B. Giannakis (2012). "Residential load control: Distributed scheduling and convergence with lost AMI messages," *IEEE Transactions on Smart Grid* **3**(2): 770–86.

Gelazanskas, L. and K. A. Gamage (2014). "Demand side management in smart grid: A review and proposals for future direction," *Sustainable Cities and Society* **11**: 22–30.

Guo, Y., M. Pan, Y. Fang and P. P. Khargonekar (2013). "Decentralized coordination of energy utilization for residential households in the smart grid," IEEE Transactions on Smart Grid **4**(3): 1341–50.

Haribabu, K., S. Prasad and M. S. Kumar (2018). "An IOT Based Smart Home Automation Using LabVIEW," *Journal of Engineering and Applied Sciences* **13**: 1421–24.

Hu, Q., F. Li, X. Fang and L. Bai (2016). "A framework of residential demand aggregation with financial incentives," *IEEE Transactions on Smart Grid* **9**(1): 497–505.

Isnen, M., S. Kurniawan and E. Garcia-Palacios (2020). "A-SEM: An adaptive smart energy management testbed for shiftable loads optimisation in the smart home," *Measurement* **152**: 107285.

Jamil, M. and S. Mittal (2017). "Building energy management system: A review," *2017 14th IEEE India Council International Conference (INDICON)*, IEEE.

Lazos, D., A. B. Sproul and M. Kay (2014). "Optimisation of energy management in commercial buildings with weather forecasting inputs: A review," *Renewable and Sustainable Energy Reviews* **39**: 587–603.

Li, W.-T., C. Yuen, N. U. Hassan, W. Tushar, C.-K. Wen, K. L. Wood, K. Hu and X. Liu (2015). "Demand response management for residential smart grid: From theory to practice," *IEEE Access* **3**: 2431–40.

Ma, J.-J., G. Du, B.-C. Xie, Z.-Y. She and W. Jiao (2015). "Energy consumption analysis on a typical office building: Case study of the Tiejian tower, Tianjin," *Energy Procedia* **75**: 2745–50.

Missaoui, R., H. Joumaa, S. Ploix and S. Bacha (2014). "Managing energy smart homes according to energy prices: Analysis of a building energy management system," *Energy and Buildings* **71**: 155–167.

Palensky, P. and D. Dietrich (2011). "Demand side management: Demand response, intelligent energy systems, and smart loads," *IEEE Transactions on Industrial Informatics* **7**(3): 381–88.

Park, K., Y. Kim, S. Kim, K. Kim, W. Lee and H. Park (2011). "Building energy management system based on smart grid," *2011 IEEE 33rd International Telecommunications Energy Conference (INTELEC)*, IEEE.

Paterakis, N. G., O. Erdinc, A. G. Bakirtzis and J. P. Catalão (2015). "Optimal household appliances scheduling under day-ahead pricing and load-shaping demand response strategies," *IEEE Transactions on Industrial Informatics* **11**(6): 1509–1519.

Safdarian, A., M. Fotuhi-Firuzabad and M. Lehtonen (2014). "A distributed algorithm for managing residential demand response in smart grids," *IEEE Transactions on Industrial Informatics* **10**(4): 2385–93.

Salehi, V., A. Mazloomzadeh, J. Fernandez, J. Parra and O. Mohammed (2011). "Design and implementation of laboratory-based smart power system," *American Society for Engineering Education Annual Conference*, ASEE.

Samadi, P., A.-H. Mohsenian-Rad, R. Schober, V. W. Wong and J. Jatskevich (2010). "Optimal real-time pricing algorithm based on utility maximization for smart grid," *2010 First IEEE International Conference on Smart Grid Communications*, IEEE.

Wang, Y., H. Lin, Y. Liu, Q. Sun and R. Wennersten (2018). "Management of household electricity consumption under price-based demand response scheme," *Journal of Cleaner Production* **204**: 926–38.

Index

Italicized pages refer to figures and **bold** pages refer to tables.

1-dB compression, 110, 124
3D printed electrode, 19–21
3D printing, 17–26

A

Accelerometer bandwidth, *88*
Accelerometer noise, 91–93
Accelerometer range, *88*
Accelerometer resolution, 91–92
Accelerometer transduction techniques
    capacitive, 90
    electromagnetic, 90
    optical, 87–103
    piezo-electric, 90, 92
    piezo-resistive, 90, 92
    resonant, 90
    thermal, 90
    tunneling, 90
Accuracy, 63, 77, 95, 230–231, 237–242
Activation function, 232, 235, 237, 241
Active tumor region, 229
Adapt, 213, 235
Advantages, 110, 167–168, 210, 213, 215–216
Ambipolarity, 145–146
American cancer society, 228
Amplification, 50–51, 56–65
    isothermal amplification, 56, 62–65, *64*
    LAMP, **55**, 62–63
    NAAT, **55**, 62–65, *64*
    RPA, **55**, 62–63, *64*
Analysis of variance test (ANOVA), 182, 186
Anisotropic network, 238
Aortic stenosis (AS), 190
Apparent optical properties (AOP's), 217
Approximate compressors, 76–77
Approximate computing, 71, 84
Approximate multipliers, 71–84
Approximate Wallace tree, *75*
Area, 4, 17, 20–21, 71–78
Area-delay product (ADP), 78
Arithmetic modules, 71
Artificial intelligence (AI) techniques, 179
Artificial neural networks (ANN), 181, *230*,
    230–231
Aspect ratio, 150
Autoencoder, 237–238
Automation, 52, 65
Average, 232

B

Band-to-band (BtB) tunneling, 138
Bandgap, 2–6, 8, 10–11, 95, 138–139, 147–148, 205
Bandwidth, 4, 88, *88*, 90–94, 100
Batch normalization, *235*, 235–236
Beam Splitter, 97
Becquerel, 204
Bias, 228–229
Bias Point, 102
Black Phosphorous (BP), 7; *see also* BP
Bonding process, 53–56
    chemical, 56
    plasma, 56
    thermal, 56, 58
Bounding box, 238
BP, *5*, 7, 10–11
Bragg Grating, 95, **96**, *99*, 99–101, *101*
Broadband photodetector, 1, 5, 11
Buck converter, 37–38, *38*, 40
Buck-boost converter, 38
Building energy management systems
    (BEMS), 246

C

Cadmium Telluride (CdTe), 208
Chamber-based PCR device, 56, 62–63
    multi, **54**, 58–61, *60*
    single, **54**, 56–57, *57*
    virtual, **54–55**, 61–62
Characteristics, 205–209, 217–218, 221
Charge controller, 30–31, 34, 40, *41*, 42–44
Circuit parameters, 71, **152**
Classification, 127–128, *205*, *214*, 214–216, 228,
    231, 231–232
CMOS technology, 137–138
Coefficient, **120**, **125**, 192, 217–219
Communication, 181
Compact model, 147
Complex, 71, 73, 76, 92, 95, 97, 190, 232
Compressor, 73, *76*, 76–77
Computer-aided, 180, 186
Concatenate, 240
Concentrated PV (CPV), 209
Conduction, 138–140, 147
Conduction Band, 4, 9, 138, 141, 146, 205, 209
Confusion matrix, 182, *183*
Consume, 71, 78, 127, 181

Continuous conduction mode (CCM), 37
Continuous wavelet transform (CWT), 190–192
Convection, 210–211
Conventional, 30, 53, 56, 138, *139*, 140, 145–147, 155, 165–166, *230*, 231
Convolution filters/kernels, 231–233, *232*
Convolve, 231–232
Coolant, 30, 210
Cooling, 204, 210–213, **212**
Copper Indium Gallium Selenide (CIGS), 208
Crtical path delay, 76, 78

D

Dadda and Wallace, 73
Dadda multiplier, 76, *76*
Damping Coefficient, 90
Data augmentation, 237–239, 241
Decoder, 237–238, 240–241
Deep learning, 71, 172–173
Deep neural networks (DNN), 230
Demand side management (DSM), 246
Demand Response (DR), 246
Deoxynucleic acid (DNA), 50
    primers, 52
    reaction mixture, 50, 62
Depth of water, 220
Detectivity, 4, 11–12, *12*
Device Characteristics, 141, *143*, 144, *145*
Dice similarity coefficient (DSC), 239
Die-area, 72, 78
Diffraction Grating, **96**, 97, 102
Diffusion current, 32–33
Digital PCR, 58, 61–62
Dilated convolution, *234*, 234–235
Direct bandgap, 2, 4, 10–11, 138–139, 147–148
Discrete Fourier transform (DFT), 191
Discrete wavelet transform (DWT), 191–193
Doping details, **141**, 147
Double-gate FDSOI, 142
Down sampling, 241
Drain Current, *143*, 145
Droplet PCR, 56, *60*, 61–62

E

Early diagnosis, 229
Edema, *229*, 229–230
Efficiency, 204, 206, 209–213
Efficient, 241, 246
Electrical equivalent circuit, 30, 32, *32*
Electrochemical sensor, 17–18, 21–26
Electrolyte, 19, 21, 25, 30–31, 34
Electrophoresis, 24, 50, 61
    agarose gel, **54–55**
    capillary electrophoresis, 61
    gel electrophoresis, **54–55**

Emergency frequency controls, 172
Encoder, 237–238, 240, 241
Energy, 2–8, 29–30, 36, 42, 47, 71, *147*, 166, 168, 197, *198–199*, 204–205, 208–209
Energy consumption, 30, 71, 246–247, 251–253
Energy Conversion, 205
Energy density, 30
Energy spectrum, 197, *198*, *199*
Energy storage systems (ESS), 30
Ensemble, 239
Environment, 203–221
Error analysis, 77
Error distance, 77
Error rate, 77, 78
Exact compressor, 76–77
External quantum efficiency (EQE), 4

F

F-measure, 182, **182**, **184**
Fabrication, 17–18, 22, 52, 53–56, 141–143
    CNC milling, 53
    laser ablation, 53, 56
    soft lithography, 53
Factors, 208–209
False positive predictions, 239
Fast Fourier transform (FFT), 191
Fault detection and diagnosis, 179
Feature maps, *233*, *234*, 239–241
Flexible plastic, 53
Float charging, 35–36
Float current, 36
Floating PV (FPV), 210
Floating system, 213–215
Fluorescence signal, 58
    CCD, *64*
    CMOS, *60*
    LED, *60*
FNET, 168, 171
Focal loss function, 239
Forced cooling, 210–213
Fourier transform (FT), 189
Free-space, 94–95, 97–98
Frequency disturbance recorders (FDR), 171
Frequency monitoring, 167–168, 171–172
Frequency stability, 165–173
Fully convolutional networks, 240–242
Fundamental Frequency, 91
Fused deposition modelling (FDM), 18

G

Gate capacitance, *144*, 145
Gate oxide thickness, 141
Gate-metal thickness, 141
Gate-Overlap Tunnel FET, 137, 140–149
Gaussian function, 193

Genes, 50, 52, 61, 63
  genetic code, 50
Global features, 234–235, 238
Gr, 2, *5*, 7–10
Graphene, 2, 20–22; *see also* Gr
Grid balancing, 166, 173
Group normalization, *235*, 235–236, 241
Guided-wave, 94, 97, 99

H

Heart valve disorders (HVD), 189
Heat dissipation chamber, 30
Heaters/heating elements, 52, 56, 58, 61, 246
  peltier, *60*, 61, 211
  resistive, 90, 92
Heating, ventilation and cooling (HVAC), 246
Hetero Junction (HJ), **208**
Heterojunction bipolar transistor (HBT), 109
Hidden layers, 230, 232
High grade glioma, 229
High-throughput, 61, 240
Hybrid power systems (HPS), 166–168, 173
Hyperparameters, 240

I

Image preprocessing, 237
Image processing applications, 71–72
Image sharpening, 78, 84
Impedance matching, 30, 32, 114, 115, 117, *117*
Incomplete charging, 42–44
Inertial Sensors, 88
Inherent optical properties (IOP'S), 217
Insulation, 209
Integration, 52, 62, 65, 95, 99, 110, 138
Intensity, 206, 211, 220, 236
Intensity Modulation, 95, **96**, 100, *101*, *102*
Inter-turn faults, 180
Interface, 5, 10, 30–31, 37, 40, 44, 46, 218, 249
Interferometry, 95, 97
  Fabry-Perot, 97, *98*, 99
  Mach-Zehnder, 95, *98*, 102
  Michelson, 95, *98*
Interlayer photoexcitation, 5–6, 10, 13
Interrogation Methods, 87, 100–103
Irradiation, 56, 204–206, 218–219

J

Jsupervisory control and data acquisition
        (SCADA) systems, 166

K

Karatsuba-Ofman Algorithm (KOA), 73

L

Lab-on-a-chip (LOC), 50
Lab-on-a-disc (LOD), 62
Labview, 62, 180, 246–248, *248*, 251
Latency, 72–73, 171, 173
Lateral BtB, 146
Lead-acid, 30
Life-threatening, 228
Limitations, 191, 193
Line of sight (LOS), 110
Lithium-ion, 30
Low-noise amplifier (LNA), 110, 112–121, *119*,
        120, *120*, *121*, *122*

M

Malignant tumor, 228
Materials, 1–3, *5*, 7, 11, 17, 26, 53, **141**, 204, 206,
        208, 216
  poly (methyl methacrylate), 53
  polycarbonate (PC), 53. *57*
  polydimethylsiloxane (PDMS), 53
  polyethylene terephthalate (PET), 57, 58
  polyimide, 53, 56
Max pooling, 232, 241
Maximum power point (MPP), 34
Maximum power point tracker (MPPT), 34–35
Mean, *235*, 235–236
Mean relative error distance, 77
Mechanical Quality Factor (Q), 91
Mechanical Sensitivity, 90, 91, 93, 98
MEMS technologies, 88
Metal Wrap Through (MWT), 208, **208**
Microchannel, 23, 50, 53, 56, 61
Microchip, 53, 56–58, *57*, 61, 64–65
Microcontroller, 31
Microfluidic device, 18, 52–56, 61–65
Microfluidics, 49–50, 61
Microgrids, 165, 167–168, 173
Micromachining, 88, 93
Miniaturization, 50, 52, 65, 88
Minimize, 73, 76, 113, 166
Mitral regurgitation (MR), 190
Mitral stenosis (MS), 190
Mitral valve prolapse (MVP), 190
Mixer, 110, 112, 125–134, *128*
Mobility, 2, 10, 13
Model Complexity, 242
Modeling, 22, 30–39, 245–254
Modeling approach, 147–148
Molecular biology, 50, 63
Mooring system, 213
Most significant portion, 73, 76–77
Motor current signature analysis (MCSA), 180
Mullis, K., 50
Multimedia, 71

N

N on P, 206, *207*, 209
Natural cooling, 210, 213, 220
Natural waters, 217
Necrotic core, 229, *229*
Nernst potential, 33
Neuroimaging modalities, 229
Nickel-cadmium, 30
Noise equivalent power (NEP), 4
Noise figure, 109–110, 112–126, **120**
Nonstationary signal, 190–191
Normalized mean error distance, 77
Novel, 230, 237, 239–241

O

Optical Circulator, 100
Optical Gradient Force, 95, **96**, 98
Optical Source, 95, 100
    Laser, **96**, 99, 102
    LED, *23*, 60, 64
Optimization, 46, 61, 241
Over fitting, 238–239, 241
Over-potential restriction, 42

P

P on N, 206, *207*, 209
Partial product generation, 72–73, *75*
Partial product reduction, 72–77
Passivated Emitter and Rear Solar Cells (PERC), 208, **208**
Passivated Emitter Rear and Totally diffused Rear-Junction (PERT_RJ), 208
Patches, 238–239, 241
Pattern recognition, 186
Performance, 238–242
Performance, 1, 3, 6, 10, 13, 40–43, 58, 72, 78, 88, 89–92, **152**
Performance benchmarking, 150–156
Perturb and observe (P&O), 34
Phase Change Material (PCM), 211
Phasor measurement units (PMU), 166
Phonocardiogram signal, 189–200
Photo-Diode, 100
Photo-generated, 2
Photo-generated-carrier, 8
Photodetector, 1–7, *8*, 10–13, *11*, *12*, 65
Photodiode, 5–7, 9–11, 95
Photoelastic Effect, 96–97
Photon flux, 4
Photon photodetector, 2–3
Photonic Crystal, 95, **96**, 97
Photonic integrated circuit, 95, 98
Phototransistor, 5–11
Photovoltaic, 3, 5–7, 9, 29, 204, 213

Planar Grating, 99

Planar Grating, 99
Point-of-care (POC), 19, 52
Polarization Controller, 94
Polymerase chain reaction (PCR), 50
    annealing, *51*, 51–52
    denaturation, *51*, 52
    extension, *51*
Pound-Drever-Hall, 103, *103*
Power, 71–78
Power conditioning unit, 213
Power consumption, 39, *39*, 73, 93, 109, 111, 113, 115, 138, 150, *154*, *155*, *156*, *157*, 210, 245
Power delay product (PDP), 78, 150
Power density, 30, 98
Power systems, 165–169
Power-delay product, 78
Precision, 182, **182**, **184**
Proof Mass, 87, 89–90, 92–95, 97–98
Propagation delay, 150, *151*, 156
Pulse width modulation (PWM), 38
PV-thermal units (PV/T), 210

Q

Qualitative, 229
Quantification, 58, 61–62, 65
    quantified, 58, 62, 94

R

Radiative transfer, 217
Random forest (RF), 181
Rate of change of frequency (RoCoF), 171
Recall, 182, **182**, **184**
Receptive field, 233, *233*, 234, *234*
Recursive multipliers, 73, *74*, *75*
Recursive-based, 72
Reflection, 215–216
Refraction, 217
Regularization, 237–238, 242
Remote fault signature analyzer (ReFSA), 180, *180*
Renewable energy sources (RES), 29, 42
Residential load management, 249, *249*, 251
Resiliency, **170**, 173
Response time, 4, 11, *12*, 30
Responsivity, 4–13, *12*
RF divider, 110, *125*, 125–126, *126*, **127**, 133–134
Ring resonator, 95, **96**, *97*, 99, 101

S

Scaling, 30, 71, 137, 192
Scalogram, 195, *195–197*, 197
Self-discharge, 32–34
Shockley, W., 208

Short-time Fourier transform (STFT), 190
Silicon, 204, 206, *207*, **208**, 208–209
Situational awareness, 173
Smart-cut, 142
Smith chart, 117, *117, 118*
Sodium-sulfide (Na-S), 30
Solar, 204–221
Solar irradiance, 34
Spatial information, 239
Spectrogram, 191, 194, *195*
Spectrum, 197, *198–199*, 219–220
Speed, 18, 40, 71–72, 111, 138, 171
Spring-mass-damper model, 89
Stack terminal voltage, 32, 34
Standard clinical procedure, 228
Standard deviation, 182, 186, *235*, 236
State estimation, 168, **170**
State of charge, 31, 33
Statistical features, 182–186, *184, 185*
Stereolithography (SLA), 18
Stiffness, 90
Stockwell transform (ST), 190, 193
Stride, 236, *236*
Sub-threshold Slope, 150
Submerged, 213, 215, 220–221
Submultiplier, 73–75, **74**
Supervisory control and data acquisition
        (SCADA) systems, 166
Support vector machines (SVM), 181
Synthesis, 13, 72, 78, 216

T

Temperature, 204, 206–210, *207*
Thermal photodetector, 2
Thermal shutdown, 46
Thermocycler, 50, 56, 58, 65
Threshold Voltage, 138
Time consuming, 24, 229
Time-frequency analysis (TFA), 189–200
TMD, 4–11, *9, 11, 12*
Tolerate, 71

Transconductance, 144–145, *145*
Transition frequency, 144
Transition metal dichalcogenide (TMD), 2, 4;
        see also TMD
Transitional frequencies *(fT)*, 109
Transmission rate, 186
Trickle current, 36
Truncation, 71, 75, 77
Tunnel field effect transistors, 137–160
Tunneling current, 93, 138

U

U-Net architecture, *240*, 240–241
Ubiquitous, 72
Underdesigned recursive, 73
Underwater (UW), 203–221
Unity-gain BW, *145*
Up sampling, 237, 241
Upgraded Metallurgical Grade (UMG), 208, **208**

V

Valance band, 4
Van der Waals, 2–13; *see also* VdW
Vanadium redox flow battery (VRFB), 30
VdW, 2–13
Vertical field, 146, 148
Visible light, 218
Voltage transfer characteristics, 150

W

Wavelet transform (WT), 190, 191–192
Wide area control, 172
Work function, 141
World Health Organization, 228
Würfel, P., 204

Z

Zigbee, *181*, 181–185

For Product Safety Concerns and Information please contact our EU
representative GPSR@taylorandfrancis.com
Taylor & Francis Verlag GmbH, Kaufingerstraße 24, 80331 München, Germany